INTRODUCTION TO
EMBRYONIC DEVELOPMENT

STEVEN B. OPPENHEIMER
GEORGE LEFEVRE, JR.
California State University, Northridge

Second Edition

INTRODUCTION TO EMBRYONIC DEVELOPMENT

ALLYN AND BACON, INC.   Boston   London   Sydney   Toronto

Library of Congress Cataloging in Publication Data

Oppenheimer, Steven B., 1944–
 Introduction to embryonic development.

 Bibliography: p.
 Includes index.
 1. Embryology. I. Lefevre, George. II. Title.
[DNLM: 1. Embryology. QS 604 0624i]
QL955.066 1984   596'.033   83-26622
ISBN 0–205–08097–9

Printed in the United States of America.
10 9 8 7 6 5 4 3 2 1   88 87 86 85 84

TABLE OF
CONTENTS

PREFACE

The first edition of this text was successful largely because of its extensive coverage of embryology, as well as of the molecular and cellular aspects of development, and because of its clarity, readability, and enthusiasm of style. In this second edition, we have preserved these important points and, in addition, completely revised the text to reflect the rapid advancements of this area of biological science.

Four completely new chapters have been added to this second edition. Chapter 1, "Developmental Biology: A Look Backward," presents the history of embryology. Chapter 2, "Reproduction, Cell Division, Mitosis and Meiosis: A Background for Developmental Biology," provides the reader with the basic information needed to understand many aspects of modern developmental biology. Chapter 3, "Genetics, DNA and Gene Regulation: A Review," introduces genetics and molecular genetics early in the text so that the exciting new developments in these areas presented later can be fully appreciated. Chapter 17, "Aging and Developmental Biology," introduces the student to current thoughts in a field that is beginning to yield exciting information and that is capturing the attention of scientists and the lay public alike.

In addition to these chapters, we have carefully updated the remainder of the text to reflect recent advances in such areas as: fertilization, cleavage, gastrulation, regeneration, gonad development, the genetics of antibody diversity, hybridoma technology and monoclonal antibodies, embryonic induction, the mechanisms of neural crest migration and morphogenesis, hormonal mechanisms in plant morphogenesis, the control of gene transcription during differentiation, intervening sequences, recombinant DNA technology, multigene families, gene amplification, gene rearrangement, levels of gene control, the commitment to differentiate, genetic engineering and the transfer of genes between species, and recent developments in cancer biology. The last include the isolation of human cancer genes and their roles in the development of human cancer, human oncogene base sequences and their chromosome localization and translocation, and the prevention of cancer. We have included many new figures to help the student grasp the subject matter and have carefully edited the entire text to make the book clear, readable, and exciting.

We hope that own enthusiasm for the subject matter will excite the curiosity of students, so that some may eventually contribute new findings to this rapidly progressing area of biological science.

We thank Jim Smith, Science Editor, who steered us and offered his friendship and expertise to make our job much more pleasant. We thank Mary Beth Finch, Production Editor, for a most meticulous job. We also thank the entire staff at Allyn and Bacon, who provided excellent assistance in all phases of the production of this book. We are also grateful to our families for their support and enthusiasm, to the many fine reviewers who helped guide us in the writing of this text, and to those who kindly provided us with excellent figures and micrographs. We encourage the readers of this text to write to us about any errors that might appear and to tell us how we can make this book even more compatible with your specific needs.

1

DEVELOPMENTAL BIOLOGY: A LOOK BACKWARD

EARLY THOUGHTS

High among the rewards in the life of any parent is the birth of a baby—a seemingly miraculous event in which mother and father join in a cooperative act of creativity. Since the beginning of time, human parents have wondered how a baby forms within the mother's body and how it develops the features and organization of a unique human child who possesses some characteristics of the father, others of the mother, yet is demonstrably different from both. Long ago, common observation established that cats have kittens, dogs have puppies, and cows have calves. Except in folklore, the maxim that "like begets like" is true without exception. The question is: What forces, mechanisms, and processes assure the development of offspring that in all significant details are replicas of their parents? Over the course of time, some answers to this question have been found, but much remains to be learned. Today, the scientific field that continues to search for answers to this question is called *developmental biology*. This book is intended to provide a description of our current knowledge of the processes of development.

Developmental biology has a long history, and it will be useful to look into the past in order to lay a foundation for understanding the field as it exists today. Although we know nothing of the earliest, prehistoric speculations, the thoughts and concepts of Aristotle and other early philosopher-scientists were written down during the "Golden Age" of ancient Greece. Important among the surviving treatises of that time is *De Generatione Animalium* by Aristotle, who lived from 384 to 322 B.C. In this treatise Aristotle questioned (among other things) whether the *embryo* is preformed and merely unfolds and grows during development, or whether it progressively *differentiates* from a formless beginning into a complex, complete individual. Aristotle clearly favored the second view, which we now call *epigenesis;* nonetheless, others held to the first view, which we now call *preformation.* This controversy remained unresolved for some 2,000 years and was not put to rest until early in the 19th century.

During the 17th century, a time of rebirth of scientific inquiry, nearly all biologists favored the preformation theory. Writing in 1669, Jan Swammerdam reported his observation that, in insects, miniature larvae are present in the "egg" and that gradual changes could be observed before a butterfly emerged. Actually, his "egg" was the pupa, or cocoon—a far cry from the "real" egg. In 1694, Antony van Leeuwenhock, an early Dutch microscopist, first observed living sperm. Until then, the semen was thought merely to stimulate or nourish the egg.

With this recognition that living cells were present in the semen, the preformation concept split in two: (1) the egg contained the preformed miniature individual, and (2) the sperm did. Thus, "ovists" and "spermists" sought evidence to support their respective views. In 1694, Nicklaas Hartsoeker drew a figure of a miniature human (*homunculus*) inside a sperm, presumably representing what he saw through the microscope. (Like the "canals" on Mars, observations such as this demonstrate that we see what we look for, not what we look at.) By 1745, however, Charles Bonnet demonstrated conclusively that unfertilized eggs sometimes develop into normal individuals *(parthenogenesis)*. This, of course, is an exceptionally strong argument that the egg, not the sperm, was the bearer of the preformed individual.

Despite these preformationist views, epigenesis had its adherents, too. As early as 1651, William Harvey, who is famous for demonstrating the continuous circulation of the blood, considered the egg to be a formless, undifferentiated mass from which the embryo developed. He is thought to have coined the term "epigenesis," following Aristotle. Despite the authority of Aristotle and Harvey, no other 17th century biologist of note supported the concept of epigenesis. More than 100 years later, Casper Friedrich Wolff (1738–1794) observed microscopically the successive stages of development of the chick egg. However, he found no evidence of a preformed chick in the egg, only progressive growth and gradual development from a simple to a complex form. The preformation theory received its final blow in 1828, when Karl von Baer published his great work, *Developmental History of Animals,* which contained his careful observations on eggs and their developmental stages showing that differentiation proceeded progressively. Preformation soon faded away as a concept of the developmental process.

Today, of course, the egg is not viewed as a completely undifferentiated, "formless" mass, nor as containing a miniature, preformed individual. In the modern view, the sperm and eggs contain a set of genetic instructions, *genes*, inherited from the parents. The genes may be thought of as the blueprints to be followed by the embryo in achieving its proper form.

NINETEENTH CENTURY EMBRYOLOGY

With von Baer's discovery of the true ova of mammals in 1828, detailed study of the embryonic stages of development began, and von Baer can properly be regarded as the "father" of modern embryology. Not only did he correctly identify the mammalian egg, but he described the primary germ layers of the embryo and the correspondence between the successive embryonic stages of higher organisms and the adult stages of more primitive organisms. Especially significant was his identification of the *notochord,* a basic embryonic structure characteristic of all animals (including ourselves) that belong to the phylum Chordata. It plays a fundamental role in establishing the organization of the early embryo. Perhaps von Baer's most significant contributions to embryology were his high standards of embryological investigation, and his introduction of comparative studies on both higher and lower organisms, the essence of *comparative embryology.*

Drawing on von Baer's work, Ernst Haeckel (1834–1919) stated his *biogenetic law,* which may be epitomized by the phrase, "ontogeny recapitulates phylogeny." This implies that a higher organism passes through a sequence of embryonic stages that reflect the adult stages of its more primitive ancestors. For example, clefts appear at an early stage in the pharyngeal (throat) region of mammalian embryos. These clefts bear a distinct resemblance to the gill slits of adult fish. However, the pharyngeal clefts close up in the mammalian embryo, and the adult shows no trace of them. Likewise, the early mammalian heart is a tubular, two-chambered structure, like that of an adult fish; later, it develops three chambers, like a frog's heart, before finally achieving the four-chambered condition seen in adult reptiles, birds, and mammals. The notochord provides structural support in adult primitive chordates, such as the lancet *Amphioxus,* but is replaced by bone in the vertebrates. It persists as a significant embryonic structure in vertebrates, though only vestiges are found in the adult.

Catchy sayings like "ontogeny recapitulates phylogeny" easily become accepted as though they possess deep meaning. However, they should not be accepted as revealed truths. A developing embryo does not truly "recapitulate" its evolutionary history; rather, the process of development requires that early stages be simple and later stages complex. A true understanding of the developmental sequences so well described by von Baer and his successors had to await the arrival of experimental embryologists, who go beyond simple description to a calculated interference with developmental activities.

THE EARLY EXPERIMENTAL EMBRYOLOGISTS

Wilhelm Roux (1850–1924) was the pioneer experimental embryologist. He was interested in the processes that determine which embryonic cells differentiate into which functional tissues. He visualized a complex of determinants already present in the fertilized egg that would be progressively parceled out during cell division to different regions of the embryo. His experiment, reported in 1888, was deceptively simple: with a red-hot needle, he punctured one of the two cells that form from the cleavage of a fertilized frog egg. The punctured cell died; the other cell gave rise to only a half-embryo. This clearly substantiated his concept of a sorting out of developmental determinants by qualitatively different cell divisions, starting from the very first division of the fertilized egg (Figure 1–1).

This early embryological experiment, with its seemingly clear-cut result, teaches us an important scientific lesson. Roux failed to provide an essential *control* for his experiment. Would the same result be obtained simply by separating the first two cells of the embryo, rather than killing one in place? (Of course, it is much easier to kill one of the first two cells than to separate them.) In 1891, Hans Driesch succeeded in separating the first two cells of sea urchin embryos, and usually two normal, though smaller, embryos—not two half-embryos—formed! Were frogs and sea urchins that different, or was one experimental result wrong? Not until 1933 was the difficult separation trick

ANALYSIS OF DEVELOPMENT

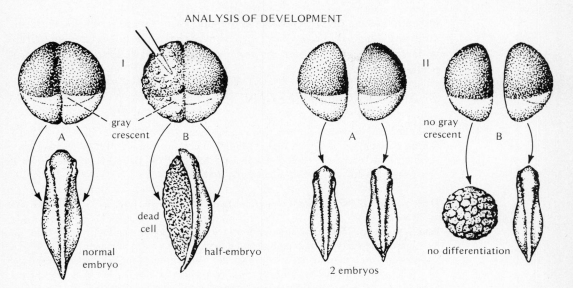

Figure 1-1. Mechanical and chemical factors in early frog development.
Ia. In the frog egg, a light area called the *gray crescent* forms on the surface diametrically opposite from the point where the sperm enters. The first cleavage furrow normally bisects the gray crescent so that each of the two blastomeres contains gray crescent material. A half-embryo develops from each blastomere.

Ib. If one of the blastomeres is killed by a needle, but is allowed to remain in place, the living cell develops into a half-embryo attached to the dead cell (Roux' experiment).

IIa. If the blastomeres are separated from one another without injury, then each gives rise to a small, but complete, embryo (Schmidt's experiment).

IIb. Sometimes the first cleavage splits the zygote so that one blastomere gets all of the gray crescent. If the two cells are then separated, the one with gray crescent develops into a complete embryo; the other divides a few times to form a ball of cells incapable of further differentiation.

These experiments clearly demonstrate that each of the first two blastomeres has the *potency* to develop into a complete embryo. However, this potentiality can be inhibited either by mechanical factors (*Ib*) or by chemical factors (*IIb*).

actually accomplished with frogs by G. A. Schmidt. In many cases, the result was two whole embryos, exactly as in sea urchins.

What was wrong with the original puncture experiment of Roux? Obviously, he did not foresee the mechanical effect of the dead cell on the movement and differentiation of cells derived from the living cell, and his experiment was not designed to recognize alternative possibilities. Nonetheless, Roux opened up the field of experimental embryology, and he can be properly called the "father" of experimental embryology, taking his place beside von Baer.

Since the time of Roux, many experimenters have tinkered with embryos, transplanting cells from one place to another, for example, or fertilizing enucleated eggs with nuclei taken from various regions of differentiating embryos. These and many other experiments have led the way to our current understanding of development processes. They will be described in appropriate places in this text.

READINGS AND REFERENCES

More detailed historical information can be found in the excellent book by Eldon J. Gardner, *History of Biology,* 3rd ed., (Minneapolis, Mn.: Burgess, 1972). Another good review of the early embryologists can be found in B. I. Balinsky, *An Introduction to Embryology,* 5th ed., (New York: Saunders, 1981).

Driesch, Hans. 1891. Entwicklungsmechanische Studien. I–II. *Zeit. wiss. Zool.* 53: 160–182.

Roux, Wilhelm. 1888. Beiträge zur Entwicklungsmechanik des Embryo. 5. Über die küntsliche Hervorbringung halber Embryonen durch Zerstörung einer der beiden ersten Furchungskugeln, sowie über die Nachentwicklung (Post-generation) der fehlenden Köperhälfte. *Virchows Arch. path. Anat. Physiol.* 64: 113–154, 246–291.

Schmidt, G. A. 1933. Schnürungs- und Durchschneidungsversuche am Amphibienkeim. *Roux Arch.* 129: 1–44.

2

REPRODUCTION, CELL DIVISION, MITOSIS, AND MEIOSIS: A BACKGROUND FOR DEVELOPMENTAL BIOLOGY

REPRODUCTION.

Various activities can be used to distinguish living from nonliving systems. Among these are growth, metabolism, responsiveness to environmental stimuli, movement, and so forth. None, however, is more fundamental and significant than *reproduction,* the ability of an existing individual to give rise to progeny like itself. If the original life form, whatever it may have been, had not been able to reproduce itself, nothing would be alive today.

Two different forms of reproduction, *asexual* reproduction and *sexual* reproduction occur in nature. Asexual reproduction occurs when one individual, by itself, engages in simple somatic cell division *(mitosis)* to produce another individual. The division of one ameba into two, for example, is a case of asexual reproduction. The original ameba can divide again and again, and its progeny in turn can divide, giving rise to a *clone*, or aggregation of *genetically identical* progeny. Most plants can produce clones of asexual, or *vegetative,* progeny by budding, by forming runners, or by rooting, among other means. In higher animals, by contrast, asexual reproduction is virtually unknown.

Relatively early in evolution, living organisms developed the capacity for sexual reproduction. Here, two different germ cells, or *gametes,* fuse together in *fertilization* to form a *zygote,* which then develops into a new individual. Typically, gametes occur in two easily distinguishable types: (1) small, motile male gametes, or *spermatozoa* (commonly called sperm), and (2) much larger, immotile *ova,* or eggs. In animals, sperm are normally produced in the *testis,* the male *gonad* or reproductive gland. Eggs are produced in the *ovary,* the female gonad. A single individual that possesses both functional ovaries and testes, such as an earthworm, is a *hermaphrodite.* When testes and ovaries are found in separate individuals, as is most often the case in higher animals, males have testes and females have ovaries.

The critical feature of sexual reproduction, as contrasted with asexual reproduction, is that the various progeny of the same parents are *not* genetically identical. They do not form a clone; they are genetically distinct from one another. The reason is that gametes are formed by *meiosis,* not mitosis. As we shall see, meiosis assures that the cells produced are genetically different one from another; whereas cells produced by mitosis are genetically identical.

Another form of asexual reproduction occurs in species that have no males. This is the parthenogenic development of unfertilized eggs. In the case of bees and ants, however, fertilized eggs develop into females; whereas unfertilized eggs develop parthenogenetically into males. This form of parthenogenesis is actually a modified type of sexual reproduction. Parthenogenetic male bees and ants are *not* genetically identical; they are as different from one another as are offspring resulting from normal reproduction, except that they are haploid rather than diploid. (As an aside, identical twins, even in humans, form a clone. One twin is asexually reproduced from the other, even though we cannot determine which twin is the "parent" and which is its asexual offspring.)

Since in this text the emphasis in our discussions of development will be largely devoted to the development of zygotes, we shall first focus on sexual reproduction, which begins with the union of gametes. The question is: can any cell of an early embryo differentiate, at the appropriate time, into a gamete, or is the capacity to form germ cells set off early in development? In other words, are *primordial germ cells* different from body *(somatic)* cells even at the earliest stages of embryonic development? In the latter part of the 19th century, August Weismann (1834–1914) and Theodor Boveri (1862–1915), among others, effectively demonstrated that the *"germ plasm"* is in fact set off from the *"somatoplasm"* early in development. In *Ascaris*, a large parasitic roundworm, Boveri observed a difference in the form of the *chromosomes* in one of the two first cells of the early embryo. One cell is larger than the other embryonic cell. Its nucleus and even its individual chromosomes are larger than those of the other embryonic cell. Through the first five successive cleavage divisions, the larger cell divides unequally, each time forming one cell like itself, and one "normal" cell. After the fifth division (32-cell stage), the larger cell, now called the *primordial germ cell*, gives rise only to cells like itself, and these eventually migrate to become incorporated in the gonads. Only from these primordial germ cells can gametes be derived. No more clear-cut example of the early separation of germ line from somatic cells can be found in any organism. The germ plasm provides the direct continuity between successive generations. Should the germ plasm not be set apart in the developing embryo, the adult will be sterile.

CELL DIVISION

When an animal or plant cell divides, two separate and distinct activities occur. First, the *nucleus* of the cell undergoes mitosis. This involves a characteristic series of changes, easily studied with a microscope, that ultimately yield two identical nuclei where only one existed before. That is to say, the nucleus reproduces itself. Second, by a process called *cytokinesis,* the *cytoplasm* of the cell divides into two parts, each containing one of the two nuclei. As a result of nuclear division and cytokinesis, two cells are formed from one. Normally, the reproduction of the nucleus and the cleavage of the cytoplasm are nicely synchronized. Under special conditions, however, either event can occur without the other.

Since animal cells are typically bounded by elastic membranes, cytokinesis can be carried out simply by constriction. A furrow appears around the middle of the cell and gradually deepens until the cell is completely divided. By contrast, plant cells possess rigid cell walls, which prevent constriction. After the nucleus has reproduced in a plant cell, a new cell wall is laid down between the daughter nuclei. At first only a thin plate, the new wall gradually thickens and hardens until the original cell is completely divided into two separate cells. Cytokinesis is never an exact process, since substances found in the cytoplasm are not evenly distributed throughout the cell. The two daughter cells, therefore, never get identical amounts of the various cytoplasmic constituents. In extreme cases, one daughter cell may receive a great deal more cytoplasm than the other *(unequal cleavage).*

The division of the nucleus is much more precise than cytokinesis. The changes that occur in the nucleus involve the appearance, exact duplication, and separation of the *chromosomes,* which bear the *genes.* The significance of nuclear division lies in the precise manner by which the chromosome activities are repeated at each cell division. This precision results in a regular and predictable distribution of the chromosomes to the daughter cells.

THE CELL CYCLE

Early embryonic development consists primarily of successive cycles of mitotic cell division, beginning with the fertilized egg. The *cell cycle,* then, consists of the activities that occur during one full cell division.

Let us begin the description of the cell cycle at the time between actual divisions—the *interphase.* Although this stage is sometimes referred to as the "resting" stage, it is actually a time of intense cellular activity. The only "resting" aspect is that the cell is not dividing. During interphase, following an initial G_1 stage, an important activity occurs in preparation for division: each individual chromosome present in the nucleus duplicates itself exactly. This *synthesis* of new chromosome material occurs during the *S* stage of interphase, and the actual division will begin after a period called the G_2 stage. The period of mitosis then follows. Mitosis takes place in four stages—*prophase, metaphase, anaphase,* and *telophase*—during which time each original chromosome and its duplicate separate from one another. As a result, the two daughter cells formed from the mitotic division of the mother cell receive exactly equivalent sets of chromosomes. After mitosis is over, each daughter cell returns to the interphase condition and remains in the G_1 stage until chromosome synthesis begins at the next *S* phase.

This cycle of activity is diagrammed in Figure 2–1. The relative lengths of the different periods can vary considerably in different kinds of cells. The overall cycle can be as short as 10 minutes in an early *Drosophila* embryo, hours at later stages or in other kinds of eggs, and days or even longer in still other kinds of cells.

MITOSIS

Mitosis is the more common of the two distinctly different kinds of nuclear division that may occur, and takes place whenever somatic, or body, cells divide, whether in animals or plants. Mitosis is frequently described as *equational division,* because each daughter cell receives exactly the same

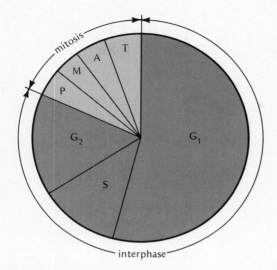

Figure 2-1. The cell cycle. In the cell cycle, G_1, S, and G_2 comprise the interphase. Mitosis consists of prophase (P), metaphase (M), anaphase (A), and telophase (T). The actual division of the cell (cytokinesis) normally occurs at telophase.

number and kind of chromosomes—hence genes—as the original cell had. Therefore, all of the cells of the body should have exactly the same chromosome and gene content as the fertilized egg, or *zygote*, because all body cells are derived from the zygote by mitosis. Mitosis is accomplished in one cycle of division. It can take place no matter what chromosomes the original cell contains, as long as the cell is viable and able to divide.

MEIOSIS

Meiosis, the only other form of nuclear division possible in eukaryotic organisms, is a specialized modification of mitosis associated with the formation of gametes. Unlike mitosis, meiosis is a *reductional division* and occurs normally only when the original cell is *diploid*—that is, having its chromosomes in *homologous pairs*. Daughter cells produced by meiosis receive only half the number of chromosomes and genes that the original cell had. More precisely, the products of meiosis contain one of each original pair of chromosomes. Meiosis occurs during the formation of gametes (sperm or eggs) in animals and during the formation of *microspores* and *megaspores* in higher plants. Gametes and spores are necessarily *haploid*, having only one chromosome of each pair.

In sexual reproduction, two gametes fuse in fertilization to form a new individual. Because haploid sperm and eggs contain only half as many chromosomes as do somatic cells, fertilization reestablishes the original diploid chromosome number in the zygote. If meiosis did not occur, the

(*Text continues on page 14.*)

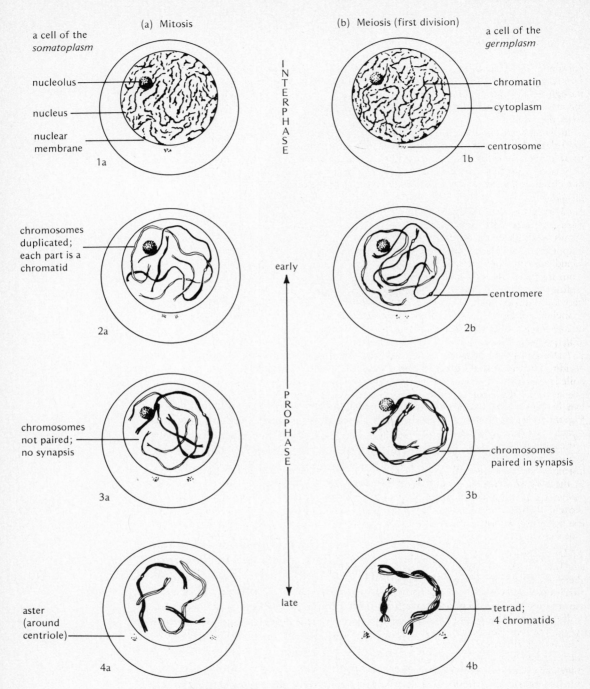

Figure 2-2(a). Cell division.

Mitosis and the First Meiotic Division

The chromosome activities that can be observed in generalized (sexless) animal cells during mitotic cell division and during the first division of meiosis are shown in the parallel series of drawings in Figures 2–2(a) and 2–2(b). These activities are divided into five stages: (1) *interphase*, (2) *prophase*, (3) *metaphase*, (4) *anaphase*, and (5) *telophase*. Even though cell division is a continuous process in which one phase blends into the next, each phase exhibits individual characteristics by which it can be recognized and distinguished from the others.

In living cells, the nucleus appears empty because the chromosomes of dividing cells are relatively transparent and hard to see. The cells, however, can be be stained with various dyes, revealing the chromosomes as dark red or black structures. In the diagrams, only two pairs of chromosomes are shown, one pair longer than the other. For contrast, one chromosome of each pair is drawn darker than the other.

Interphase

1a (*Mitosis*) and 1b (*Meiosis*). During interphase, the nucleus of a cell about to divide by mitosis is essentially identical to that of a cell ready to divide by meiosis. The nucleus shows no visible evidence of chromosomes. However, the nucleus is not as structureless as it appears to be. The chromosomes are present as submicroscopically fine strands, and a round *nucleolus* (some cells have more than one) is also present. The nucleus is separated from the *cytoplasm* by a *nuclear membrane*. (The cytoplasm contains many structures in addition to the *centrosome* shown, but since they do not pertain directly to cell division, they are not illustrated.) Chromosome duplication takes place during the *S* period. Even though the duplicated chromosomes are essentially invisible in interphase, each consists of 2 *chromatids*, held together at the centromere region.

Early Prophase

2a (*Mitosis*) and 2b (*Meiosis*). In mitosis, the early prophase is much like that in cells dividing by meiosis. As division begins, the long, thin chromosome threads become coiled, so that they become visible under the microscope as elongated structures. In most preparations, the chromosomes appear as single strands; however, in favorable cases they can be seen to be doubled, as shown here. The duplicate *chromatids* are held close together. The chromosome behaves as a single unit as long as the two chromatids remain attached at their centromere regions. During early meiotic prophase, the cell grows noticeably in size. The prophase chromosome activities of the first meiotic division are more complex than those of mitotic prophase. Five meiotic prophase substages are normally recognized. (1) In the *leptotene* stage, the chromosomes are unpaired, as in 2b. (2) In the *zygotene* stage, homologous chromosomes become *synapsed*, as in 3b. (3) In the *pachytene* stage, the paired chromosomes, or *tetrads*, become thicker and *crossing over* (not shown) occurs, as in 4b. (4) In the *diplotene* stage, tetrads "open up" as synapsis becomes less tight (not shown). (5) In *diakinesis*, the tetrads reach a state of maximum compactness (not shown).

Middle Prophase

3a (*Mitosis*). As the chromosomes progressively shorten and thicken by tighter coiling, their double nature becomes more evident. In mitosis, each individual chromosome carries out its own activities without reference to the behavior of its partner. Indeed, chromosomes do not have to be in pairs in order for normal mitotic division to occur.

3b (*Meiosis*). Shortly after the visible appearance of the chromosomes (in the leptotene stage), the two members of each chromosome pair attract one another and come to lie side-by-side in *synapsis*. This is accomplished by a submicroscopic *synaptonemal complex* (not shown) that develops between the duplicated chromosomes of each pair. At this stage (zygotene) the chromosomes are so intimately synapsed that, in actual cells, it is usually impossible to see that, in fact, two duplicated chromosomes, rather than just one, are present in each *tetrad*.

Late Prophase

4a (*Mitosis*). As prophase continues, the chromosomes further shorten and thicken as each chromatid becomes more tightly coiled. Each chromosome now visibly consists of two identical chromatids connected to each other only at the region of the centromere. The nucleolus disappears during late prophase, and in the cytoplasm, the *centrioles* move apart as an *aster* forms around each.

4b (Meiosis). The chromatids (and hence the chromosomes) shorten, as in mitosis. The number of tetrads equals the number of chromosome pairs present. In real cells, it is often difficult to distinguish the 4 individual chromatids of a tetrad.

When the nuclear membrane ruptures at the end of prophase, fibers, actually microtubules, extend between the separating centrioles of the two asters to form the *spindle,* which provides the mechanism for separating the chromosomes at anaphase. Its position determines the plane of cleavage and whether the cytoplasm will divide equally or unequally.

(Text continued from page 11.)

chromosome number would double every generation. Two complete cycles of division are required for meiosis; thus, the *first meiotic division* must be distinguished from the *second meiotic division.*

We now present a diagrammatic representation of the chromosome activities associated with mitosis and meiosis in Figures 2–2(a), 2–2(b), and 2–3. These diagrams show, side-by-side, how the chromosome activities differ in mitosis and meiosis. Mitosis produces two daughter cells, each of which has a chromosome constitution exactly identical to that of the mother cell; whereas, meiosis produces (after two successive divisions) four meiotic products, each of which contains only one chromosome of each pair originally present in the mother cell. However, the diagrams represent generalized, sexless cells and, for simplicity, omit some details of meiosis that will be treated in Chapter 4. In particular, Chapter 4 will deal with the differences between the processes producing sperm and eggs.

Although most higher animals have their chromosomes in homologous pairs, some individuals have an abnormal chromosome constitution. People suffering from Down's Syndrome, for example, have an extra, unpaired chromosome. Nonetheless, mitosis successfully produces new cells, each exactly duplicating the abnormal condition. In mitosis, each individual chromosome goes through its sequence of activities without regard to the activities of any other chromosome, including its homologous partner.

A normal meiotic division, on the other hand, *requires* that the chromosomes be present in the diploid, paired condition, and always produces haploid products. "Haploid" is not literally synonymous with "half"; it is a very special half composed specifically of one chromosome of each pair. Consider an organism having only two pairs of chromosomes, one longer than the other. The diploid condition would consist of two long and two short chromosomes, four altogether. This is conventionally symbolized as $2n = 4$ (*read:* the diploid chromosome number is 4) and for the haploid condition $n = 2$ (the haploid chromosome number is 2). However, two long or two short chromosomes do not comprise a haploid set, even though each condi-

(Text continues on page 16.)

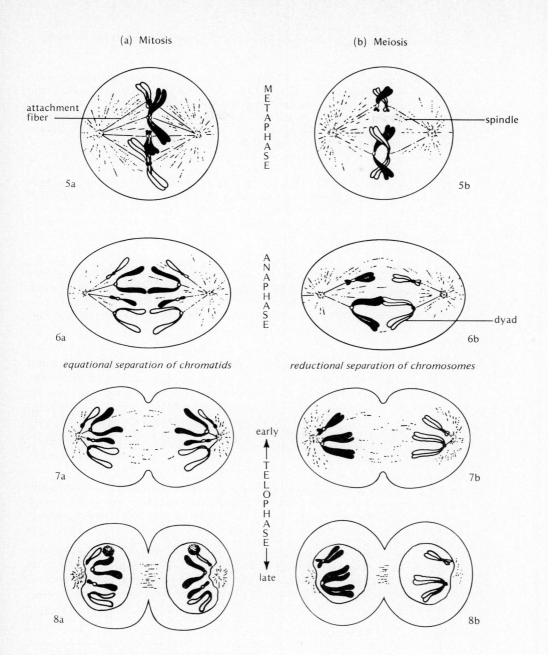

(a) Mitosis   (b) Meiosis

METAPHASE

attachment fiber

spindle

5a   5b

ANAPHASE

dyad

6a   6b

equational separation of chromatids   *reductional separation of chromosomes*

early

TELOPHASE

late

7a   7b

8a   8b

Figure 2–2(b). Cell division, continued.

Metaphase

5a (Mitosis). At the end of prophase, the chromosomes reach their shortest, most compact state and move to the center of the newly formed spindle at the *equatorial plate*. At this time, the double nature of the centromeres of each chromosome becomes evident. Spindle fibers attach to the centromere of each chromatid, connecting them to the opposite poles of the spindle.

5b (Meiosis). The chromosomes move onto the spindle to form the equatorial plate, but

15

remain together as tetrads. Moreover, the sister centromeres of each chromosome act as though they were single, rather than duplicate as they really are. Thus, for each tetrad, the fiber from one pole of the spindle is attached to the functionally single centromere of an entire duplicated chromosome and the fiber from the opposite pole is attached to the centromere of the homologous chromosome. The two chromosomes of each metaphase tetrad, in turn, tend to repel one another, forming X-shaped or circular bodies on the equatorial plate.

Anaphase
6a (Mitosis). During anaphase, the two chromatids that constituted each metaphase chromosome undergo *equational separation* and are drawn toward opposite ends of the spindle by their attachment fibers. The complement of chromosomes going to one pole of the spindle is exactly equal to that going to the other pole. Furthermore, the complement going to either end is exactly like that of the original cell. The chromatids, once separated, are called chromosomes.

6b (Meiosis). Because sister centromeres do not separate during the first meiotic division, the two synapsed chromosomes that formed each metaphase tetrad undergo a *reductional separation* during anaphase. One (duplicated) chromosome of each original pair goes to one pole of the spindle; the other to the opposite pole. Each of these separating bodies, which consist of a duplicated chromosome with effectively one centromere, is called a *dyad*; the daughter cells are *haploid*. (Note that the spindle elongates during anaphase.)

Early Telophase
7a (Mitosis). The chromosomes form a compact *diploid* group at each pole of the spindle. Cytokinesis begins with a furrow in the cell membrane. In plant cells with rigid cell walls, cytokinesis begins as a thin plate forms between the two groups of chromosomes.

7b (Meiosis). The dyads form a *haploid* group at each pole of the spindle. As drawn, the two dark chromosomes and the two light chromosomes are shown together. This need not be the case, as equally often, the orientation of the tetrads on the equatorial plate will be such as to distribute one light and one dark chromosome to each pole. Cytokinesis occurs as in mitosis.

Late Telophase
8a (Mitosis). As cytokinesis nears completion, the chromosome coils loosen, so that the chromosomes appear large and fuzzy. A nuclear membrane forms around the chromosomes in each daughter cell, the nucleolus reappears, and the chromosomes gradually return to the interphase condition. As a result of this equational division, each new cell has exactly the same chromosome and gene content as the original cell.

8b (Meiosis). As in mitosis, the chromosomes become less distinct in late telophase, and a new nucleus forms in each cell. However, in some species, the chromosomes do not return to a true interphase condition; instead, each cell promptly begins the second meiotic division. The period between the first and second meiotic divisions is usually called *interkinesis,* rather than interphase. At the end of telophase, each daughter cell contains only one chromosome of each original pair. Thus, the first meiotic division is reductional, and the daughter cells contain the haploid number of chromosomes. However, each of these chromosomes has already duplicated in preparation for the second meiotic division.

(Text continued from page 14.)

tion has two chromosomes. The true haploid condition must be composed of one short and one long chromosome.

The most important consequence of meiosis is that an individual who develops as a result of the union of male and female meiotic products has a

(Text continues on page 18.)

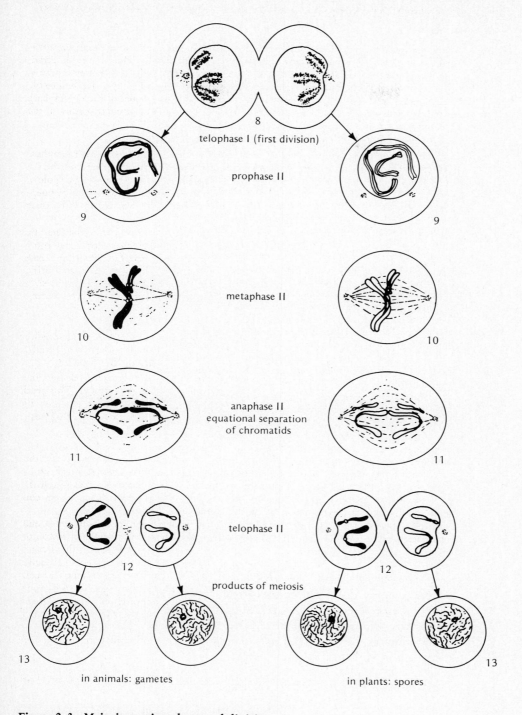

telophase I (first division)

prophase II

metaphase II

anaphase II
equational separation
of chromatids

telophase II

products of meiosis

in animals: gametes

in plants: spores

Figure 2–3. Meiosis continued: second division.

The Second Meiotic Division

In meiosis, two successive divisions take place, but chromosome duplication occurs only

17

once. Hence, the chromosome number in the final daughter cells is only half that of the original cell. The first, reductional meiotic division produces haploid cells from diploid ones, as illustrated in Figures 2–2(a) and 2–2(b). By contrast, the second division is, in effect, equational, like mitosis. It is customary to use the terms prophase I, metaphase I, anaphase I, and so on when referring to the first meiotic division, and to use the terms prophase II, metaphase II, and so on when referring to the second meiotic division. Figure 2–3 illustrates the significant phases of the second meiotic division in a generalized animal cell. (The numbering in Figure 2–3 continues from Figure 2–2(b).)

8. *Telophase I.* (Drawing 8b of Figure 2–2(b) redrawn.) This drawing represents both interkinesis and early prophase II. The two dark chromosomes are shown in one cell and the two light ones in the other, which need not be true.

9. *Prophase II.* The two daughter cells are shown in the middle prophase stage of meiosis II. Note that the two chromatids are still held together by their sister centromeres. In real cells, each chromosome would be called a dyad, but the constituent chromatids are often difficult to distinguish. (There is no cell growth after telophase I.)

10. *Metaphase II.* At the end of prophase II, the nuclear membrane disappears, and the spindle forms as in meiosis I. The chromosomes form an equatorial plate and now, at last, attachment fibers connect with the two centromeres, just as in the metaphase of mitosis. (Compare Drawings 10 and 11 with Drawings 5a and 6a in Figure 2–2(b).) In mitosis, both members of each chromosome pair are present, and the duplicated chromatids of each chromosome separate. In metaphase II, only *one* member of each chromosome pair is present in each cell, and its duplicate parts, chromatids, will separate at anaphase II. The similarity between meiosis II and mitosis is evident; however, no chromosome duplication takes place during interkinesis, and the chromosome condition is haploid, not diploid.

11. *Anaphase II.* An *equational separation* of the chromatids occurs in anaphase II. However, the chromatids are now called chromosomes, as each is independent and has its own centromere.

12. *Telophase II.* Four haploid meiotic products have formed from one original diploid cell. Insofar as chromosomes are concerned, however, only two kinds of cells have actually been produced, one containing "dark" short and long chromosomes and one containing "light" short and long chromosomes. Other cells entering meiosis will produce cells that contain the other two possible combinations of "dark" and "light" short and long chromosomes.

13. Following telophase II, the cells separate and the nuclei return to an interphase condition. In plants, the 4 haploid cells form spores, which later divide by mitosis to form the haploid phase (*gametophyte*) of the plant life cycle. Meiosis in a male animal results in 4 haploid cells, which, without further cell division, transform into 4 functional male gametes, or sperm. Meiosis in a female animal results in only one functional gamete, the *ovum* or egg, and 3 small, nonfunctional polar bodies, which will be described in detail in Chapter 4.

(*Text continued from page 16.*)

genetic content distinctly different from that of either parent. His or her genetic heritage comes half from the father, half from the mother, and is necessarily different from that of either parent alone. An asexually reproduced individual, by contrast, has exactly the same chromosome and gene composition as that of its single parent.

READINGS AND REFERENCES

A good presentation of the chromosome activities during mitosis and meiosis can be found in M. W. Farnsworth, *Genetics* (New York: Harper and Row).

3

GENETICS, DNA, AND GENE REGULATION: A REVIEW

An understanding of cell activities during the period of development requires a knowledge of the basic facts of genetics: gene transmission, the nature of the gene, the mechanism of mutation, the manner of gene action, and the regulation of gene activity. Students who have already taken a course in general genetics can use this chapter for review. Others should study this material to prepare them for the material in subsequent chapters.

CLASSICAL GENETICS: GENE TRANSMISSION

The basic laws of inheritance were first published by Gregor Mendel in 1866. Mendel was an Austrian monk working in a monastery located in Brünn (now Brno in Czechoslovakia), and he raised peas in his monastery garden. Performing very simple experiments that involved crossing one true-breeding variety of garden peas (*Pisum sativum*) with another variety differing in some striking way from the first (for example, *tall* versus *dwarf*), he identified the following genetic properties:

1. The first generation, or F_1, progeny resembled one parental (P_1) type. For example, all of the progeny of the *tall* by *dwarf* cross were *tall*. From this and similar crosses that gave the same kind of result, he concluded that traits like *tall* are *dominant*, others like *dwarf* are *recessive* (Figure 3–1).

2. Upon inbreeding F_1 progeny to produce a second generation (F_2) of progeny, he observed not only that both *tall* and *dwarf* F_2 progeny appeared, but also that the two forms appeared in particular and predictable ratios. In the case of the *tall* and *dwarf* F_2 progeny, three-quarters (75%) were *tall* and one-quarter (25%) were *dwarf*, a 3 : 1 ratio. The F_2 *dwarf* progeny were identical to the original parental *dwarf* progeny (Figure 3–2); however, the *tall* F_2 progeny did not all behave alike when crossed with *dwarf* plants. About two-thirds of the *tall* F_2 plants produced *tall* and *dwarf* progeny in equal proportions (a 1 : 1 ratio) when crossed with *dwarfs*. The other third of the *tall* F_2 progeny behaved like the *tall* parental plants and yielded only *tall* progeny when crossed with *dwarf* plants (Figure 3–3).

To explain these basic facts, Mendel concluded that:

3(a). The appearance of traits such as *tall* and *dwarf* depends on the

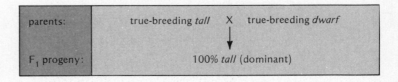

Figure 3–1. A cross of true-breeding *tall* and *dwarf* pea varieties. Note that no *dwarf* progeny are produced in this cross—that is, *dwarf* is recessive.

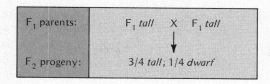

Figure 3–2. A cross of two *tall* F₁ plants. Note that F₁ *tall* plants cannot be genetically the same as parental, true-breeding *tall* plants. Inbreeding the latter would produce only *tall* progeny.

activity of hereditary factors (now called *genes*) that can occur in alternative forms (now called *alleles*). The dominant allele of the gene governing plant growth habit produces *tall* plants; the recessive allele, *dwarf* plants.

3(b). These hereditary factors occur in *pairs*. A gene pair can be composed of two identical alleles (both *tall*, for example, or both *dwarf*) or of two unlike alleles (one *tall*, one *dwarf*). In the latter case, neither allele is modified by the presence of the other—alleles do not fuse with or contaminate one another. However, the dominant allele can express itself even in the presence of the recessive allele, but the recessive allele is masked and cannot be detected by simply observing the plant. Recessive alleles are expressed only when *both* alleles of the pair are recessive.

3(c). At the time of reproduction, only one allele of a pair is transmitted to an offspring by each parent. This separation of allele pairs is called *segregation*, and Mendel's first genetic principle is called *The Law of Segregation*. Its significance is that a gamete receives only one gene of a pair, not both, and that this is true of both sperm and eggs. Fertilization, then, restores the paired gene condition; for every gene pair, one is inherited from the father, the other from the mother.

3(d). Further experiments by Mendel involved crosses where the origi-

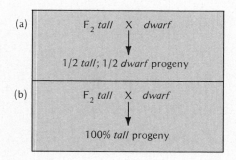

Figure 3–3. Testcrosses of F₂ *tall* with *dwarf* plants. Two-thirds of all F₂ *tall* plants tested followed pattern (a); one-third followed pattern (b). Therefore, the *tall* F₂ plants are not all genetically alike; one-third are like the parental (true-breeding) *tall* plants, and two-thirds are like the F₁ *tall* plants (not true-breeding).

nal parents differed from each other in two or more distinct ways. For example, one parent was *tall* and produced *purple* flowers, the other was *dwarf* and produced *white* flowers (Figure 3–4). Crosses such as these demonstrated that each different gene pair segregates *independently*. That is, when the dominant *tall* allele enters a gamete, it is just as likely to be associated with the dominant *purple* allele of the flower-color gene pair as with the recessive *white* allele. This property of different pairs of genes is Mendel's second genetic principle: *The Law of Independent Assortment*.

3(e). To explain the numerical ratios that he found, Mendel concluded that the union of gametes is not affected by the gene content of the gametes. An egg containing a dominant allele, for example, is fertilized equally well by a sperm carrying a recessive allele as by a sperm carrying a dominant allele: *fertilization is random*.

Mendel recognized (as no doubt you do) the difficulty of describing his crosses in words. To solve that problem, he invented a convention, followed to this day, for symbolizing the different kinds of alleles. Dominant alleles are symbolized by capital letters, recessive alleles by small letters. Thus, the *tall* allele is represented by *T*, the recessive allele by *t* (not *d*, for *dwarf*). Alleles of the same gene *must* be represented by different forms of the same letter. True-breeding parental types are then *TT* and *tt* for the *tall* and *dwarf* parents, respectively. The F_1 progeny, inheriting one allele from each parent, are *Tt*. We characterize the true-breeding parental types (having identical alleles) as being *homozygous* and the F_1 type (*Tt*) as *heterozygous*. Homozygous parents (*TT* or *tt*) produce only one genetic kind of gamete (*T* or *t*), but heterozygous parents (*Tt*) form two different classes of gametes (*T* and *t*) in equal numbers. The genetic constitution of an individual (*TT*, *Tt* or *tt*) is its *genotype*; the physical appearance of an individual (*tall* or *dwarf*) is its *phenotype*.

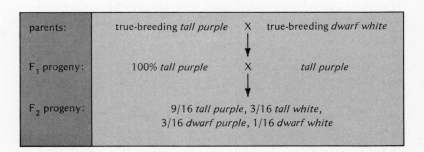

Figure 3-4. A cross of *tall purple* with *dwarf white* pea varieties. Note that the F_2 progeny include both *tall white* and *dwarf purple* plants, although neither combination was present in the parental or F_1 generations.

As an example, let us symbolize the first Mendelian cross involving the contrasting traits, *tall versus dwarf*. This is a *monohybrid* (one pair of genes) cross (Figure 3–5). Note that the three different genotypes in the F_2 generation are present in a 1 : 2 : 1 ratio (1 *TT*, 2 *Tt*, 1 *tt*), but the two phenotypes are in a 3 : 1 ratio (3 *tall*, 1 *dwarf*). Inspection alone cannot distinguish the homozygous (*TT*) *tall* from the heterozygous (*Tt*) *tall* F_2 progeny. The two types of *tall* F_2 progeny, however, can be differentiated by a *testcross* to a recessive *dwarf* plant (Figure 3–6). Two of every three *tall* F_2 plants yield both *tall* and *dwarf* progeny in a 1 : 1 ratio; these F_2 plants are heterozygous *Tt*. One of every three *tall* F_2 plants yields only *tall* progeny; these plants are homozygous *TT*.

Figure 3–7 is a diagram of a *dihybrid* cross (two pairs of genes). Note that in the F_2 generation, nine different genotypes appear in a 1 : 2 : 1 : 2 : 4 : 2 : 1 : 2 : 1 ratio (1 *TT PP*, 2 *TT Pp*, 1 *TT pp*, 2 *Tt PP*, 4 *Tt Pp*, 2 *Tt pp*, 1 *tt PP*, 2 *tt Pp*, 1 *tt pp*). However, only four phenotypes appear in a 9 : 3 : 3 : 1 ratio (9 *tall purple*, 3 *tall white*, 3 *dwarf purple*, 1 *dwarf white*).

A dihybrid cross consists simply of two monohybrid crosses occurring simultaneously. Note that if we ignore the flower-color alleles in the *tall purple* × *dwarf white* cross, the F_2 genotypes are 4 *TT*, 8 *Tt*, and 4 *tt* (a 1 : 2 : 1 genotypic ratio) and the F_2 phenotypes are 12 *tall*, 4 *dwarf* (a 3 : 1 phenotypic ratio). The same is true if we consider only the flower-color alleles and ignore the plant-height alleles.

Any dihybrid (or higher) cross can be analyzed by identifying the monohybrid components and *multiplying* the monohybrid results together to

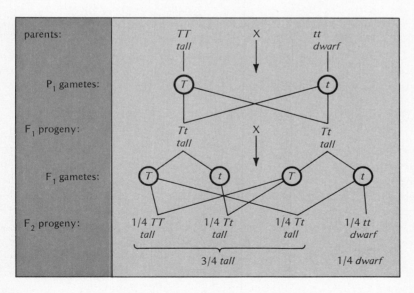

Figure 3–5. A diagram of the cross between true-breeding *tall* and *dwarf* pea varieties. Compare this figure with Figures 3–1 and 3–2.

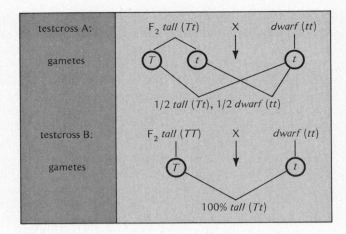

Figure 3–6. A diagram of the testcross of *tall* F$_2$ progeny. Two-thirds of all the *tall* F$_2$ progeny are heterozygous *Tt*. When they are testcrossed with *dwarf* (*tt*) plants (a), half of the testcross progeny are *tall* and half are *dwarf*. One-third of the *tall* F$_2$ plants are homozygous *TT* (b). When they are testcrossed with *dwarf* (*tt*) plants, all of the testcross progeny are *tall* (*Tt*).
Compare this figure with Figure 3–3.

give the dihybrid results. For example, in the *tall purple* × *dwarf white* cross, three-quarters of the F$_2$ progeny are *tall* and one-quarter are *dwarf*. Likewise, three-quarters of the F$_2$ progeny are *purple* and one-quarter are *white*. Thus, $^3/_4$ × $^3/_4$ = $^9/_{16}$ will be *tall purple*, $^3/_4$ *tall* × $^1/_4$ *white* = $^3/_{16}$ *tall white*, $^1/_4$ *dwarf* × $^3/_4$ *purple* = $^3/_{16}$ *dwarf white*, and $^1/_4$ *dwarf* × $^1/_4$ *white* = $^1/_{16}$ *dwarf white*. If we crossed *Aa Bb cc Dd* × *Aa bb Cc dd*, whatever the traits are that each gene pair controls, we would find that progeny showing the dominant phenotype for each trait would constitute $^3/_4$ *A* × $^1/_2$ *B* × $^1/_2$ *C* × $^1/_2$ *D*, or $^3/_{32}$ of the total.

NEO-MENDELIAN GENETICS

Mendel's original paper was essentially ignored by the biological community, and his contributions were not appreciated until the year 1900. In that year, his paper was "rediscovered" by a trio of investigators, each of whom essentially repeated Mendel's work and drew the same conclusions, but, to their chagrin, 34 years too late. With the rediscovery of Mendelism, the science of genetics took off, and the basic Mendelian principles were soon verified in many different kinds of plants and animals, including *Homo sapiens*.

Before long, however, results occurred that seemed to deny the universality of Mendelian principles. For example, not all crosses gave offspring resembling one parent but not the other—that is, dominance was not always true. In zinnias, crosses of *red-flowered* with *white-flowered* plants yield *pink*

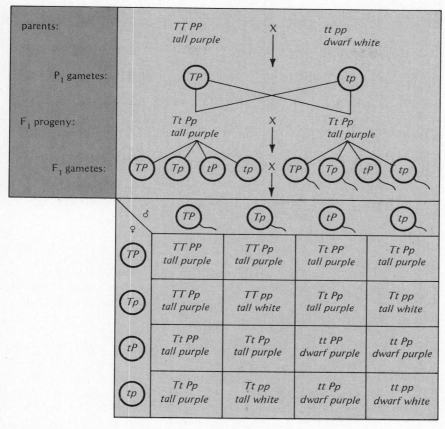

Phenotypic and genotypic summaries of F$_2$ progeny:

9/16 *tall purple*, 3/16 *tall white*, 3/16 *dwarf purple*, 1/16 *dwarf white*
1/16 *TT PP*, 2/16 *TT Pp*, 1/16 *TT pp*, 2/16 *Tt PP*, 4/16 *Tt Pp*, 2/16 *Tt pp*,
1/16 *tt PP*, 2/16 *tt Pp*, 1/16 *tt pp*
Compare this figure with Figure 3-4.

Figure 3–7. A diagram of a cross between true-breeding *tall purple*-flowered and *dwarf white*-flowered pea varieties.

offspring, not *red* or *white* (Figure 3–8). Here the heterozygote has a distinctly different phenotype from that of either homozygous parent, a relationship that has been variously described as *incomplete dominance, lack of dominance,* or *intermediate inheritance.* Many examples of this type of interaction between the two alleles of a pair are known; complete dominance is not universal.

Two further exceptions to basic Mendelism appeared in the study of the inheritance of the human ABO blood types. First, the number of different

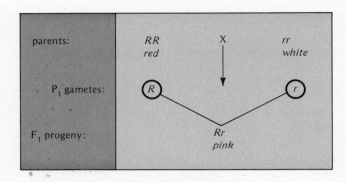

Figure 3–8. A cross of true-breeding *red*-flowered with true-breeding *white*-flowered zinnias. Note that each genotype (*RR*, *Rr*, and *rr*) has a distinctive phenotype (*red*, *pink*, and *white*). No testcross is necessary to determine the genotype of any individual plant.

kinds of alleles is not limited to two, one dominant and one recessive. In order to explain the inheritance of the four basic ABO blood types—Type A, Type B, Type AB, and Type O—a system of *three* alleles is needed. These are symbolized I^A, I^B, and i, and form a *multiple allelic series*. Since any individual can possess only two alleles, there are six different two-allele combinations of the three alleles: (1) $I^A I^A$, (2) $I^A i$, (3) $I^B I^B$, (4) $I^B i$, (5) $I^A I^B$, and (6) ii. Combinations (1) and (2) produce the Type A phenotype; (3) and (4), Type B; (5), Type AB; and (6), Type O. These dominance relationships are unusual. Although alleles I^A and I^B are both dominant to the recessive i allele, I^A and I^B are codominant—that is, *both* are fully expressed in the blood type of the heterozygous Type AB individual.

To Mendel, the characters or traits produced by different genes were unitary—that is, each different individual phenotypic trait was determined by a separate gene. After 1900, however, many examples appeared in which several genes were involved in determining a single trait—a process known as *gene interaction*. One particularly good example is the inheritance of eye color in the fruit fly *Drosophila*. The normal eye color, which is a dull brick-red, is produced by the deposit of two separate and chemically different eye pigments, one red and the other yellow or brownish. The recessive allele for *white* eye color, in the homozygous state, eliminates both pigments, leaving the eye white; a nonallelic recessive gene for *vermilion* eye color eliminates only the brown pigment, leaving the eye bright red; and yet another recessive gene eliminates only the red pigment, leaving the eye *brown*. In addition, many other genes reduce the concentration of one eye pigment or the other, or both. As a result of all this, the eye color in different fruit flies can range from pigmentless (white) through every conceivable shade of red and brown and combinations thereof. At least three dozen, and probably more, separate pairs of genes interact to establish the particular shade of eye color expressed by a particular fly.

Still another type of interaction is evidenced by the *white* eye-color gene. The dominant allele of this gene provides the ability to produce pigment. However, when the recessive *white* allele is homozygous, no other eye-color gene can express itself because the eye will be pigmentless. The *albino* gene in mammals is similar; albinism is a recessive trait in which no eye color, hair color, or skin color pigment (*melanin*) can form. Only when pigment can form are other genes able to determine the shade of eye color, hair color, or skin color. In these examples, the absence of pigment is *epistatic* to the expression of any shade of color. Through epistasis, one pair of genes can mask or prevent the expression of other, nonallelic pairs of genes. Epistatic interactions are not limited to pigmentation genes. Baldness is epistatic to curly or straight hair, for instance, as well as to any shade of hair color. Epistasis, in short, describes the same sort of relationship between different pairs of nonallelic genes that dominance describes between alleles of a pair.

CYTOGENETICS

During the period from about 1880 to 1900, the manufacture of microscope lenses reached almost the level of sophistication found in contemporary bright-field, research-grade microscopes. Along with improved lenses came progress in techniques of fixing tissues, slicing them exquisitely thin, and differentially staining them to bring out different cell constituents. These new techniques rapidly matured into a new field, *chromosome cytology*. During this period, the chromosome activities of mitosis and meiosis were worked out. It became apparent that chromosomes were typically in diploid pairs in somatic cells, but in the haploid state in gametes.

In 1903, W. S. Sutton drew the very reasonable conclusion that the activities of chromosomes provide the physical basis for the transmission and behavior of genes. He suggested that the two genes of a pair are located one on each chromosome of a pair. This pair of chromosomes separates during meiosis, in the process segregating the genes. Not only does this explain how the invisibly small genes segregate, it also explains how different pairs of genes segregate independently: they need only to be on different chromosome pairs.

Although Sutton's suggestion was not immediately accepted by many early geneticists, by 1916 conclusive proof that genes were physically borne by chromosomes was offered by C. B. Bridges. With this investigation, the science of *cytogenetics* came into existence: a study of the correlation of chromosome behavior with gene transmission—and, more importantly, the effect of chromosome abnormalities on the distribution of genes.

Bridges knew that a chromosome difference existed between males and females of most animals. In males, one particular pair of chromosomes consists of homologs that do not match in size or shape—they can easily be told apart. This pair is composed of *sex chromosomes*, so called to distinguish them from all other chromosomes, which are the *autosomes*. In human males, the sex-chromosome pair consists of a large X chromosome and a much smaller Y chromosome. In females, however, two X chromosomes make up the sex-chromosome pair; females have no Y chromosome at all. Bridges realized that this chromosome condition could provide the genetic explanation for sex determination. Also, if a phenotypic trait could be found whose alleles were distributed differently between the sexes, it could also provide proof that genes were physically located on chromosomes. In birds and Lepidoptera (butterflies and moths), the sex-chromosome condition is reversed: females have the unlike XY pair, and males have the matching XX pair.

Bridges already knew that some traits are differently expressed between the two sexes. For instance, human males are much more frequently afflicted with color blindness than are females. A similar situation is found with the *white*-eyed trait in fruit flies. Traits such as these are said to be *sex linked*. As we now know, sex linkage occurs when the gene involved is physically located on the X chromosome. In flies, when a *white*-eyed male is crossed with a normal *red*-eyed female, all of the F_1 progeny are *red*-eyed; in the F_2 generation, one-quarter of the progeny are *white*-eyed as they should be, but every *white*-eyed F_2 fly is *male*. This can be explained by the inheritance of the sex chromosomes if the gene for *white* eyes is on the X chromosome and the Y

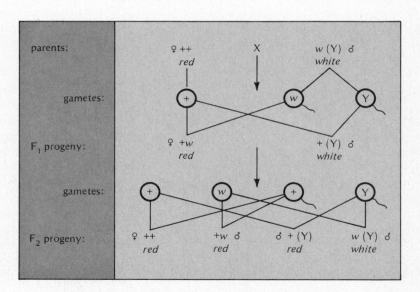

Figure 3-9. Sex-linked inheritance, as illustrated by a cross of a *red*-eyed female and a *white*-eyed male *Drosophila melanogaster*. Because the Y chromosome has no sex-linked genes, males do not have sex-linked genes in pairs: sex-linked genes in males are inherited exclusively from their mothers.

chromosome bears no eye-color alleles (Figure 3–9). (Note that in Figure 3–9, a "+" sign is used to symbolize the dominant or *wild-type* allele for *red* eyes, rather than a capital *W*. This convention is used by *Drosophila* geneticists, but not by botanical geneticists.)

The proof, however, that genes are physically located on the chromosomes awaited Bridges' discovery that a cross between a *white*-eyed male and a *red*-eyed female sometimes produces an unexpected *white*-eyed son, and a cross between a *white*-eyed female and a *red*-eyed male occasionally produces an unexpected *white*-eyed daughter. Such *exceptional* progeny cannot occur if the sex-chromosome condition is normal. Bridges showed that all excep-

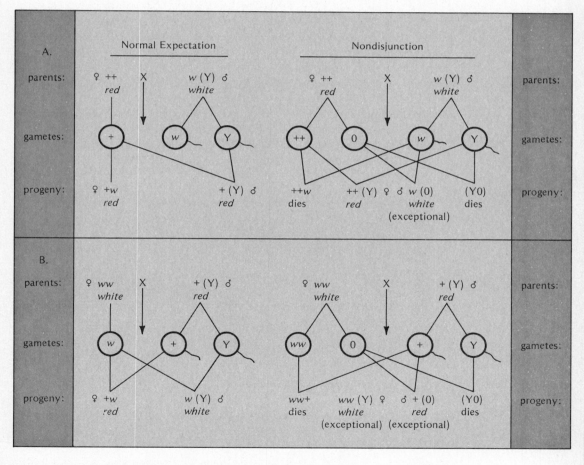

Figure 3–10. X-chromosome nondisjunction and the formation of exceptional progeny. The failure of the two X chromosomes to separate during oogenesis results in viable XXY and XO progeny that possess unexpected (exceptional) phenotypes.

tional progeny have abnormal sex-chromosome constitutions that result from *nondisjunction*, the failure of the partner chromosomes to separate at anaphase I (see Figure 3-10). This constitutes experimental proof that the sex-linked genes are on the X chromosome. In addition, Bridges recognized that meiosis is not a mechanically perfect mechanism; nondisjunction also accounts for the occurrence of Down's syndrome, for example. People with Down's syndrome have an extra nonsex, or autosomal, chromosome. (Many different kinds of abnormal chromosome constitutions are known in humans and other organisms; almost invariably they are associated with developmental abnormalities.)

LINKAGE AND CROSSING OVER

In 1905, two English geneticists, W. Bateson and R. C. Punnett, studied the sweet pea, *Lathyrus odoratus*. They came upon an unexpected result when they followed the inheritance of pollen shape (dominant *long* versus recessive *round*) and flower color (dominant *blue* versus recessive *red*). Each pair of genes showed a typical 3 : 1 ratio in monohybrid F_2 generations. However, in the F_2 of a dihybrid cross of *blue long* by *red round*, they recorded 1528 *blue long*, 106 *blue round*, 117 *red long*, and 381 *red round* progeny rather than the expected 9 : 3 : 3 : 1 ratio. A "perfect" 9 : 3 : 3 : 1 ratio would have been close to 1199, 400, 400, and 133. The difference was far too great to be accounted for by chance; yet, the ratio of both *blue* to *red* and *long* to *round* was fairly close to 3 : 1. This result is shown in Figure 3-11.

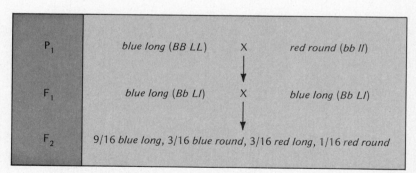

The actual F_2 results were:

   1528 *blue long*, 106 *blue round*, 117 *red long*, 381 *red round*

These numbers cannot possibly represent a 9:3:3:1 ratio. Yet, in the F_2 generation, the ratio of *blue* to *red* was 1634 to 498 and the ratio of *long* to *round* was 1645 to 487. Both of these ratios are fairly close to 3:1, showing that the two pairs of genes segregated properly, but they did not segregate *independently*.

Figure 3-11. A diagram of the phenotypic expectations in a cross of homozygous *blue*-flowered, *long* pollen and *red*-flowered, *round* pollen sweet peas (*Lathyrus odoratus*).

Bateson and Punnett then carried out a *testcross* of the same characters—that is, they bred the doubly heterozygous F_1 parent with a doubly recessive individual (testcross parent). Four phenotypic classes should have been found in a 1 : 1 : 1 : 1 ratio, but the actual numbers approximated a 7 : 1 : 1 : 7 ratio. In general, they noted that the combination of characters that enter the cross together—in this case *blue long* and *red round*—reappear together in F_2 or testcross progeny in much greater proportions than would be expected on the basis of independent assortment; while the new combinations of characters (*blue round* and *red long*) appear together much less frequently than expected. Bateson and Punnett called this nonrandom association of the parental combinations of characters "gametic coupling." Their attempts to explain this phenomenon, however, were completely unavailing, probably because at that time they were not prepared to accept the view that genes were located on chromosomes.

It remained for T. H. Morgan and his students, C. B. Bridges, H. J. Muller, and A. H. Sturtevant, to clear up the problem. Morgan had initiated genetic studies on *Drosophila melanogaster* at Columbia University in the early years of the century. Between 1910 and 1915, many inheritable traits were discovered. These included eye-color mutations such as *white* (w), *brown* (bw), and *vermilion* (v); body color mutations such as *yellow* (y) and *ebony* (e); wing mutations such as *miniature* (m) and *vestigial* (vg); bristle mutations such as *forked* (f) and *singed* (sn), and many others. Since *D. melanogaster* has only four pairs of chromosomes, it was only a matter of time before crosses carried out to study the interrelationships between these new mutants involved cases where the genes were located on the same chromosome. Obviously, crosses of any two sex-linked mutants necessarily involve genes on the same chromosome—specifically, the X chromosome. Since most of the mutants found in the early days were sex linked (they are more easily detected than autosomal mutants), their study soon led not only to a proof that genes were physically located on chromosomes, but also to the identification of a new basic principle of inheritance: linkage.

The various genes on the same chromosome form a *linkage group* and are transmitted together in inheritance. This simple, intuitively obvious statement embodies the concept of linkage. Our knowledge of meiosis with its orderly disjunction of chromosomes at anaphase I assures that genes on the same chromosome cannot assort independently, but must inherit together. For example, if, in a heterozygote, dominant alleles A and B are located on one chromosome of a pair and the recessive alleles a and b on the other, then the gamete that, by chance, receives the chromosome containing allele A must also receive allele B, not b. Similarly, a gamete receiving a must also receive b. With complete linkage, only two classes of gametes will be produced: AB and ab. No gametic combinations of A with b, or a with B, should be possible. This complete linkage of A and B is shown in Figure 3–12.

If this were the full story of linkage, nothing would be simpler. The

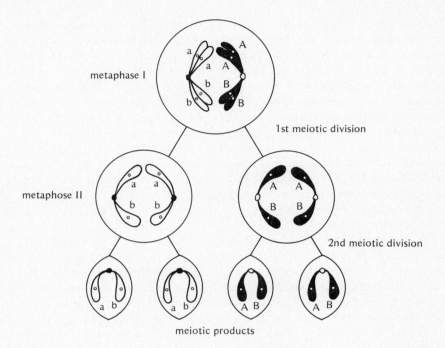

metaphase I

1st meiotic division

metaphose II

2nd meiotic division

meiotic products

Figure 3–12. A diagram of the hypothetical distribution of two completely linked pairs of genes, Aa and Bb, to the four meiotic products. Note that they cannot assort independently.

particular combination of genes present on a given chromosome would pass through successive generations in that same invariant combination.

In fact, however, some new combinations are formed even when genes are linked. In a real case where alleles *A* and *B* are on one chromosome and alleles *a* and *b* are on the other (see Figure 3–12), some gametes with *Ab* and *aB* will be produced in addition to those with *AB* and *ab*. These "escapes" from linkage, or *crossovers*, cause many complications in the study of inheritance. Linkage is reflected in the tendency of genes on the same chromosome to be inherited together; crossing over is the process by which linked genes become separated from one another in inheritance.

Because of crossing over, the inheritance of linked genes cannot be described simply. No general mathematical ratios, such as 3:1, 9:3:3:1, or 1:1:1:1, will be encountered repeatedly in crosses involving different pairs of linked genes. The results of linkage crosses will differ both according to the particular genes involved and to how the characters are combined in the parental generation. For example, in sweet peas, the cross of *blue long* by *red round* (described above) gives a result different from that of a cross of *blue round* by *red long*. Furthermore, even if we determine what these two results are, we cannot predict what the result would be if two entirely different linked characters were crossed. Each result must be determined empirically because the "strength" of linkage between two particular linked genes is likely to be different from that between any other two linked genes.

To gain insight into the process of crossing over, let us follow the inheritance of two sex-linked genes in *Drosophila*: eye color and wing size (Figure 3–13). If a *red*-eyed female with *normal* wings is crossed with a *white*-eyed *miniature*-winged male (*w m*), the F₁ daughters will be doubly

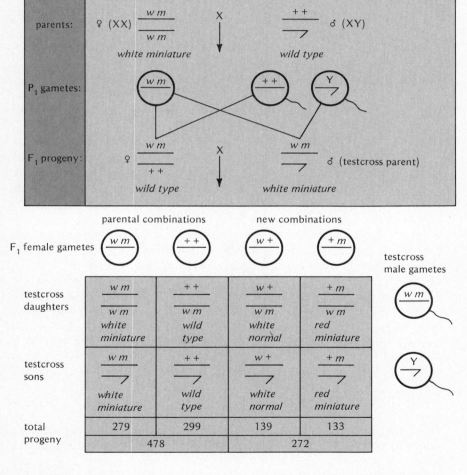

Figure 3–13. A diagram of a cross between a homozygous *white-miniature D. melanogaster* female and a *red-eyed normal-winged (wild-type)* male, together with a testcross of the F₁ female progeny.

heterozygous ($w\,m/++$). Now let us testcross them with *white miniature* males like their father. Because of linkage, there should be equal numbers of *red normal* and *white miniature* male and female offspring, the parental combination of traits. In addition, there should be two new combinations of traits (also in equal, but smaller, numbers) *red miniature* and *white normal*, caused by crossing over.

In a real cross, the four classes of testcross progeny might consist of 279 *white miniature*, 299 *wild-type*, 139 *white normal*, and 133 *red miniature* sons and daughters. Of the 850 total progeny, 272 (32.0%) represent the new combinations—that is to say, 32% of the progeny are *recombinant*, or there was 32% crossing over.

Before interpreting the results of the *white miniature* × *wild-type* cross diagrammed in Figure 3–13, let us examine the results of a cross involving a homozygous *white* female and a *miniature* male, followed by testcrossing the F_1 females with *white miniature* males (see Figure 3–14). The numerical results might well be 88 *white miniature*, 96 *wild-type*, 181 *white normal*, and 176 *red miniature* testcross progeny—that is, 34% recombination. The percentage of new combinations, or recombinants, in the two testcrosses is similar: 32% *versus* 34%.

What is significantly different in the two crosses is the relative numbers of *white miniature* and *wild-type* progeny, as compared with the numbers of *white* and *miniature* progeny. In the first cross (*white miniature* × *wild-type*), *white miniature* and *wild-type* testcross progeny occurred in greater numbers because, in that cross, those were the *parental combinations*. In the second cross, *white* × *miniature*, the *white* and *miniature* testcross progeny occurred in greater numbers for the same reason. In either case, new combinations are less frequent than parental combinations, but in the two crosses these combinations are just reversed. Had the genes not been linked, the four classes of testcross progeny would have been equally numerous in both crosses.

The explanation for the new combinations, or *recombinants*, is that they arise as a result of *crossing over*. This activity takes place during the pachytene stage of prophase I, when two *nonsister* chromatids of a tetrad rupture and *exchange* homologous chromosome segments.

THE CYTOLOGICAL BASIS OF CROSSING OVER

During meiotic prophase I (following the pachytene stage) the synapsed chromosomes relax their tight attraction as the synaptonemal complex is shed. At the ensuing diplotene stage, the *tetrads*, or *bivalent chromosomes*, open up and exhibit shapes such as crosses, rings, and figure-eights (Figure 3–15). Close inspection of these diplotene tetrads shows that the chromatids exhibit an exchange of pairing partners at the point where they cross one another

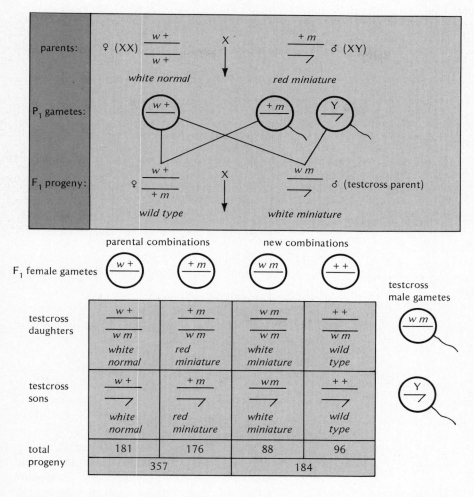

Figure 3–14. A diagram of a cross between a homozygous *white D. melanogaster* female and a *miniature* male, together with a testcross of the F₁ female progeny.

(see Figure 3–16a). This point is called a *chiasma* (plural: chiasmata). A tetrad can have one, two, or more chiasmata.

Two interpretations of chiasmata are possible: (1) The nonsister chromatids simply overlap one another (Figure 3–16b). (2) There has been a physical break and rejoining of the two chromatids (Figure 3–16c). The second interpretation is the correct one. During the pachytene stage (when the four chromatids of the tetrad are tightly synapsed and cannot really be distinguished individually), two nonsister chromatids literally break at exactly the same point along the length of the tetrad and quickly join back

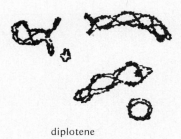

diplotene

Figure 3–15. Grasshopper tetrads at the full diplotene stage of meiosis I in the male. The two chromatids of each homolog are visible; the sites of genetic exchange are the chiasmata, or cross-shaped intersections. (Drawing after a micrograph by B. John)

together in opposite orientation (Figure 3–17). This produces an *exchange* of homologous segments, or *crossover*, between the two chromosomes of the pair. When the tetrad opens up at the diplotene stage, chiasmata become visible because the sister chromatids stay tightly together, while the homolo-

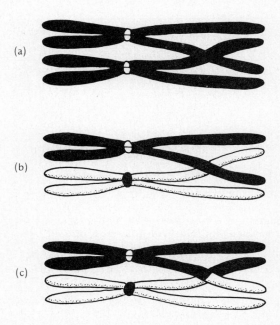

Figure 3–16. The interpretation of a chiasma. (a) A diagram of a chiasma as seen under the microscope. (b) By drawing the two chromosomes of the pair in different colors, the chiasma can be interpreted. One interpretation is that two nonsister chromatids simply overlap. (c) The correct interpretation is that two nonsister chromatids have broken and exchanged connections, and that the sister chromatids have remained together at all points.

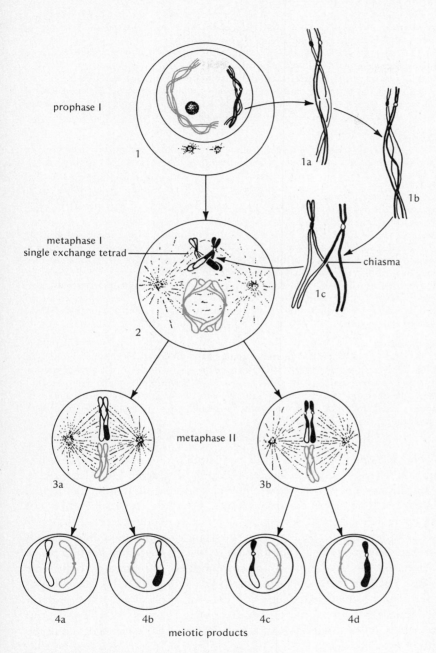

prophase I

metaphase I
single exchange tetrad

chiasma

metaphase II

meiotic products

Figure 3–17. Crossing over: single exchange.

Prophase I
1. In this part, one tetrad is illustrated as it would appear at the pachytene stage, after

chromosome duplication and synapsis. (A second tetrad, representative of the others in the cell, is shown faintly in outline.)

1a. This part shows the tetrad enlarged at a slightly later stage. Two nonsister chromatids have ruptured at identical levels.

1b. The two ruptured nonsister chromatids rejoin to produce an *exchange*.

1c. This part shows the exchange tetrad as it would appear at the diplotene stage. Under the microscope, a tetrad at this stage exhibits a *chiasma*, a visible indication that an exchange of genetic material—a *crossover*—has taken place.

Metaphase I
2. The tetrad has positioned itself in the middle of the spindle, and the two functionally single centromeres become connected with attachment fibers leading to opposite poles.

Metaphase II
3a and *3b.* Two daughter cells are produced as a result of the first meiotic division. At metaphase II, the dyads are oriented on the metaphase spindle. Note that the centromeres have divided and that each dyad is composed of one normal and one *crossover* chromatid.

An inspection of the dyads in 3a and 3b should convince you that reduction was not fully accomplished by the first meiotic division. In diagrams such as these, reduction is represented by the separation of black from white chromosomes. It is apparent that crossing over resulted in the failure of part of the chromosomes to separate reductionally at the first meiotic division. Only in the neighborhood of the centromere is reduction complete; distal to the point of exchange, the chromosomes have separated in an equational manner. For such regions, the second meiotic division will be reductional (*second division segregation*).

Meiotic Products
4a, 4b, 4c, and *4d.* Four haploid meiotic products are produced by the second meiotic division. Each has one chromatid derived from the original tetrad: 4a and 4d possess normal, *noncrossover* strands, but 4b and 4c possess *crossover* strands. Note that from a tetrad in which one exchange occurs, only two of the four resulting meiotic products contain crossover strands. In females, only one of the four meiotic products becomes a functional egg; the other three are polar bodies.

---

gous chromosomes appear to repel one another. A chiasma is the visible indication that a physical exchange of chromatid segments has taken place.

In addition, chiasmata prevent homologous chromosomes from simply falling apart. If no chiasmata are produced, the two chromosomes of a pair come apart and arrive at the metaphase I plate independently, and their subsequent movement to the poles at anaphase I is at random. The result is exactly same as if the chromosomes had never synapsed in the first place. At anaphase I, then, two unsynapsed chromosomes of a pair pass to the same pole just as often as to opposite poles. The failure of effective synapsis (preventing chromatid exchange and chiasma formation) is the primary cause of nondisjunction. Since nondisjunction is relatively rare, however, it follows that at least one chiasma normally forms at some point along the length of every tetrad.

More important, even, than its role in chromosome disjunction, the physical exchange of homologous chromatid segments at pachytene provides the basis for crossing over. If genes are placed on the chromosomes, as in

Figure 3–18(b), we can see that an exchange *between* the genes generates two chromatids with a new combination of alleles; the other two chromatids retain the parental combinations. Suppose, however, that the exchange happens at a position along the chromosome *not* between the genes, as in Figure 3–18(a). All four chromatids will then retain the parental combination

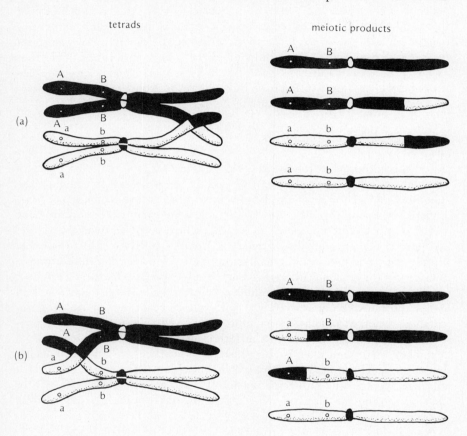

Figure 3–18. The relationship between exchange (crossing over) and gene recombination.

In part (a), exchange *is not* between the loci of genes *A* and *B*; therefore, no new combinations of genes are produced.

In part (b), exchange *is* between genes *A* and *B*; therefore, two new combinations of genes are produced.

Although an *exchange* occurred in both (A) and (B), only in (B) is a *new combination* (recombination) of genes produced. Recombination requires that an exchange, or crossover, occur during the pachytene stage of Prophase I. However, exchange does not guarantee that a new combination of genes will occur, because recombination requires that the exchange take place *between* the loci of the genes involved.

of genes and none will show the new combinations, even though an exchange has occurred in the tetrad. Since an exchange can occur anywhere along the chromosome, only some fraction of all exchanges will occur between two particular genes on the same chromosome. The fraction will be large if the genes are physically far apart, small if they are close together. Each different gene occupies a specific position, or *locus* (pl. loci), on the chromosome; therefore, the frequency of recombination—that is, the percentage of cross-over combinations—reflects the physical distance between any two genes on the same chromosome. The results of many linkage testcrosses have been used to produce linkage maps that indicate the sequence and relative separation of the genes constituting each linkage group. A linkage map for *Drosophila melanogaster* is shown in Figure 3–19.

MOLECULAR GENETICS

All of the additions to genetic knowledge that followed the rediscovery of Mendelism in 1900 took place and were fully comprehended by about 1920. The major unsolved problem, however, was the nature of the gene. Most geneticists during the 1920s and 1930s—and even early 1940s—would have bet that genes were composed of *proteins*, which were known to be large, ubiquitous macromolecules formed by linking together many relatively simple *amino acid* molecules, one after another. Living organisms form proteins from, altogether, 20 different amino acids. Genes certainly possess great *specificity*, because so many different traits are controlled by so many different genes. Proteins exhibit specificity. Different *enzymes*, which are proteins, catalyze different metabolic reactions. Different *antibodies*, which are proteins, are needed to react with different *antigens*. Furthermore, Friedrich Miescher, an organic chemist, demonstrated as early as 1871 that cell nuclei, including sperm, contained only two chemical substances in any significant amounts: protein and "nuclein," an acidic substance containing phosphorus that, according to Miescher, was "not comparable with any other group known at present." Today, we call nuclein *nucleic acid*. Thus, only two choices were possible: (1) genes are protein, or (2) genes are nucleic acid.

The problem was that genes, whatever they are made of, must be able to replicate themselves exactly, but no one could visualize a mechanism that would permit proteins to do this. On the other hand, most people believed in the early days that nucleic acid did not have sufficient specificity to serve as the gene. Impasse!

Until 1944, no clear choice could be made. Much was learned about the actual chemical composition of nucleic acid in the intervening years, and two different classes of nucleic acid were identified: deoxyribonucleic acid (*DNA*) and ribonucleic acid (*RNA*). DNA was mostly confined to the nucleus, but RNA (and proteins) were more prevalent in the cytoplasm,

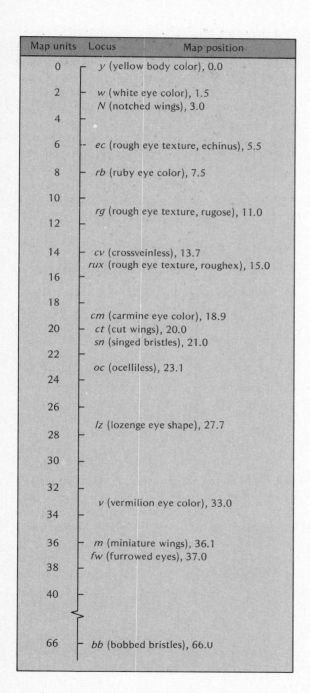

Map units	Locus	Map position
0	*y* (yellow body color), 0.0	
2	*w* (white eye color), 1.5	
	N (notched wings), 3.0	
4		
6	*ec* (rough eye texture, echinus), 5.5	
8	*rb* (ruby eye color), 7.5	
10		
	rg (rough eye texture, rugose), 11.0	
12		
14	*cv* (crossveinless), 13.7	
	rux (rough eye texture, roughex), 15.0	
16		
18		
	cm (carmine eye color), 18.9	
20	*ct* (cut wings), 20.0	
	sn (singed bristles), 21.0	
22		
	oc (ocelliless), 23.1	
24		
26		
	lz (lozenge eye shape), 27.7	
28		
30		
32		
	v (vermilion eye color), 33.0	
34		
36	*m* (miniature wings), 36.1	
	fw (furrowed eyes), 37.0	
38		
40		
66	*bb* (bobbed bristles), 66.0	

Figure 3-19. A linkage map of the *Drosophila melanogaster* X chromosome.

though certainly not excluded from the nucleus. In 1944, however, a team of three investigators, O. T. Avery, C. M. McLeod, and M. McCarty, working at the Rockefeller Institute in New York City, reported their studies on the penumonia bacterium, *Diplococcus pneumoniae*. Their work explained the earlier discovery of F. Griffith (1928) that (1) live avirulent, noncapsulated strains of *D. pneumoniae* had no effect when injected into mice; (2) live virulent, capsulated strains killed mice; (3) heat-killed virulent, capsulated bacteria had no effect on mice; but (4) the simultaneous injection of live avirulent, noncapsulated and dead heat-killed virulent, capsulated bacteria yielded virulent, capsulated bacteria that killed mice. These virulent bacteria did not form because the virulent cells had somehow survived the heat treatment, but because the live nonvirulent forms were *transformed* into virulent cells by some "principle" released from the dead virulent cells that entered the live avirulent cells.

The difference between the two forms of the bacteria was known to be a genetic difference. The question therefore was: What was the transforming principle? Avery, MacLeod, and McCarty extracted all of the different types of substances that could be extracted from dead virulent bacteria—proteins, carbohydrates, lipids, DNA, and RNA in particular—and showed that only DNA was able to transform avirulent into virulent bacteria.

This demonstration that DNA was the genetic material did not win instant acceptance by the genetic community; but time proved it correct. Still, the questions remained: How did DNA replicate itself? How could DNA occur in enough specifically different forms to serve as the genetic material?

THE CHEMICAL NATURE OF DNA

Even before 1944, biochemists were interested in working out the chemical nature of DNA. During the 1930s, for instance, P. A. Levene showed that isolated nucleic acid molecules could be broken down into small subunits called *nucleotides*. Each nucleotide consisted of a 5-carbon sugar (*deoxyribose* in DNA; *ribose* in RNA), a phosphate group, and an organic base containing nitrogen. A nucleic acid molecule, however, contained only four different organic bases, two of them purines (*adenine* and *guanine*, or A and G) and the other two smaller pyrimidines (*thymine* and *cytosine*, or T and C, in DNA; *uracil* and *cytosine*, or U and C, in RNA). These nucleotide structures are shown in Figure 3–20.

In his studies on DNA, Levene was misled into suggesting that one each of the four different DNA nucleotides were linked together to form a *tetranucleotide*, and that numerous tetranucleotides then joined together to form complete DNA molecules. In hindsight, we can see that such a simple, invariant organization could not possibly account for the tremendous specificity required by the genetic material.

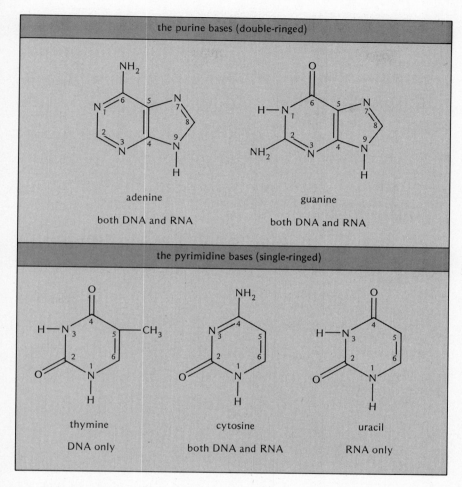

Figure 3-20. Chemical formulas for the five different nitrogenous organic bases found in DNA and RNA. Note that thymine is found only in DNA; uracil only in RNA. The purine bases are adenine and guanine; the pyrimidine bases are cytosine and thymine in DNA and uracil in RNA.

Not until the 1940s was the true picture recognized: The four nucleotides in DNA were not in equal proportions, 25% each. E. Chargaff, among others, clearly showed that the amounts of adenine and thymine were equally frequent (A = T), as were those of guanine and cytosine (G = C). However, samples of DNA from most sources (including humans) had more adenine *plus* thymine than guanine *plus* cytosine (AT rich); whereas, some samples (mostly bacteria, such as *Sarcina lutea*) were just the reverse (GC rich).

The four different kinds of DNA nucleotides are shown in Figure 3-21,

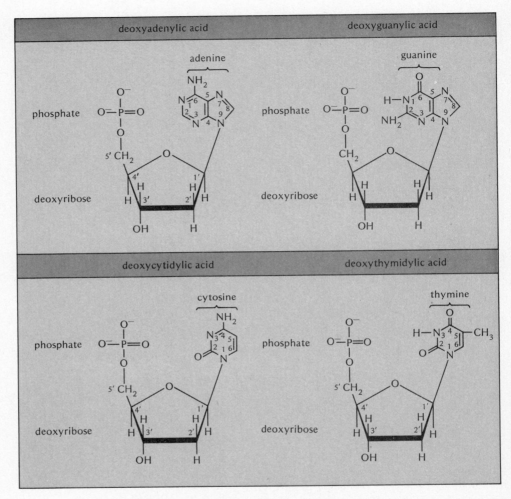

Figure 3–21. The four different DNA nucleotides. Note the numbering system used to indicate the ring positions in the bases and in the deoxyribose formulas. The sugar position numbers include primes to distinguish them from the base position numbers.

and the manner in which adjacent nucleotides are connected together is shown in Figure 3–22.

THE DOUBLE HELIX

Even though the nucleotide structure of DNA had been determined, this did not explain how DNA could provide the specificity needed for genes. After all, there are only four different nucleotides; whereas there are twenty different amino acids that can be linked together in any sequence to form

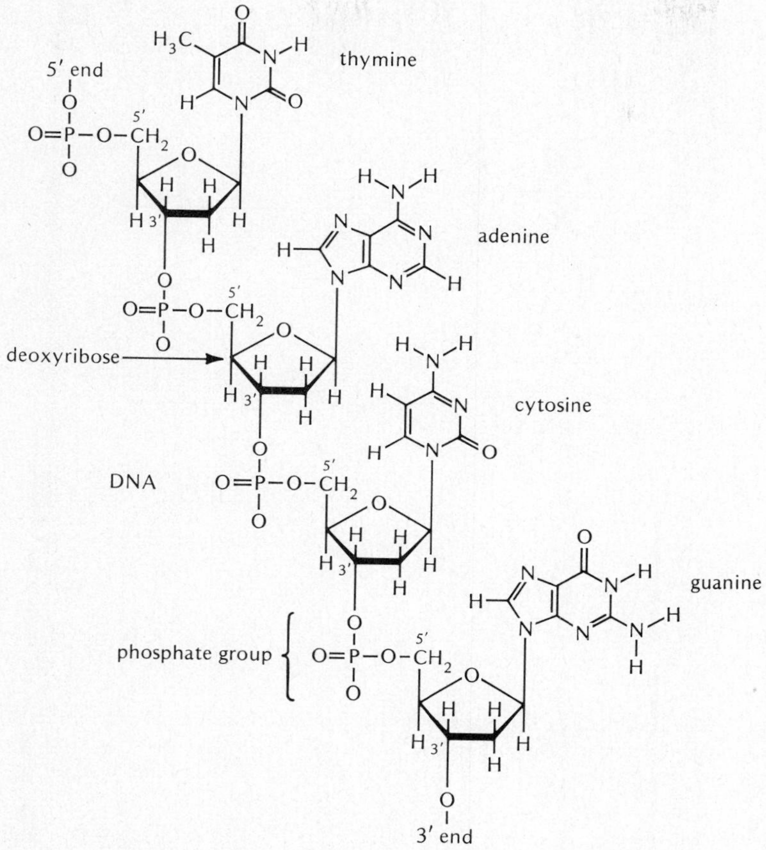

Figure 3–22. The linkage of nucleotides in DNA, showing the sugar-phosphate "backbone." The bases extend laterally. Note that the connection between adjacent bases is always from the 3′ carbon of one deoxyribose molecule through the phosphate group to the 5′ carbon of the deoxyribose molecule of the next nucleotide.

proteins. Moreover, the nucleotide structure provided no insight into how DNA could replicate itself, which is an absolute requirement for the genetic material. The same, however, could be said of proteins.

The solution was not forthcoming until the publication, in 1953, of a paper by James D. Watson and Francis C. Crick in which they proposed the *double helix* model of DNA structure. According to their model, DNA is composed of two adjacent, complementary nucleotide strands spiraled about one another. These strands are held together by specific *hydrogen bonds* between adenine bases in one strand and thymine bases in the other and between guanine bases in one strand and cytosine bases in the other (Figure 3–23). Individual DNA double-helix molecules were found to be extremely

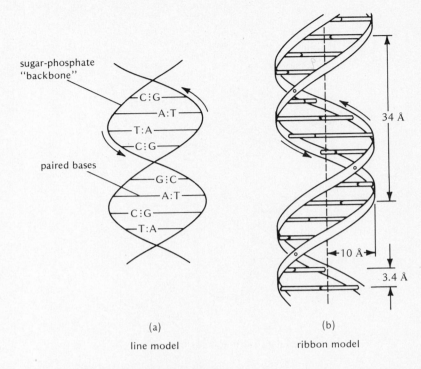

(a)

line model

(b)

ribbon model

Figure 3–23. Diagram of the double helix. Two antiparallel polynucleotide strands are held together by hydrogen bonding between adenine and thymine, and between guanine and cytosine. (*Note:* Two hydrogen bonds hold A to T; three hold G to C.)

long. In some cases, they contained as many as 10^8 nucleotide pairs, occasionally even more, and single DNA molecules can stretch for several centimeters if they are straightened out.

DNA REPLICATION

The double-helix model of DNA was rapidly accepted by the geneticists because, for the first time, it provided a clear answer to the question of how the genetic material could duplicate itself. The answer, suggested by Watson and Crick, was experimentally proven by M. Meselson and F. W. Stahl in 1958. At the time of DNA duplication (the S phase of interphase), each DNA molecule progressively "unwinds" by releasing the hydrogen bonds that hold the two strands together. This is carried out by specific "unwinding" proteins. The bases exposed by the separation can then bond with the appropriate *complementary* bases carried by the free nucleotides present in the cell, A with T, and G with C (Figure 3–24). The final result is two double-helix molecules where there was but one before. Each new molecule

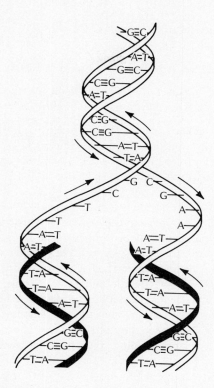

Figure 3–24. DNA replication. Hydrogen bonds holding the two strands of the double helix together are released, and each of the two separate strands serves as a template for the assembly of a new double helix. (The newly formed DNA strands are shown in solid color.)

is identical to the original one in nucleotide sequence, and each is composed of one "old" and one newly synthesized strand. This is a *semiconservative* replication process. A conservative replication, by contrast, would be one in which the whole "old" double-helix molecule would serve as a model for synthesizing an entirely new double helix. However, as Meselson and Stahl proved, the true manner of DNA replication is semiconservative.

RNA

After Watson and Crick reported the structure of DNA, the genetic community turned its attention to RNA. What was the alternative form of nucleic acid good for? RNA was already known to differ from DNA in several major respects:

1. Most importantly, RNA is *not* in double-helix form, and there are no rules restricting the relative amounts of the four different RNA nucleotides.

2. The thymine of DNA is replaced by uracil in RNA.

3. The deoxyribose sugar of DNA is replaced by ribose (having one more oxygen atom than deoxyribose).

4. RNA is found throughout the cell—most plentifully in the cytoplasm—and is not confined to the nucleus as is DNA. (Actually, some DNA is found in mitochondria and chloroplasts and accounts for cases of non-Mendelian inheritance, especially in plants.)

As early as the 1940s, a correlation had been noticed between the protein production of cells and the amount of cytoplasmic RNA that they contained. Cells that produce large amounts of protein, such as liver, pancreas, and silk-gland cells, contain a relatively large amount of cytoplasmic RNA; cells that produce little protein, such as kidney, heart, and lung cells, contain much less. Furthermore, in 1941 G. W. Beadle and E. L. Tatum, after studying *Neurospora crassa* mutants that could not synthesize specific *enzymes*, suggested that specific genes control the production of specific enzymes. Indeed, their thesis was epitomized by another catchy phrase: "one gene, one enzyme." Thus, a relationship between genes and protein synthesis seemed self-evident.

There was one significant problem, however. With minor exceptions, DNA is confined to the nucleus (in fact to the chromosomes), whereas protein synthesis occurs in the cytoplasm, in association with submicroscopic organelles called *ribosomes*. Thus, if DNA controls protein production, there *must* be some sort of intermediary to transport genetic information from DNA to the cytoplasm where the information can be used to assemble protein molecules. The intermediary turned out to be RNA.

Unlike DNA, RNA occurs in three different forms, each involved in one way or another with protein synthesis. The three kinds of RNA are: (1) *messenger* RNA (mRNA), (2) *ribosomal* RNA (rRNA), and (3) *transfer* RNA (tRNA).

Compared with DNA molecules, all RNA molecules are relatively small. Transfer RNA molecules contain only 80 or so nucleotides, even though mRNA and rRNA molecules can be composed of a few thousand nucleotides.

All RNA molecules are produced by copying from DNA; they cannot replicate themselves. (Certain viruses, however, use RNA as their genetic material.) First, a section of the DNA double helix unwinds, as it does in preparation for DNA duplication. In this case, however, the exposed bases of one of the two DNA strands, the *sense* strand, hydrogen-bond with complementary bases of RNA nucleotides present in the cell. The base uracil (U) in

RNA substitutes for thymine (T) of DNA; thus, the complementary base pairs in RNA synthesis are AU and GC rather than AT and GC (Figure 3–25).

If the stretch of DNA copied in this manner is a regular (*structural*) gene, such as one for an eye-color pigment, the resulting RNA molecule is messenger RNA. It contains (although in complementary RNA language) the same information that was present in the gene it copied. When completely synthesized, the mRNA molecule is released from the DNA and moves from the nucleus into the cytoplasm. The other two types of RNA are copied from specific DNA regions that serve only as the templates for rRNA and tRNA synthesis. When released from the DNA, rRNA enters the nucleolus, where it accumulates and (together with various proteins) forms *ribosomes*. (All protein synthesis occurs on the surface of ribosomes.) Usually, three different size classes of rRNA molecules are formed and distribute themselves in various proportions among the large and small subunits that together form each ribosome (Figure 3–26). The final type of RNA, transfer RNA, occurs in about 40 different varieties. tRNA molecules can combine with and "carry" specific amino acids that are used in assembling proteins. tRNA genes are distributed throughout the genome.

In general, the role of the three types of RNA is to assemble appropriate amino acids into specific polypeptide chains on the basis of the instructions encoded in DNA.

THE GENETIC CODE

By the late 1950s, the fact that genetic information in DNA is used to assure the formation of specific proteins was clearly recognized, as was the obvious similarity between DNA structure (a sequence of many nucleotides) and protein structure (a sequence of many amino acids). The question, then, was: How is the genetic information of DNA encoded, and how does the cell decode and use that information to construct proteins?

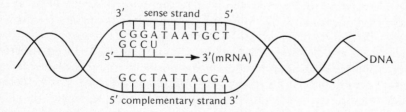

The mRNA is synthesized in a 5' ⟶ 3' direction. Note that its sequence is complementary to the DNA sense strand, and the same as the complementary strand of the DNA (with the exception of uracil).

Figure 3–25. mRNA synthesis.

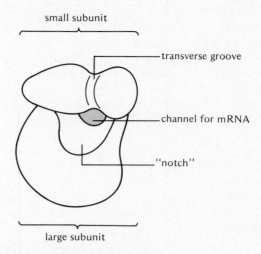

small subunit

transverse groove

channel for mRNA

"notch"

large subunit

Figure 3–26. Eukaryotic ribosome structure.

Since the messages copied in mRNA form are not readily understood by students of English (or Chinese, for that matter), they must exist in some coded form: the genetic code. Our problem is to decipher the code in order to learn how the cell uses mRNA, rRNA, and tRNA to make protein molecules.

The most obvious approach was to assume that somehow the linear sequence of DNA nucleotides in a structural gene corresponds directly to the linear sequence of amino acids in the protein encoded by the structural gene. However, two major difficulties with this were readily apparent: (1) each DNA double helix has two different nucleotide sequences, and (2) DNA is composed of only 4 different nucleotides, but proteins are composed of 20 different amino acids in proteins.

The first difficulty was overcome by recognizing that only one of the two DNA strands (the *sense* strand) contains meaningful genetic information; the other, complementary strand is there simply to allow the DNA double helix to replicate itself. The second difficulty was overcome by realizing that, although there could not be a one-to-one correspondence between 4 nucleotides and 20 amino acids, a given amino acid might be encoded by a short sequence of nucleotides. A 2-nucleotide sequence would still be insufficient, because only 16 different two-nucleotide sequences can be formed from four different nucleotides. A 3-nucleotide sequence, on the other hand, would be redundant—64 different 3-nucleotide sequences can be formed from 4 different nucleotides, far more than are needed to encode 20 amino acids.

Nonetheless, 3-letter sequences of nucleotides (*codons*) were proved to encode the 20 amino acids. Many of the 3-letter codons, however, are synonymous. There is more than one correct way to "spell" most of the amino acids, even as in English either "harbor" (American) or "harbour" (British) spellings are perfectly equivalent and never lead to misunderstanding.

The breakthrough in deciphering the genetic code came in 1961, when M. W. Nirenberg and J. H. Matthei reported that the amino acid phenylalanine was encoded by the three-letter codon UUU in mRNA. That, of course, would correspond to the sequence AAA in the DNA strand from which the mRNA molecule was copied.

In relatively short order, all 64 of the RNA codons were identified, as shown in Figure 3–27. Note that four of the codons carry special significance: (1) AUG identifies the starting, or *initiation*, point in constructing a protein molecule (at other positions, AUG identifies the amino acid methionine). (2) UAA, UAG, and UGA all act as "periods" and identify the termination of the amino acid chain. None of these correspond to an amino acid.

first letter		second letter				third letter
		U	C	A	G	
U		UUU ⎫ phe UUC ⎬ UUA ⎫ leu UUG ⎭	UCU ⎫ UCC ⎬ ser UCA ⎪ UCG ⎭	UAU ⎫ tyr UAC ⎬ UAA ⎫ stop UAG ⎭	UGU ⎫ cys UGC ⎬ UGA stop UGG try	U C A G
C		CUU ⎫ CUC ⎬ leu CUA ⎪ CUG ⎭	CCU ⎫ CCC ⎬ pro CCA ⎪ CCG ⎭	CAU ⎫ his CAC ⎬ CAA ⎫ gln CAG ⎭	CGU ⎫ CGC ⎬ arg CGA ⎪ CGG ⎭	U C A G
A		AUU ⎫ AUC ⎬ ileu AUA ⎭ AUG* met	ACU ⎫ ACC ⎬ thr ACA ⎪ ACG ⎭	AAU ⎫ asn AAC ⎬ AAA ⎫ lys AAG ⎭	AGU ⎫ ser AGC ⎬ AGA ⎫ arg AGG ⎭	U C A G
G		GUU ⎫ GUC ⎬ val GUA ⎪ GUG ⎭	GCU ⎫ GCC ⎬ ala GCA ⎪ GCG ⎭	GAU ⎫ asp GAC ⎬ GAA ⎫ glu GAG ⎭	GGU ⎫ GGC ⎬ gly GGA ⎪ GGG ⎭	U C A G

The abbreviated names of amino acids are as follows: ala = alanine, arg = arginine, asn = asparagine, asp = aspartic acid, cys = cysteine, gln = glutamine, glu = glutamic acid, gly = glycine, his = histidine, ileu = isoleucine, leu = leucine, lys = lysine, met = methionine, phe = phenylalanine, pro = proline, ser = serine, thr = threonine, try = tryptophan, tyr = tyrosine, val = valine, stop = termination codon.

*AUG also serves as the initiation codon.

Figure 3–27. The genetic code dictionary in RNA "language."

PROTEIN SYNTHESIS

Working out the genetic dictionary made possible a full understanding of the mechanism by which the coded information in DNA is first *transcribed* into mRNA, and then (in cooperation with ribosomes and tRNA) *translated* into protein. The secret lies principally in the nature of tRNA molecules.

There are at least 20 (and possibly 40 or so) different types of tRNA molecules. Each can attach to, or "carry," a specific amino acid. Each different tRNA molecule is formed from a different sequence of about 80 nucleotides linked together, but all contain a specific sequence of three nucleotides (CCA) at their 3′ ends, to which the particular amino acid attaches. The tRNA molecules fold themselves into a cloverleaf shape, held together by hydrogen bonds between complementary bases. The middle leaf contains a distinctive sequence of three nucleotides, the *anticodon*. This sequence is complementary to the codon for the particular amino acid carried by the tRNA molecule (Figure 3–28).

Figure 3–29 shows the formation of a particular amino acid chain, or *polypeptide*. A mRNA molecule first attaches to a dissociated small ribosomal subunit in such a way that the AUG codon falls at a specific *tRNA binding site* on the ribosomal subunit. A special tRNA molecule with a UAC anticodon will bond to this site, carrying methionine. This complex is then joined by a large ribosomal subunit, holding the mRNA molecule in place between the two subunits much as a magnetic tape is held at the read head of a tape recorder. On the large subunit are two adjacent binding sites, P and A. As the mRNA AUG codon with its methionine-carrying tRNA molecule comes into position at the P site, a tRNA with an anticodon complementary to the mRNA codon exposed at the A site will also bond to the mRNA molecule. The two amino acids, now adjacent to each other, join together under the influence of a specific enzyme present in the large ribosomal subunit and form a *dipeptide*. The original tRNA molecule is released, and the mRNA molecule moves a distance of one codon relative to the ribosome. This takes the second tRNA molecule, now holding both amino acids, from the A site to the P site. The next mRNA codon is now exposed at the empty A site, where a third tRNA molecule having the appropriate anticodon will bring a third amino acid into position for linkage with the first two. (Most often, the original methionine will be clipped off, because few proteins begin with methionine.)

This process continues until the last exposed A site is filled by a termination codon, for which no tRNA has an anticodon. This triggers the release of the entire polypeptide and the dissociation of the two ribosomal subunits. However, the leading (5′) end of the original mRNA molecule will have usually become associated with other ribosomes long before the synthesis of the first polypeptide is completed. The association of one mRNA

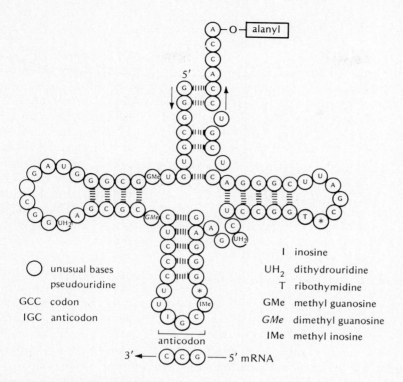

Figure 3–28. A tRNA molecule. Reprinted by permission from Watson, James D., *Molecular Biology of the Gene*, 3rd ed. (Menlo Park, CA: The Benjamin/Cummings Publishing Company, Inc., 1977), p. 306, fig. 12–1. Cited in R. D. Dyson, *Cell Biology: A Molecular Approach.* 2nd ed. (Boston: Allyn and Bacon, Inc., 1978), Fig. 8.13.

molecule with numerous ribosomes at the same time—a *polysome* complex—permits the sequential synthesis of several identical polypeptide chains, one from each ribosome.

GENE REGULATION

The orderly occurrence of mitosis should ensure the presence of identically the same chromosomes and genes in every cell of the organism, since all cells are derived from the zygote by mitosis. Yet cells of the body occur in many different types: blood, bone, muscle, fat, skin, brain, and so on. Each type is distinguished by its own characteristic form and function. If the genes that guide cellular differentiation are identical in all cells, how can the cells become different?

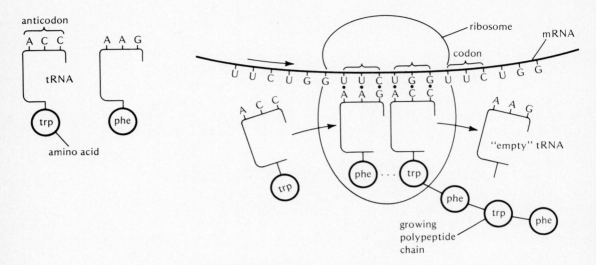

Figure 3–29. Protein synthesis: an outline. (Only the tryptophan (Trp) and phenylalanine (Phe) tRNA are used.)

One might visualize various mechanisms that could produce different genetic activity in cells originally of the same genotype—for example, gene loss from some cells and gene duplication in others. The most reasonable explanation, however, lies in mechanisms of *gene regulation* by which particular genes are "turned on" in some cells and "turned off" in others.

The best-understood cases of gene regulation occur in bacteria such as *Escherichia coli*, which in large part form the microflora of the gut in organisms such as ourselves. *E. coli* is not a green plant and therefore must rely on outside sources for its energy. Glucose, a digestion product of many sugars and starches, provides its usual source of energy, and *E. coli* possesses the enzymes necessary to metabolize glucose. Consider, however, the crisis *E. coli* would face if its host were unable to ingest anything but milk, like a baby. Milk sugar, *lactose*, is not acted on by the same enzymes that metabolize glucose. Different enzymes are needed, especially *beta-galactosidase*, the enzyme that can cleave lactose into its constituent monosaccharides, galactose and glucose.

Nonetheless, *E. coli*, confronted by lactose as its sole sugar source, promptly begins to synthesize beta-galactosidase and two other enzymes needed to deal with lactose rather than glucose. Lactose can thus *induce* the production of the appropriate enzymes, which are then said to be *inducible*. This is so even though any given enzyme (being a protein) can be produced only if the cell has a gene that carries the code for it and if that gene is transcribed by mRNA and then translated into the enzyme.

A model for the coordinate control of enzyme synthesis was proposed in 1961 by F. Jacob and J. Monod to explain the appearance of inducible enzymes such as beta-galactosidase. Briefly, they proposed that the DNA of *E. coli* is organized into functional units called *operons*. Each operon consists of a group of functionally related *structural* genes immediately adjacent to one another, preceded by a controlling region called the *operator*. The operator, in turn, is immediately adjacent to a *promoter*, a region in front of every gene that the RNA polymerase responsible for mRNA synthesis recognizes and at which it begins transcription (see Figure 3–30).

Separated from the operon is a *regulator gene*, commonly symbolized *i*, whose protein product is a *repressor* that specifically binds with and blocks the operator (*o*) region. In this condition, mRNA synthesis of the structural genes (*z*, *y*, and *a*) cannot occur because the RNA polymerase cannot move through the operator region. As a result, no lactose enzymes are produced: the lactose operon is *repressed*. When lactose appears, however, binding of the operator is released because the repressor protein has an even greater affinity for lactose than it has for the operator: the operon is *derepressed*. In this situation, the lactose operon is turned on, the lactose genes are transcribed and translated, and the three lactose enzymes, including beta-galactosidase, are produced as

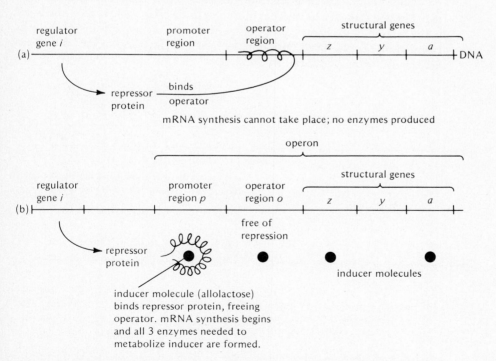

Figure 3–30. An inducible operon. (a) A repressed inducible *lac* operon. (b) A derepressed inducible *lac* operon.

needed to metabolize the inducer, lactose. When lactose is again replaced by glucose, the system reverts to its *repressed* state until such time as lactose may reappear to *derepress* the lactose operon again.

Many systems like this are known by which microorganisms coordinate the production of enzymes needed to deal with specific metabolites. The genes themselves are always present, but their activity is controlled by the presence of specific substances that enter the cell.

In addition to inducible operons, such as the lactose operon, *repressible* operons are also known. Here, the end-product of the enzyme activity stops the activity of the genes in the operon. For example, the amino acid tryptophan is produced as a result of the sequential activity of at least five adjacent structural genes in *Salmonella typhimurium*. An operator and a regulator gene are present, but the protein product of the regulator gene is not the active repressor; rather, it is in an inactive state, an *aporepressor*. Thus, the operon is on and the end product, tryptophan, is produced. If excess tryptophan enters the cell so that cellular synthesis of tryptophan is no longer needed, the excess tryptophan combines with the aporepressor. This complex becomes the active repressor, binding the operator and turning the tryptophan operon off (Figure 3–31).

Gene regulation in higher organisms is much more complicated than this. For one thing, certain genes must be turned on in some cells of the body, while at the same time those genes must be turned off in other cells. Hemoglobin is produced only in red-blood-cell-forming tissues, not in any other cells of the body, and so on.

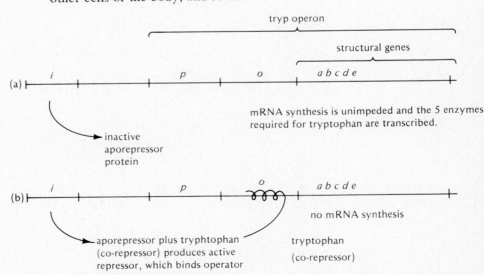

Figure 3–31. A repressible operon: the tryptophan operon. (a) Derepressed (b) Repressed

The operon form of organization does not seem to occur in higher organisms. For example, the two genes needed to produce the two polypeptide chains present in every hemoglobin molecule, the alpha and beta chains, are not even on the same chromosome, much less immediately next to each other; yet, their activity must be regulated.

Although much remains to be learned about gene regulation in higher organisms, some clues are readily apparent. DNA in higher organisms (*eukaryotes*), unlike DNA in lower organisms such as bacteria (*prokaryotes*), is always associated with protein. In fact, chromosomes contain about half DNA and half protein. Protein may be the key factor in differentiation, being able to repress some genes while allowing others to remain active. There are two major types of chromosomal proteins: (1) *histone* proteins and (2) *nonhistone* proteins. Histones are low molecular weight (10,000–20,000) proteins that are rich in basic amino acids, such as lysine and arginine. Histones from different organisms and different cell types appear to be structurally similar. Nonhistone proteins associated with DNA are not low in molecular weight and are structurally much more heterogeneous than are histones. Most of them are phosphorylated and are acidic, rather than basic.

Some experimental results suggest that histone proteins bind rather nonspecifically to the DNA of the chromosomes, causing compaction of the DNA and preventing binding of the RNA polymerase that allows RNA synthesis (transcription) to occur. However, the nonhistone proteins may protect regions of DNA from histone binding or may remove histones from specific regions of DNA so that RNA synthesis can proceed at these regions. It may be that the acidic, negatively charged nonhistone proteins bind to the positively charged histones, removing them from specific regions of the DNA. In this way, particular genes may be activated. Hormones and other factors are known to activate specific genes in specific tissues during differentiation. Hormone complexes may in some way "turn on" genes by binding to nonhistone protein on the chromosomes.

Recent work on chromosome structure has led to the finding that eukaryotic chromatin is composed of particles called *nucleosomes*. Although the structure of these units is still being investigated, evidence suggests that nucleosomes are 70–100 Å in diameter and contain 2 each of 4 different histone molecules (H-2A, H-2B, H-3, and H-4) and about 140 base-pairs of DNA. Nucleosomes are like beads on a string; between the beads, the piece of string consists of 30–60 base-pairs of DNA. The DNA double helix is wrapped over and around the histones rather than being embedded within the nucleosomes. As a result, nucleosome DNA is accessible to chemical reagents and can be transcribed. A fifth type of histone (H-1, which is rich in lysine) is not present within the nucleosome particles; rather, H-1 is associated with the short 30–60 nucleotide strands of DNA between the nucleosomes. The other four histones, which are less rich in lysine than H-1, are present in the nucleosomes themselves (Figure 3–32).

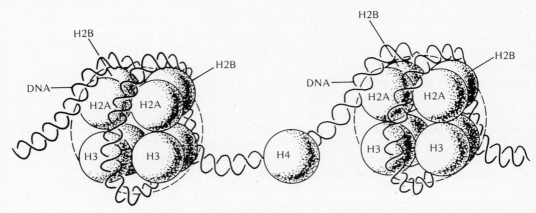

Figure 3-32. The structure of nucleosomes as proposed by R. D. Kornberg. An $(H-2A)_2\,(H-2B)_2\,(H-3)_2\,(H-4)_2$ octamer forms the protein core of the nucleosome. DNA is wrapped around the outside of the histone core, in a coil of 1-¼ to 1-½ turns. The entire structure is 10-11 nm in diameter. Histone H-4 is associated with the DNA strand between successive nucleosomes.

Although the ratio of histone protein to DNA is relatively constant from cell type to cell type, the amount of nonhistone chromosomal protein is quite variable. Cells with very active genes tend to have more nonhistone protein than cells with inactive genes. In addition, there are many types of nonhistone proteins, not just five. This suggests that nonhistone proteins play a specific role in gene activation. For example, brain chromatin can be dissociated into its major components—DNA, histone, and nonhistone proteins. If it is then reassembled using the nonhistone protein fraction from red blood cells rather than from the brain, this hybrid chromatin can produce globin messenger RNA. Globin is the protein of hemoglobin; it is normally produced only by red blood cell precursors, not by brain cells. Here, however, adding nonhistone protein from red blood cells allows brain chromatin to produce globin mRNA. Although controversial, this experiment suggests that nonhistone proteins can activate specific genes, possibly by altering the histones that mask a given gene.

Before we leave the eukaryotic chromosome, we should mention that each structural gene in the cell, such as that for globin, probably is associated with a series of control genes that can turn it on and off. The products of these control genes might be the specific nonhistone proteins. A given activator such as a hormone might bind to the specific nonhistone protein, and this complex might then bind directly to the gene. In other cases, the hormone might bind to the cell surface and cause reactions that result in the accumulation of small secondary molecules, such as cyclic AMP. These molecules, in turn, might bind to the nonhistone proteins, activating the gene.

Several systems have been found in which hormones act in this way. For example, steroid sex hormones in mammals penetrate the cell membrane and bind to specific protein receptors in the cytoplasm. The receptors are absent in cell types that are not stimulated by the given hormone. The activated receptor, in turn, penetrates the nucleus and interacts with the chromatin to activate specific genes. Specific receptors for estrogens, progesterones, androgens, and cortisol have been isolated. The estrogen receptor, for example, is a protein of about 80,000 daltons that binds estrogen, but not similar molecules without estrogen activity. This topic will be discussed further in Chapter 13.

RECOMBINANT DNA

During the past few years, remarkable technological advances have made possible a new type of genetic analysis. Individual genes can be isolated from DNA extracts, captured in virus-like plasmids, placed in bacterial cells, cloned in large numbers, and then analyzed to determine the specific DNA sequences of the gene. The difference between the gene and its various mutant alleles can be described at the nucleotide level. Furthermore, genes from one organism can be combined with genes of virtually any other organism, no matter how distantly related. In short, *genetic engineering* is with us today, for good or evil.

All of this has come about because it is now possible to manipulate individual DNA and RNA strands in a test tube. Moreover, by measuring the ability of strands from different origins to *reanneal*—that is, to hydrogen-bond together—the genetic relatedness of two DNA samples can be determined. The more similar the nucleotide sequences are, the more completely will double-helix DNA molecules form, and the more closely related the organisms are that provided the DNA samples.

Even DNA and RNA sequences can be compared. A recent, unexpected finding is that, in higher organisms, the mRNA is often much shorter in length than the DNA region from which it was copied. The DNA contains internal stretches of nucleotides (*introns*) that are not found in the active mRNA molecules transcribed from it. These introns are transcribed along with the rest of the gene to form a large *heterogeneous RNA* (hnRNA) molecule. However, before the transcript is released into the cytoplasm to begin translation, the molecule is *processed* to remove the copied intron regions. The functional mRNA molecule is then ready to go to work. (Specific experiments are described in Chapter 13.)

MUTATION

Some intron-like sequences, called *transposons*, can apparently transpose themselves from one gene to another. After being inserted in a gene, such transposons often change that gene's expression. This is a kind of *mutation* that is quite different from the usual sort. Normally, one DNA base substitutes for another, converting one codon into another and thus modifying the protein product by substituting one amino acid for another.

Exactly this kind of change occurred in the mutation from normal hemoglobin to sickle-cell hemoglobin: The amino acid at the sixth position from the beginning of the beta chain changed from the normal glutamic acid (codon GUA or GUG) to valine (codon GAA or GAG). This change of only one nucleotide in a glutamic acid codon (U to A) changes exactly 1 amino acid out of the 146 in the beta chain (and none out of the 141 amino acids of the alpha chain). Even so, this small change significantly reduces the oxygen-carrying capacity of sickle-cell hemoglobin as compared with normal hemoglobin.

Today, rapid progress is being made in determining the exact DNA nucleotide sequences of gene after gene, normal and mutant, even in higher organisms. One day, we may well be able to write down in full detail the genetic formula for a human being! Such is the future of molecular genetics.

READINGS AND REFERENCES

For the more classical topics in genetics, the comprehensive treatment by M. W. Strickberger in *Genetics*, 2nd ed. (New York: Macmillan, 1976) is most useful as a general reference. For the molecular aspects of genetics, an excellent reference is *Principles of Genetics* by J. W. Fristrom and P. T. Spieth (New York: Chiron Press, 1980).

Avery, O. T., MacLeod, C. M., and McCarty, M. 1944. Studies on the chemical nature of the substance inducing transformation of pneumococcal types. Induction of transformation by a deoxyribonucleic acid fraction from pneumococcus type III. *Jour. Exp. Med.* 79: 137–58.

Bateson, W., and Punnett, R. C. 1905. Experimental studies in the physiology of heredity. *Reports to the Evolution Committee Royal Society*. II. London: Harrison and Sons.

Beadle, G. W., and Tatum, E. L. 1941. Genetic control of biochemical reactions in *Neurospora. Proc. Natl. Acad. Sci. U.S.,* 27: 499–506.

Bridges, C. B. 1916. Nondisjunction as proof of the chromosome theory of heredity. *Genetics.* 1: 1–52, 107–63.

Chargaff, E. 1950. Chemical specificity of nucleic acids and mechanism of their enzymatic degradation. *Experientia.* 6: 201–09.

Griffith, F. 1928. The significance of pneumococcal types. *Jour. Hygiene.* 27: 113–59.

Jacob, F., and Monod, J. 1961. Genetic regulatory mechanisms in the synthesis of proteins. *Jour. Mol. Biol.* 3: 318–56.

Levene, P. A., and Bass, L. W. 1931. *Nucleic acids*. New York: Chemical Catalog Co.

Meselson, M., and Stahl, F. W. 1958. The replication of DNA in *Escherichia coli. Proc. Natl. Acad. Sci. U.S.* 44: 671–82.

Miescher, F. 1871. On the chemical composition of pus cells. *Hoppe-Seyler's Med.-Chem. Untersuch.* 4: 441–60.

Nirenberg, M. W., and Matthei, J. H. 1961. The dependence of cell-free protein synthesis in *E. coli* upon naturally occurring or synthetic polyribonucleotides. *Proc. Natl. Acad. Sci. U.S.* 47: 1588–1602.

Sturtevant, A. H. 1913. The linear arrangement of six sex-linked factors in *Drosophila* as shown by their mode of association. *Jour. Exp. Zool.* 14: 43–59.

Sutton, W. S. 1903. The chromosomes in heredity. *Biol. Bulletin.* 4: 213–51.

Watson, J. D., and Crick, F. H. C. 1953. Molecular structure of nucleic acids. A structure for deoxyribose nucleic acids. *Nature.* 171: 737–38.

4

GAMETOGENESIS

We could introduce our serious discussion of embryonic development by describing the fertilized egg and its early activities. It is fitting, however, to begin by describing the formation of the gametes—sperm and eggs—that must unite in order to produce the fertilized egg. Gametogenesis is concerned with just that process. It is subdivided into (1) *spermatogenesis*, the formation of sperm, and (2) *oogenesis*, the formation of eggs, or ova.

In Chapter 2, we outlined the activities of the chromosomes during meiosis, the form of nuclear division associated with gametogenesis that reduces the chromosome number from diploid to haploid in both sperm and eggs. In that discussion, however, we did not describe the significant differences between the formation of a minute, motile sperm and that of a large, yolk-laden egg. Here, we shall describe in detail the cellular activities that distinguish spermatogenesis from oogenesis, remembering that the chromosomal activities per se are identical in the two processes.

SPERMATOGENESIS

First, we shall outline the events involved in spermatogenesis to provide a framework for discussing some of the more interesting, recently discovered aspects of the process. The area in the testis where sperm are produced, the *seminiferous tubules*, consists of two important types of cells: the germ cells, which are in various stages of meiosis, and the *Sertoli cells*, which support and nourish the developing sperm cells. The Sertoli cells probably supply the developing sperm cells with specific factors that cause growth, division, and differentiation (Figure 4–1).

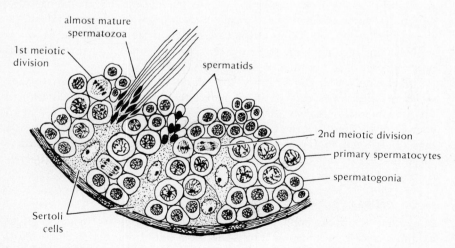

Figure 4–1. Section of a mammalian seminiferous tubule, showing cells undergoing meiosis. From B. I. Balinsky, *Introduction to Embryology*, 4th edition (Philadelphia: W. B. Saunders Co., 1975), p. 19.

Primordial germ cells do not originate in the gonad, but migrate or are carried by the blood to the gonad from other regions of the body. Most evidence suggests that the sperm and egg cells are derived from the primordial germ cells that take residence in the gonads. In Chapter 10 we shall examine the origin of primordial germ cells in more detail.

The primordial germ cells in the sperm-forming (seminiferous) tubules of the testis give rise to the *spermatogonia*, the cells that eventually develop into sperm. In many vertebrates, the spermatogonia are located in the outer region of the seminiferous tubules. As the spermatogonia develop and mature, they move toward the inner region where the *lumen* or canal of the tubule is located (Figure 4–1).

Large numbers of spermatogonia accumulate through mitotic cell division in the testes of most vertebrate males. As the male reaches sexual maturity, some of the spermatogonia continue to divide, but others undergo a noticeable increase in size. These cells are committed to divide meiotically and become sperm. These enlarged cells are now called *primary spermatocytes*. Just as in mitosis, each chromosome duplicates itself in the last premeiotic interphase so that each primary spermatocyte chromosome is formed of two identical chromatids with identical DNA content, joined at their centromere regions. The chromosomes synapse in pairs to form tetrads during the zygotene stage of prophase I (see Figure 2–2a).

At the following pachytene stage, the process of *crossing over* takes place. Crossing over (which was not illustrated in Figure 2–2a) is shown in Figure 3–17. Homologous chromosome segments are exchanged by a process of rupturing two nonsister chromatids at the same level and subsequently rejoining the four broken ends in opposite orientation. This is the basis for genetic recombination between genes that are physically located (*linked*) on the same chromosome. After an exchange, two of the four meiotic products will contain recombinant, or crossover, chromosomes, and the other two noncrossover chromosomes (see Figure 3–18, p. 39).

At telophase I, the primary spermatocyte divides into two smaller *secondary spermatocytes*, each of which undergoes the second meiotic division to form even smaller *spermatids*. These are normal-looking, round cells with round nuclei. Each spermatid then differentiates without further division into a functional, motile sperm. This cytoplasmic transformation of spermatids into sperm is called *spermiogenesis*, not to be confused with spermatogenesis. Figure 4–2 summarizes the process of spermatogenesis.

MECHANISMS OF SPERMATOGENESIS

Now that we understand the basic steps of forming sperm cells, let us examine some experimental results that shed light upon the mechanisms of spermatogenesis. In addition, we shall look at results that may lead to safe and effective male contraceptives. Basic developmental biology, as will be seen, has many medically and socially promising applications. We shall concen-

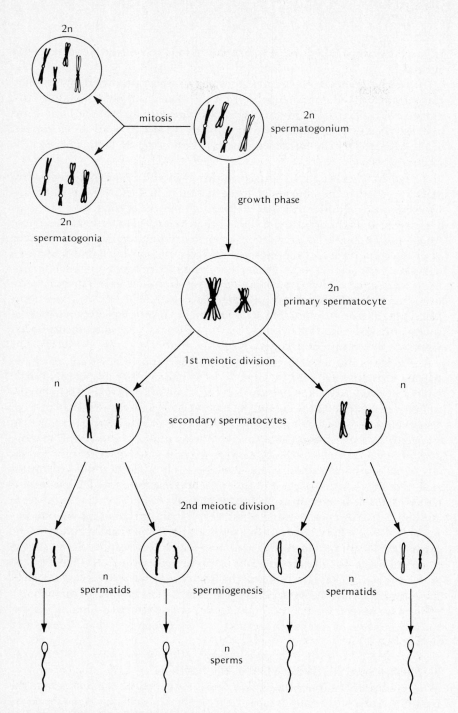

Figure 4-2. Summary of spermatogenesis.

trate on mammalian development, the system of direct importance to humans.

Spermatogenesis is controlled by two hormones produced by the anterior pituitary gland, luteinizing hormone (LH) and follicle-stimulating hormone (FSH). These hormones (especially luteinizing hormone) stimulate interstitial cells in the testis (Leydig cells) to synthesize male sex hormones called *androgens*. These hormones appear to be among the factors needed for sperm maturation.

Over 100 years ago, La Valette noticed that the developing male germ cells appeared to be connected to each other. In 1970, Fawcett and co-workers, using the electron microscope, demonstrated that the spermatocytes were in fact connected to each other by cytoplasmic bridges. These connections seem to be incompletely pinched-off cell membranes left over after the cell division of spermatogonia. They amount to channels between daughter cells (Figure 4–3). Many developing spermatocytes are thus connected together, allowing molecules to pass from cell to cell. This may be an important factor in insuring that many sperm can differentiate at one time. Many sperm must be mature at specific times to insure sufficient numbers of sperm for successful fertilization. Sertoli cells may provide molecules that stimulate sperm differentiation.

At this point, let us digress for a moment to consider some phases in spermatogenesis that may be vulnerable to interference by potential male contraceptives. Fawcett suggests that the stimulation of gametogenesis by follicle-stimulating hormone and the separation of joined sperm may be steps that eventually could be interrupted by specific drugs. Also, mature sperm traveling in the collecting ducts of the testis must be vulnerable to drug action. Several drugs are being studied that directly affect mature sperm. These include α-chlorohydrin and 1-amino, 3-chlor, 2-propanol hydrochloride. The site of action of these drugs on the sperm is not as well understood as that of other new substances. Several studies have indicated that both sperm and eggs have sugar-containing molecules on their surfaces. Certain substances called *lectins* (carbohydrate-binding proteins) specifically attach to the sugars on the cell surfaces of the sperm and eggs. It is possible that fertilization could be prevented by treating sperm or eggs with lectins that bind to surface sugars. Oikawa, Yanigamachi, and Nicolson have found that one of these lectins (wheat germ agglutinin) blocks the fertilization of mammalian eggs by sperm. Such blockage might occur because the lectin binding inhibits sperm motility or prevents sperm from becoming attached to egg cell surfaces.

MECHANISMS OF SPERM DIFFERENTIATION

We have briefly examined meiosis, but have stopped short of describing the transformation of spermatids into mature sperm. We call this transformation phase spermiogenesis. In the next chapter on fertilization we shall investigate

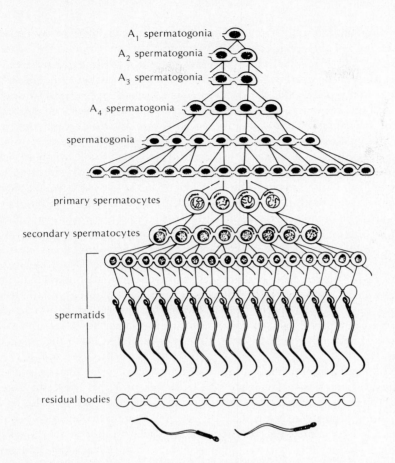

A₁ spermatogonia

A₂ spermatogonia

A₃ spermatogonia

A₄ spermatogonia

spermatogonia

primary spermatocytes

secondary spermatocytes

spermatids

residual bodies

Figure 4-3. Interconnections of mammalian sperm. From D. Fawcett, "Gameto-genesis in the Male: Prospects for Its Control," in C. L. Markert and J. Papaconstantinou, eds., *The Developmental Biology of Reproduction* (New York: Academic Press, 1975), p. 38.

the structure of sperm and the function of its various parts. Here, let us look at the process by which the round, nondescript spermatids are transformed into highly differentiated, mature sperm cells.

During the primary spermatocyte stage (before any sperm differentia-tion occurs), the chromosome that confers "maleness" (in humans, the Y chromosome) is active in synthesizing messenger RNA. This was demon-strated by using a technique called autoradiography to observe the appear-ance of RNA on the loop of the Y chromosome. The RNA was previously labeled using radioactive ^{3}H-uridine, which is specifically incorporated into newly formed RNA. A photographic emulsion is then placed over the specimen, and radioactivity in specific regions of the specimen (such as the Y chromosome) is observed as black dots. The researchers used the fruit fly,

Drosophila, but the results are probably relevant to many systems. What is the importance of the synthesis of RNA on the Y chromosome in the primary spermatocyte? First of all, this Y chromosome RNA in mammals presumably codes for specific proteins that are responsible for "maleness." In *Drosophila*, these proteins are needed for differentiating spermatids into mature sperm. Remember, however, that spermatids contain the haploid chromosome number. Each spermatid contains *either* an X or a Y chromosome, not both. Thus, any messenger RNA formed by the Y chromosome for spermatid differentiation must be made before the first meiotic division, when the X and Y chromosomes are separated into different cells. Apparently, the important Y chromosome messenger RNA is produced in the primary spermatocytes and distributed to all their progeny. Thus, all secondary spermatocytes and spermatids have the genetic information from the Y chromosome needed for synthesizing the specific proteins required for forming mature spermatozoa.

The differentiation of spermatids into spermatozoa involves transforming the Golgi apparatus into a structure called the *acrosome*. This structure will be described in the next chapter. Its main function is to contact the egg surface and to aid the sperm in penetrating the outer egg coats during fertilization. Not all sperm possess acrosomes; those that do not have to penetrate complex egg coats do not need them. Acrosomes contain a variety of substances, including enzymes, that help to break down the outer egg coats.

Transforming spermatids into sperm also involves compacting the nucleus and associating the nuclear DNA with basic proteins. The basic proteins may help compress the DNA into the small space of the sperm head. At the same time, a sperm tail is formed by assembling proteins called microtubules, beginning at a centriole of the spermatid. The structure of sperm will be examined in the next chapter. Figure 4–4 shows a summary of spermiogenesis. Figure 4–5 shows some of the varying forms that mature sperm take in different organisms.

OOGENESIS

SUMMARY OF EVENTS

Before looking at the development of female gametes, let us summarize the process of oogenesis. In this way, we shall have a framework for discussing the findings that follow.

1. Primordial germ cells, as indicated in our discussion on spermatogenesis, do not originate in the gonad. Instead, they migrate or are carried to the gonad from other regions of the body. (This will be discussed in more detail in Chapter 11.)

2. The primordial germ cells proliferate and give rise to the oogonia, the cells that develop into eggs.

3. Some of the oogonia grow. In some species, such as birds and reptiles,

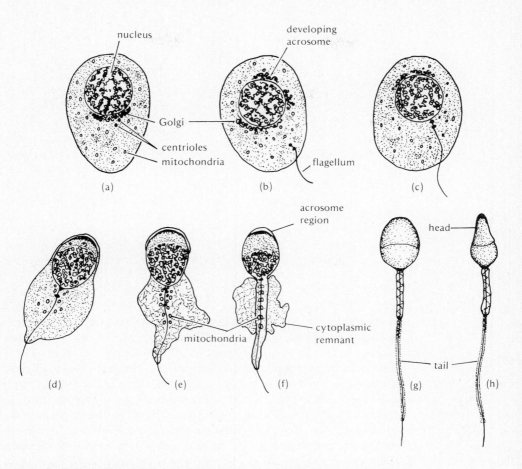

Figure 4-4. Transformation of spermatid into mature sperm (chronological sequence). After Gatenby and Beams, 1935.

this growth can be quite striking. The volume might increase to 100,000 times the original volume. This extensive growth obviously provides the egg cell with the materials needed for the embryo to develop.

4. These growing cells are called primary oocytes. While oogonia have the same amount of chromosomal DNA as normal body cells, the primary oocytes (like primary spermatocytes) possess twice as much chromosomal DNA as normal body cells, because the DNA duplicated itself during the last premeiotic interphase.

5. The primary oocytes, like primary spermatocytes, undergo two meiotic divisions. By the end of these divisions, the cells have only half as many chromosomes as normal body cells. Thus, haploid eggs can combine with haploid sperm to form diploid fertilized eggs.

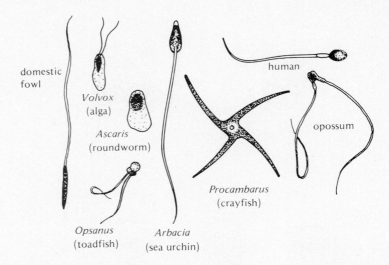

Figure 4–5. Variety of sperm forms.

In oogenesis and spermatogenesis, the meiotic stages are exactly the same in terms of crossing over and the distribution of the chromosomes to daughter cells (Figure 4–6). One major difference, however, involves cytoplasmic division. One primary oocyte divides (in meiosis I) to form one large secondary oocyte and one tiny polar body (that may or may not divide again). In meiosis II, the secondary oocyte divides to form one large mature egg and another tiny polar body (Figures 4–6 and 4–7). (The mechanisms that cause such unequal cytoplasmic division are discussed in Chapter 7.) Such unequal division preserves for one cell the vast stores of materials produced in the oocyte growth phase. If division were equal, each cell would have much less of the stored material needed for the successful development of the embryo. The only other major difference between oogenesis and spermatogenesis is that spermatids must go through a differentiation phase to form mature sperm. The egg formed after the second meiotic division is already prepared for its role in supporting the new embryo (Figure 4–7).

As in spermatogenesis, hormones control oogenesis in most, if not all, higher organisms. Different hormones act in different systems. Some hormones act at the surface of the oocyte cell to stimulate the completion of meiosis. In mammals (including humans) a variety of hormones influence the maturation of the oocyte, the development of the follicle that surrounds the oocyte, and the rupture of the follicle, releasing the egg (Figure 4–8). The hormones that control oocyte development in mammals are rather well understood. The oral contraceptive, or birth control pill, is an example of how knowledge of the hormonal control of oogenesis has been applied to an important social problem. These pills contain specific hormones (such as

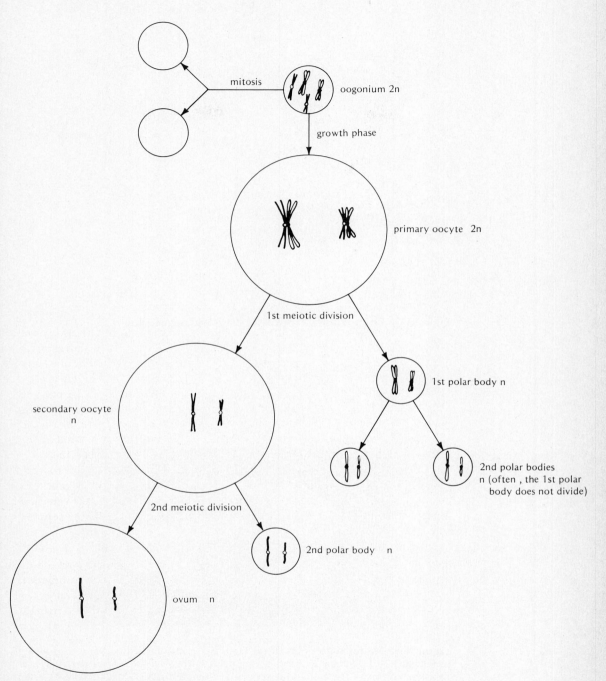

Figure 4-6. Summary of oogenesis.

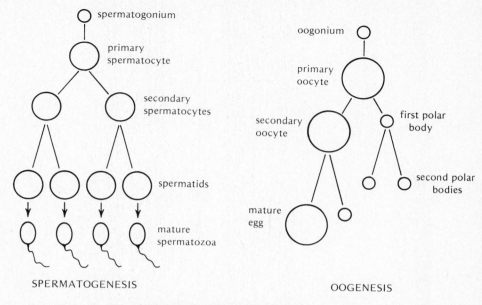

Figure 4-7. Comparison of spermatogenesis and oogenesis.

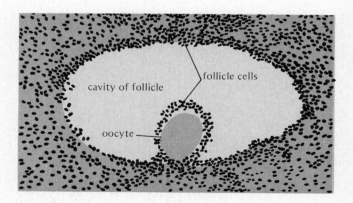

Figure 4-8. Mammalian oocytes. (a) Drawing of mammalian oocyte in its follicle, surrounded by follicle cells. From J. D. Ebert and I. M. Sussex, *Interacting Systems in Development* (New York: Holt, Rinehart and Winston, 1975), p. 17. (b) Photo of cat oocyte in its follicle. Photo by Richard L. C. Chao.

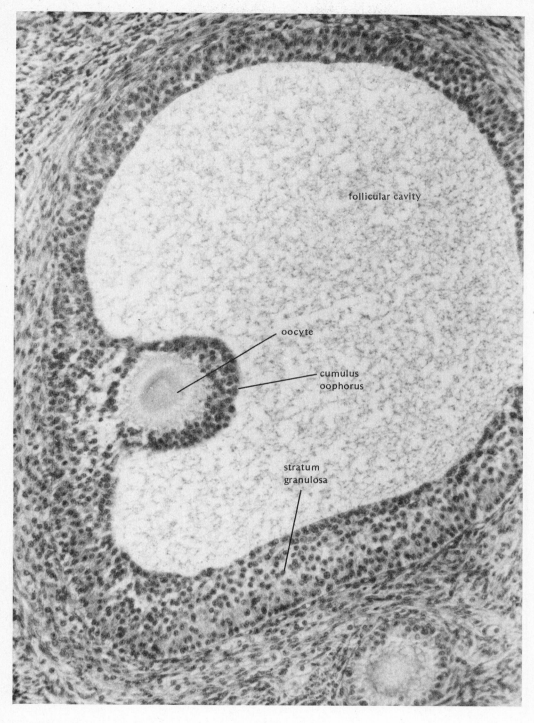

follicular cavity

oocyte

cumulus oophorus

stratum granulosa

(b)

estrogen and progesterone) that control oocyte and follicle development. The concentration of these hormones in the pills is chosen so as to prevent the development and release of eggs.

OOCYTE GROWTH AND DEVELOPMENT

Unlike sperm cells, egg cells must possess extensive stores of materials to nourish the embryo and control its development. One key aspect of oogenesis, therefore, is how the cell accumulates the necessary stores of nucleic acids, proteins, and other substances needed for embryonic development. Let us now examine this rather remarkable process of oocyte growth.

NUCLEIC ACIDS. The primary oocytes enter the first meiotic prophase and go through the same phases as described previously for spermatocytes: leptotene, zygotene, pachytene, and diplotene (Figure 4–9). During spermatogenesis, the chromosomes condense and prepare to separate from each other in diplotene. In oogenesis, however, the chromosomes become greatly

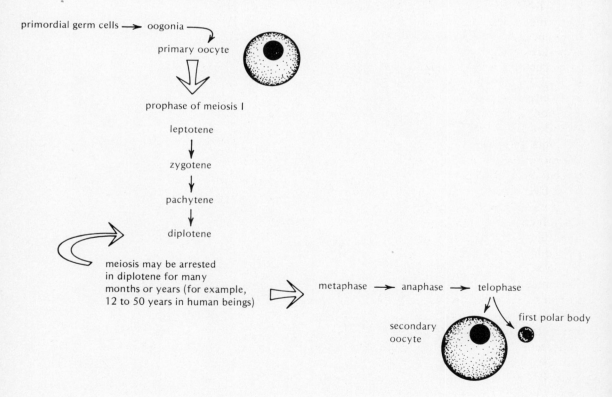

Figure 4–9. Meiosis I in vertebrate oogenesis.

extended during diplotene, forming the so-called lampbrush structure, which are particularly prominent in amphibian oocytès. These chromosomes resemble lampbrushes because thin threads or loops develop perpendicular to the long axis of the chromosomes proper (Figure 4–10). Lampbrush chromosomes are actively involved in synthesizing the RNA needed for the development of the embryo.

Primary oocytes often remain in the diplotene stage for months or even years. In the human female, oocytes remain in the diplotene stage for up to 50 years. During this time, only one oocyte usually matures each month in response to hormonal stimulation. It is well known that birth defects are more likely to occur in children born to older than to younger women. The reason may be that the oocytes that give rise to these children have remained in diplotene for decades, and that this increases the probability of chromosomal damage.

The loops of the lampbrush chromosomes actively synthesize RNA. This can be shown by incubating oocytè chromosomes with radioactive uridine which RNA contains (one of the building blocks of RNA). Many of these experiments are done using amphibian oocytes because of their large size, large nuclei, and easily obtainable lampbrush chromosomes. Isolated oocyte nuclei are broken up on a slide, releasing the lampbrush chromosomes. These are incubated with radioactive uridine and then washed to remove the uridine. Any newly synthesized RNA incorporates the radioactive uridine; thus, only newly synthesized RNA is radioactively labeled. A photographic emulsion is placed over the specimen. Radioactive RNA is then revealed as black dots in the emulsion over the radioactive regions of the specimen. This technique of autoradiography has shown that the loops of the lampbrush chromosomes are actively engaged in synthesizing RNA. Only about five percent of the total genome is in the form of loops at any one time, and only about five percent of the genome is transcribed at the lampbrush stage.

Some of the RNA synthesized during the lampbrush stage is messenger RNA. It codes for proteins that are needed during early embryonic development. Thus, the egg is preprogrammed with messenger RNA that will be needed later on.

In addition to messenger RNA, oocytes synthesize a large amount of ribosomal RNA. This RNA is needed to produce the many ribosomes where protein synthesis is carried out. In many oocytes, the production of ribosomal RNA occurs by a process called selective gene amplification. The DNA segments of the chromosomes that code for ribosomal RNA are selectively replicated, while other genes are not. Thus, the ribosomal RNA genes, and only the ribosomal RNA genes, are reproduced thousands of times. Ribosomal RNA is then formed (transcribed) in these nucleoli from DNA segments. A single oocyte can contain over 1,000 nucleoli. It can be shown (again by using radioactive uridine) that the nucleoli are actively synthesizing ribosomal RNA. It has been estimated that it would take about 500 years to

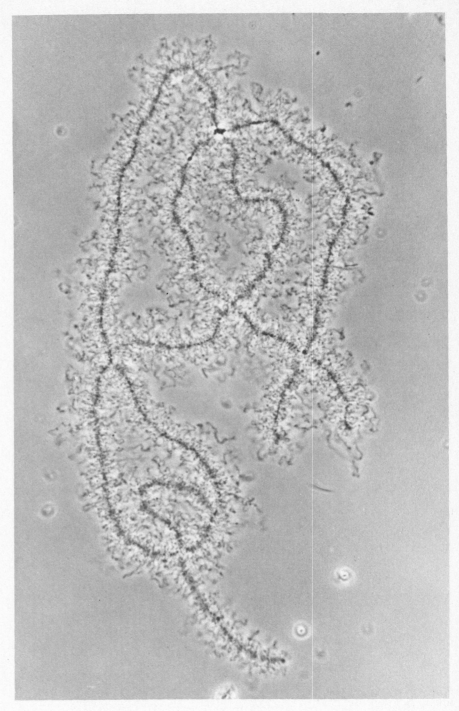

Figure 4-10. Photomicrograph showing lampbrush chromosomes. Courtesy of J. G. Gall.

synthesize all of the ribosomal RNA needed for early development if gene amplification did not occur. Instead, a frog can synthesize the needed ribosomal RNA in only a few months.

Not all species possess a lampbrush stage. In species without lampbrush chromosomes, the oocytes receive RNA (and sometimes DNA) from cells that surround the oocytes. A *Drosophila* (fruit fly) oogonium divides four times, giving rise to 16 cells. Only one of these becomes the oocyte; the others become nurse cells that nourish the oocyte. Thus, the nurse cells and oocytes are derived from oogonia and are very closely associated. There are, in fact, direct cytoplasmic bridges between the oocyte and the nurse cells. The nurse cells synthesize large amounts of RNA, ribosomes, and proteins, and pour this material into the oocyte through the connecting cytoplasmic bridges. Autoradiography has been used to show that nucleic acid flows from the nurse cells into the cytoplasm of the oocyte in the house fly, *Musca domestica* (Figure 4–11).

PROTEIN AND OTHER MOLECULES. Two key aspects of oocyte growth are the synthesis of large amounts of RNA and the accumulation of proteins. Now that we have outlined the RNA story, we move on to proteins. Many of the proteins that accumulate in oocytes for use in embryonic development are produced outside the ovary and brought to it by the bloodstream. Experiments with identifiable proteins (either radioactively labeled or detected with specific antibodies) have shown that proteins do enter oocytes from the outside.

How do protein molecules get to the oocytes? What kind of molecules

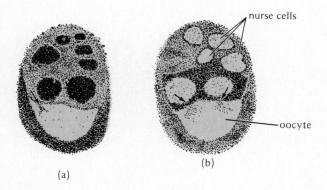

(a) (b)

Figure 4–11. Passage of nucleic acids from nurse cells to oocyte of the fly. (a) Follicle (oocyte and nurse cells) one hour after injection of ³H cytidine. (b) Five hours after injection. Arrows show radioactive nucleic acid streaming from nurse cells into oocyte. After K. Bier, *Roux Arch.* 154 (1963): 552–75.

are they, and what is their function? One important substance that accumulates in eggs is yolk, which is a major food reserve for many developing embryos. There are different types of yolk, and many methods of yolk production. Some yolk is mainly protein with some lipid (protein yolk). Other types of yolk consist mainly of phospholipid and fat and possibly some protein (fatty yolk). In vertebrates, yolk is synthesized in the liver, dissolved in the bloodstream, and carried to the ovaries. Once in the ovaries, it is picked up and transferred to the oocyte by follicle cells that surround the oocyte. Unlike nurse cells, the follicle cells are not formed from oogonia, but from ovarian epithelium, the surface layer of the ovary. Follicle cells are not connected by cytoplasmic channels to the oocytes. Instead, fine projections called microvilli extrude from the follicle cell surface and intertwine with microvilli on the oocyte cell surface (Figure 4–12). The microvilli of the oocyte either absorb yolk by pinching off tiny portions of the membranes of the follicle cell microvilli and engulfing the yolk material in these vesicles, or they absorb yolk previously released by the follicle cell microvilli. This process of cell drinking is termed pinocytosis or micropinocytosis. In animals without circulatory systems, yolk is sometimes synthesized in the oocyte itself on ribosomes on the endoplasmic reticulum associated with the Golgi apparatus.

Even in organisms in which yolk is transported to the oocytes from the liver, the yolk often undergoes packaging into platelets in the oocyte. Some yolk platelets are probably formed inside the mitochondria of fish, amphibian, and snail oocytes (Figure 4–13). A mitochondrial enzyme called protein kinase appears to play an important role in yolk platelet formation. This enzyme adds a phosphate group to one type of soluble yolk (phosvitin), causing the yolk to become insoluble. This causes the yolk to crystallize out of solution, forming the platelets, or yolk granules, needed to nourish an embryo. Yolk platelets are also formed in the cytoplasm within the vesicles formed by micropinocytosis.

In amphibians, oocytes mature under the control of hormones. Seasonal environmental factors cause the hypothalamus to stimulate the pituitary to secrete gonadotropin. Gonadotropin, in turn, stimulates ovarian follicle cells to synthesize estrogen. The liver synthesizes the yolk precursor, *vitellogenin*, in response to estrogen. Vitellogenin is transported to oocytes by the circulatory system. Recent work suggests that in amphibians vitellogenin is not transferred by follicle cells to oocytes. Instead, it leaves the blood capillaries in the ovary and flows through spaces between the follicle cells to the surface of the oocyte. There, the yolk precursor is engulfed by micropinocytosis. In oocytes, vitellogenin is converted into two kinds of yolk: phosvitin (a phosphorylated protein yolk) and lipovitellin (a lipoprotein yolk). These molecules form large, crystalline yolk platelets.

Many other molecules besides yolk accumulate in oocytes for use in early embryonic development. These substances include glycogen, an impor-

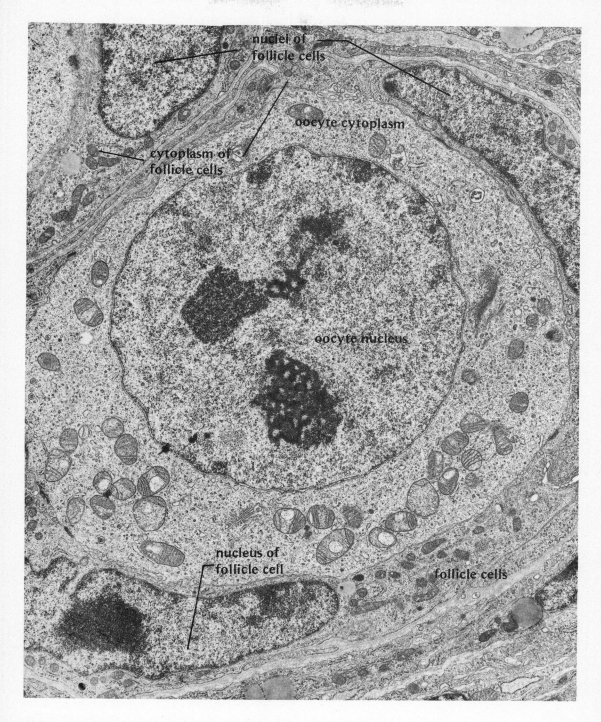

Figure 4-12. Young mouse oocyte surrounded by follicle cells. Courtesy of
Dr. E. Anderson, Harvard Medical School.

79

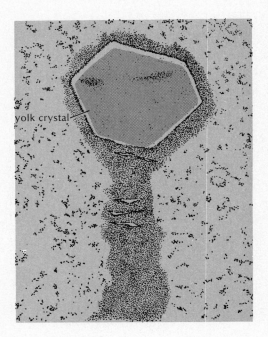

yolk crystal

Figure 4–13.   Yolk platelet inside mitochondrion of frog oocyte. After R. T. Ward, *J. Cell Bio.* 14 (1962): 309–41.

tant energy-rich carbohydrate storage molecule, lipid, used for membrane synthesis and energy supply, and a variety of proteins. Some of the proteins are subunits for the cytoplasmic contractile system (Chapter 13); others are enzymes.

To summarize: In many organisms, yolk is synthesized in the liver and brought by the bloodstream to the follicle cells surrounding the oocytes. The follicle cells transfer this material to the oocytes, where it is packaged into yolk platelets. Glycogen, lipid, and nonyolk proteins also accumulate in oocytes to help prepare these cells for embryonic development. Let us turn next to the larger structures that are essential components of oocytes.

HIGHER ORDERS OF STRUCTURE. We have examined how oocytes accumulate large stores of RNA, proteins, and other molecules that are required for the development of the embryo. Oocytes, however, are prepared for fertilization and development by additional activities other than accumulating and synthesizing molecules. They also develop new cellular organelles (cell components with specific functions) that are needed for fertilization and development. These organelles are composed of many, many molecules that fit together to form more complex structures.

In the next chapter, we shall see that many kinds of eggs contain

FROG

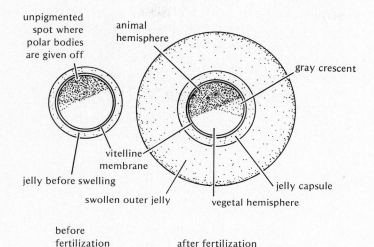

unpigmented
spot where
polar bodies
are given off

animal
hemisphere

gray crescent

vitelline
membrane

jelly before swelling

swollen outer jelly

jelly capsule

vegetal hemisphere

before
fertilization

after fertilization

HEN

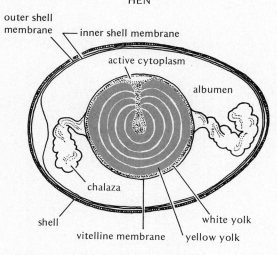

outer shell
membrane

inner shell membrane

active cytoplasm

albumen

chalaza

shell

vitelline membrane

white yolk

yellow yolk

Figure 4–14. Frog's and hen's eggs. From I. B. Balinsky, *An Introduction to Embryology*, 4th ed. (Philadelphia: W. B. Saunders, 1975). © 1975 by W. B. Saunders Co., p. 65.

structures called cortical granules. These structures, located in the surface cytoplasm just below the plasma membrane of the egg, play a major role in the fertilization reaction. Cortical granules form from the oocyte Golgi membrane complex and move to the periphery of the oocyte. These granules contain glycoproteins (proteins with attached sugar chains) that play a role in the fertilization reaction described in the next chapter.

Oocytes develop a variety of surface coats that protect the cell and probably help assure that only the right type of sperm will stick to the egg. The coats also often help form the fertilization membrane (see Chapter 5). Many eggs develop special coats in the tight space between the oocyte cell membrane and the follicle cell membranes. The follicle cells (and possibly the oocyte itself) secrete mucoproteins and fibrous proteins into this space. The coat formed in this way is called the vitelline membrane in molluscs, insects, amphibians, and birds, the chorion in fishes and tunicates, and the zona pellucida in mammals. In addition to these coats produced between oocytes and follicle cells, other coats are formed by various glands as certain kinds of eggs pass through the oviducts. These coats include the frog egg jelly coat and the egg white and shell layers of the bird egg (Figure 4–14).

SUMMARY

Gametogenesis results in the formation of haploid gametes. Genetic information is mixed in gametes by the process of crossing over and the random distribution of maternal and paternal chromosomes during meiosis. Male gametes become highly specialized as a result of the differentiation of spermatids. During the growth phase of the first meiotic prophase, female gametes accumulate vast stores of the RNA, protein, and other substances essential for development. The first and second meiotic divisions preserve these stores by unequal cytoplasmic divisions. Thus, only one large, functional egg is formed from each primary oocyte.

The next step in our study of embryology is fertilization. In this process, male and female gametes join to form the fertilized egg or zygote. We shall see how the relatively inactive egg is suddenly turned on to produce the multitude of materials required for the rapid development of a newly formed being.

READINGS AND REFERENCES

Fawcett, D. W. 1975. Gametogenesis in the Male: Prospects for Its Control. In C. L. Markert and J. Papaconstantinou, eds. *The Developmental Biology of Reproduction*, pp. 25–53. New York: Academic Press.

Fawcett, D. W. 1958. The Structure of the Mammalian Spermatozoon. *Int. Rev. Cyt.* 7: 195–234.

Fawcett, D. W. 1970. A comparative view of sperm ultrastructure. *Biol. Reprod.* supplement 2: 90–127.

Gall, J. G., and Callan, H. G. 1962. ^{3}H-uridine incorporation in lampbrush chromosomes. *Proc. Natl. Acad. Sci. U.S.* 48: 562–70.

Hadek, R. 1965. The Structure of the Mammalian Egg. *Int. Rev. Cyt.* 18: 29–71.

MacGregor, H. C. 1972. The nucleolus and its genes in amphibian oogenesis. *Biol. Rev.* 47: 177–210.

Oppenheimer, S. B. 1977. Interactions of lectins with embryonic cell surfaces. In A. A.

Moscona and A. Monroy, eds., *Current Topics in Developmental Biology*, Vol. II, pp. 1–16.

Raven, C. P. 1961. *Oogenesis: The Storage of Developmental Information*. Pergamon.

Richards, J. S. 1979. Hormonal control of ovarian follicular development. *Rec. Prog. Hormone Res.* 35: 343–73.

Roth, T. F., and Porter, K. R. 1964. Yolk protein uptake in the oocyte of the mosquito. *Aedes aegypti L. J. Cell. Biol.* 20: 313.

Wallace, R. A., and Dumont, J. N. 1968. The induced synthesis and transport of yolk proteins and their accumulation by the oocyte in *Xenopus laevis. J. Cell. Physiol.* 72 (supplement I): 73.

Wallace, R. A. 1978. Oocyte growth in nonmammalian vertebrates. In R. E. Jones, ed., *The Vertebrate Ovary*, pp. 469–502. New York: Plenum.

5
FERTILIZATION

What turns on the complex series of metabolic reactions that occur in eggs immediately after fertilization? How can a tiny sperm that fuses with only 0.0002 percent of the egg surface trigger the multitude of changes that occur in the new zygote? The answers to these questions are becoming better understood, and the program of events that comprise the fertilization process is becoming well established. We shall examine the fertilization reaction from a variety of approaches so as to answer, in part, the questions raised above. We shall look at: the ultrastructural aspects of fertilization—the aspects that we can visualize with the electron microscope; the biochemical and physiological program of events occurring during fertilization; and the molecular aspects of sperm-egg recognition, or how a sperm gets to and sticks to an egg. No attempt will be made to examine the fertilization reaction in all types of organisms. Instead, we shall discuss representative systems that for one reason or another have yielded unusually important results in the field.

ULTRASTRUCTURAL ASPECTS OF FERTILIZATION

The structural basis of fertilization can be found by examining the sperm, the eggs, the sperm-egg interaction, and the resulting changes at the egg surface with the aid of the electron microscope.

SPERM STRUCTURE

A typical sperm cell is shown in Figure 5-1, a diagram of the sperm of an annelid worm, *Hydroides*, drawn from an electron micrograph. The head of the sperm includes a nucleus, which holds the genetic information, and components of the acrosome, the anterior tip of the sperm head. The

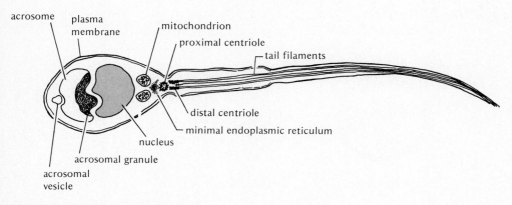

Figure 5-1. *Hydroides* **sperm.**

acrosome aids the sperm in penetrating the outer egg coats and establishing connection with the egg cytoplasm.

Vacquier and his colleagues isolated a molecule from the acrosomal granule of sea urchin sperm. This molecule is a protein called *bindin* that apparently enables the sperm to recognize and adhere to the egg surface. In addition to bindin, the acrosomal granule contains enzymes called lysins that aid the sperm in penetrating the outer egg coats.

Directly behind the nucleus, in the neck region of the sperm, mitochondria and centrioles are found. The mitochondria help supply the sperm with energy. The proximal centriole helps form asters during the division of the fertilized egg; the distal centriole provides an aster for attaching the tail fibers. The tail contains fibers composed of microtubules (rod-shaped proteins) that contract to propel the sperm cell.

EGG SURFACE STRUCTURE

Before examining the ultrastructural aspects of the fertilization reaction, we shall briefly look at some representative eggs and their surfaces. The cell surface plays a major role in sperm-egg interaction. Three types of eggs are shown in Figure 5-2.

Different types of eggs have different types of surface coats. All, however, contain a similar true membrane, the plasma membrane, bounding the cytoplasm. The coats can be cellular in nature (as in the tunicate) or proteinaceous (as in the sea urchin), but the true plasma membrane has a well-defined structure. The sperm plasma membrane consists of the same structure. It is generally agreed that the plasma membrane consists of a lipid bilayer in which protein or glycoprotein molecules are imbedded, as shown in Figure 5-3.

This "fluid-mosaic" model of the cell membrane was developed by Singer and Nicolson. The proteins and glycoproteins can move in the plane of

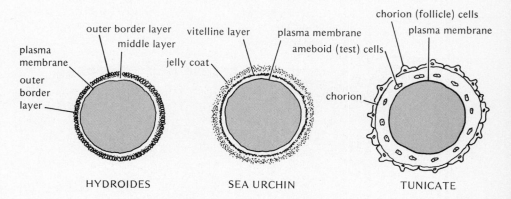

Figure 5-2. Surfaces of three eggs.

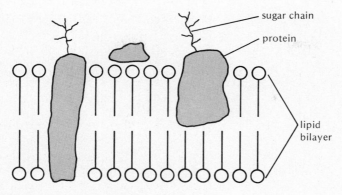

Figure 5-3. Current model of plasma membrane.

the membrane like floating islands in a sea of lipid. This movement is sometimes restricted by rod-shaped protein elements (microtubules or microfilaments) attached to the inner membrane surface. The control of membrane protein mobility by these rod-shaped proteins is not well understood but is the topic of a great deal of recent research. As can be seen in Figure 5-3, the outermost boundary of the plasma membrane consists of sugar chains attached to the protein or lipid. These sugars may be important in mediating the initial contact between the sperm and egg plasma membranes during fertilization.

SPERM-EGG INTERACTION

Let us now consider the ultrastructural aspects of sperm-egg interaction. What does the fertilization process look like under the electron microscope? Figures 5-4 and 5-5 are elegant photos of sperm attaching to the egg surface, as seen with the scanning electron microscope.

Mature spermatozoa are not always able to fertilize eggs. For example, freshly ejaculated human sperm cannot fertilize eggs because the acrosome reaction does not occur in such sperm. The process by which sperm become capable of fertilizing eggs is called *capacitation*. Human sperm must be in the female reproductive tract for about seven hours before they can fertilize eggs. Capacitation time for rabbits is about six hours and for mice, one hour. Capacitation apparently involves removing or deactivating a so-called decapacitating factor that binds to sperm as they pass through the male reproductive tract. This factor apparently blocks the acrosome reaction. Sperm in the decapacitated state live longer than capacitated sperm and (unlike capacitated sperm) cannot penetrate the linings of the male and female reproductive tracts, preventing damage to these linings.

To show what happens between the sperm and egg membranes during fertilization, the transmission electron microscope must be used. The transmission electron microscope is used to view ultrathin sections through the

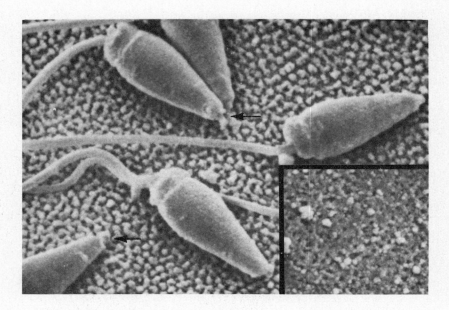

Figure 5–4. Scanning electron micrographs of fertilizing sperm on the egg surface.
Inset shows inner vitelline layer. Courtesy of Dr. Charles Glabe and Victor Vacquier.

sperm and eggs, thus revealing the interior parts of the membranes, coats, and cytoplasm.

A classic study of fertilization in *Hydroides* was performed by Colwin and Colwin using the transmission electron microscope. Figure 5–6 is a diagram of their results.

When the sperm head encounters the outer layer of the egg, the acrosomal vesicle bursts and the acrosomal membrane becomes continuous with the sperm plasma membrane. The acrosomal granule dumps lysins onto the egg. These enzymes help the sperm penetrate the outer egg coats. When the sperm head is approximately halfway through the outer egg layers, microvilli form at the base of the acrosome and also on the egg plasma membrane. The sperm and egg microvilli eventually fuse, the sperm plasma membrane and the egg plasma membrane become continuous, and the nucleus of the sperm and its other contents move into the egg cytoplasm. Thus, fertilization represents a fusion of the sperm and egg.

EGG CORTICAL REACTION

What happens, at an ultrastructural level, in the egg at fertilization? We shall use the egg of the sea urchin as an example. The sea urchin has become a model system in the study of fertilization because gametes can be extracted from them in massive numbers. Furthermore, the entire fertilization reaction can be easily observed in the laboratory in plain sea water. Figure 5–7 shows

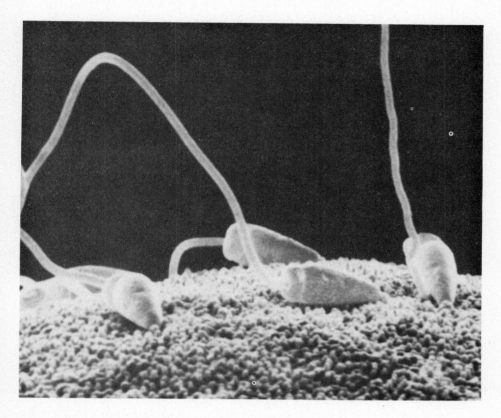

Figure 5-5. Scanning electron micrograph of sperm bound to the egg surface.
Courtesy of David Epel.

that upon fertilization in the sea urchin egg, directly below the plasma membrane, tiny cortical granules fuse with the plasma membrane. Each cortical granule is about one micrometer (10^{-3} millimeter) in diameter; each egg contains about 15,000 of these tiny structures. The cortical granules release some of their contents into the space between the plasma membrane and the vitelline layer. Cortical granules contain enzymes, structural proteins, and sugar-protein complexes. Carroll and Epel have shown that one of the cortical granule enzymes alters sperm-receptor proteins on the vitelline layer, preventing additional sperm from attaching. This enzyme is thus an important part of the so-called block to polyspermy, the prevention of fertilization by more than one sperm. This is not the whole story, however; even earlier in the fertilization reaction (about one second after sperm attachment) sodium ions flow into the cell, causing a brief voltage change that appears to prevent additional sperm from entering the egg.

Another cortical granule enzyme disconnects the vitelline layer from

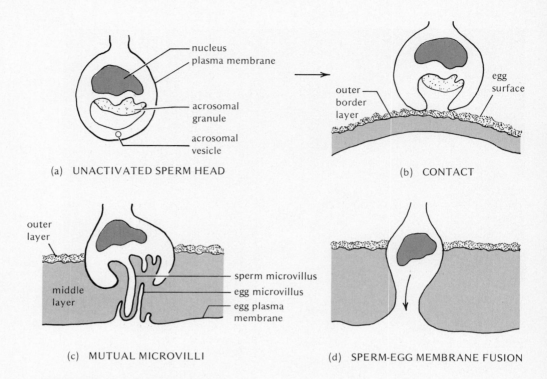

Figure 5–6. Fertilization in *Hydroides*.

the plasma membrane. It then lifts away from the surface of the egg, forming the fertilization membrane. The fertilization membrane, therefore, is a combination of the vitelline layer and certain structural proteins released from the cortical granules. The fertilization membrane is an effective additional means of blocking polyspermy. Other material derived from the cortical granules sticks to the plasma membrane, forming a clear surface coat called the hyaline layer (Figure 5–7). This layer helps the sea urchin embryo cells stay together during the cleavage or cell division stage of development.

About twenty minutes after the sperm touches the sea urchin egg, the sperm nucleus fuses with the egg nucleus. The first cleavage division occurs shortly thereafter. Fertilization is complete, and the embryo begins to develop. Some of the events occurring during sea urchin fertilization are shown in Figs. 5–7 through 5–10.

FATE OF SPERM MITOCHONDRIA

Before turning to the biochemical and physiological events occurring during fertilization, let us briefly consider one additional question. What happens to the sperm mitochondria at fertilization? Mitochondria supply cells with energy and contain DNA. Sperm mitochondria appear to have a different

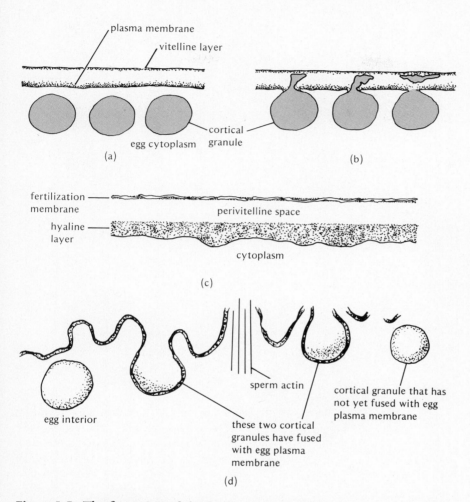

Figure 5-7. The formation of the fertilization membrane in a sea urchin egg, showing the cortical reaction.

fate in different organisms. In insects, for example, sperm mitochondria enter the egg. In the rat, they also enter but disintegrate in about 30 minutes. In *Hydroides*, sperm mitochondria enter the egg at fertilization and remain intact at least through the fifth cleavage division. In tunicates, however, an elegant study by Ursprung and Schabtach using the electron microscope showed that the sperm mitochondrion did not enter the egg. Instead, it was knocked off the sperm as the sperm began to enter the outer egg coats (see Figure 5-11). As the sperm squeezes between the chorion cells at the outer boundary of the egg, the mitochondrion can not make it through and separates from the rest of the sperm.

Not only is the fate of mitochondria different in different species, so is

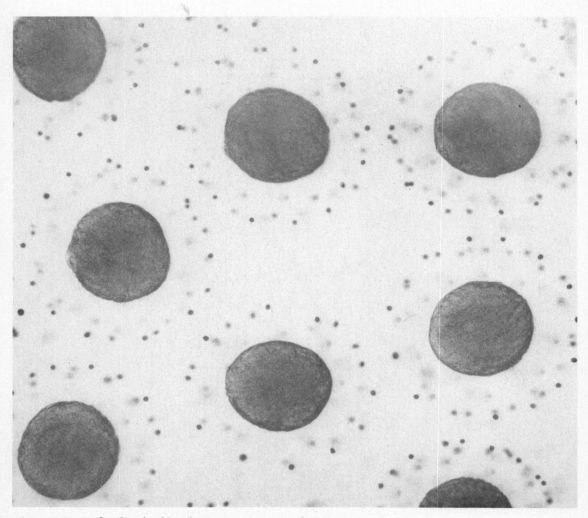

Figure 5-8. Unfertilized echinoderm egg. Courtesy of Victor Vacquier.

the number of sperm that enter the egg. We mentioned a block to polyspermy that prevents more than one sperm from entering the egg. This, however, is not the case in all organisms. For example, many sperm can enter the eggs of some molluscs, selachians, birds, reptiles, and urodeles. All but one, however, eventually degenerate in the egg cytoplasm. Thus, these eggs allow more than one sperm to enter, but eliminate all but one of them. This type of fertilization is called physiological polyspermy.

SUMMARY
Different things occur during fertilization in different organisms, and very few mechanisms are exactly the same in all organisms. Many successful

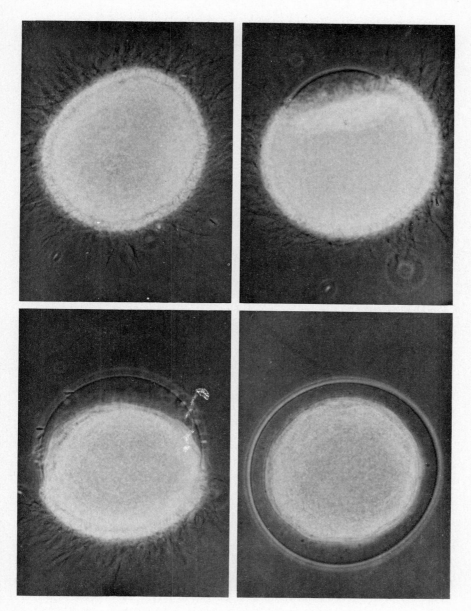

Figure 5–9. Sea urchin fertilization. Note the progressive formation of the fertilization membrane. Courtesy of Victor Vacquier.

components of individual fertilization processes have evolved. In this section we have examined sperm structure, egg surface structure, sperm-egg interaction at the structural level, the egg cortical reaction, the block to polyspermy, the fate of sperm mitochondria.

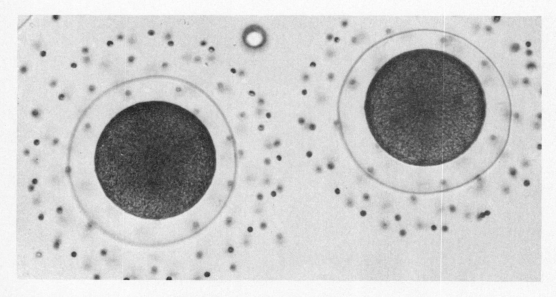

Figure 5-10. Fertilized sea urchin egg. The fertilization membrane is complete. Courtesy of Victor Vacquier.

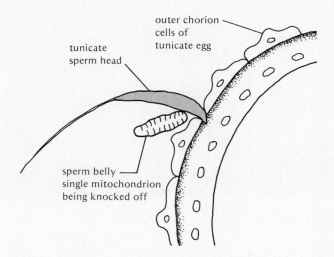

Figure 5-11. Fate of the sperm mitochondrion in the tunicate. Based on experiments by H. Ursprung and E. Schabtach, *J. Exp. Zool.* 159 (1965): 379–84.

With this background in the larger structural aspects of fertilization, we can examine the fertilization reaction at the molecular level. The story at the molecular level is still incomplete, but a body of fascinating information is beginning to emerge.

THE BIOCHEMICAL AND PHYSIOLOGICAL EVENTS OF FERTILIZATION

At the beginning of this chapter we asked: What turns on the complex series of metabolic events that occur in eggs after the initial sperm contact? Some of the mysteries clouding this question have recently been solved. Let us now examine some of the physiological and biochemical events that occur during fertilization and egg activation.

ACTIVATION OF EGGS BY AN ACROSOMAL PROTEIN

How do sperm activate eggs? Recently, Cross, Hornedo, and Gould–Somero prepared a fraction of acrosomal material from *Urechis* sperm. The fraction contained a basic protein (consisting of 50% lysine and arginine and less than 0.1% carbohydrate) that binds to and agglutinates *Urechis* eggs, sea urchin eggs, and zygotes of the alga *Pelvetia*. *Urechis* eggs treated with this "binding protein" were activated and began to develop. A surface coat elevated, the germinal vesicle (nucleus) broke down, and polar bodies formed. A transient, positive shift in egg membrane potential (similar to the depolarization that occurs at fertilization) also occurred in eggs treated with binding protein. These workers may have discovered the sperm protein that activates the egg during the normal fertilization reaction.

TIMETABLE OF EVENTS FOLLOWING FERTILIZATION

Figure 5–12 gives the timetable of events that follow the fertilization of a representative system, the egg of the sea urchin *Strongylocentrotus purpuratus*. As mentioned previously, much of the work in this area has been done with the sea urchin, because massive numbers of eggs and sperm can be removed from these animals in the laboratory and because all of the early events of fertilization can easily be observed in the laboratory, in sea water.

The sequence of events that follows the contact between the sperm and egg is as follows :

1. influx of sodium ions
2. liberation of calcium ions from intracellular depots

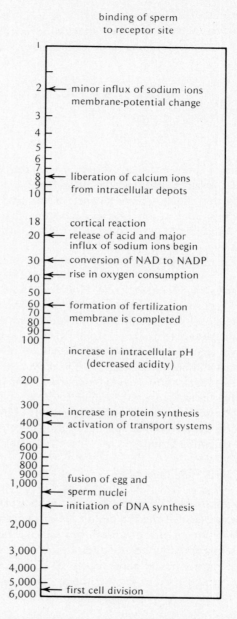

Figure 5–12. Timetable of events following fertilization of the sea urchin egg. The time scale shows seconds elapsed after sperm binds to egg. From D. Epel, "The Program of Fertilization," *Scientific American* 237 (1977): 129 Copyright © 1977 by Scientific American, Inc. All rights reserved. Used with permission.

3. cortical granule reaction, release of acid, and another major influx of sodium ions

4. conversion of NAD to NADP

5. rise in oxygen consumption

6. completion of fertilization membrane

7. increase in intracellular pH (decrease in acidity)

8. increase in protein synthesis

9. activation of transport systems

10. fusion of egg and sperm nuclei

11. initiation of DNA synthesis

12. first cleavage division of zygote

Experiments from many laboratories have provided the sequence of events shown in Figure 5–12 (reviewed by Epel, 1977). One example is the elegant set of experiments by Ridgway, Gilkey, and Jaffe showing that calcium ions are released into the egg cytoplasm early in the fertilization process, as first suggested by Mazia in 1937. They used a protein, extracted from jellyfish, that glows in the presence of free calcium ions. This protein, called aequorin, was injected into the large unfertilized eggs of a fish, the Japanese medaka (*Oryzias latipes*). The eggs glowed slightly. Upon fertilization, however, there was a 10,000-fold increase in luminescence, indicating that a large amount of calcium was released from a bound state in the egg. To determine if the rise in calcium plays a key role in egg activation, Steinhardt, Epel, Chambers, Pressman, and Rose used a substance called ionophore A23187, which causes the release of calcium ions in cells. In the absence of sperm, this ionophore caused an activation response resembling that accompanying fertilization in sea urchin eggs. Epel, Carroll, Yanagimachi, and Steinhardt found that the ionophore activated many types of eggs, including those of amphibians, mammals, tunicates, and mollusks. Thus the release of calcium seems to be an important early event in the fertilization reaction that may directly trigger some of the other events involved in the activation of the egg.

As can be seen in Figure 5–12, an enzyme, NAD kinase, is activated about 30 seconds after the sperm has bound to the egg. This enzyme catalyzes the transfer of a phosphate group from ATP to NAD, forming NADP. NAD and NADP are coenzymes. An increase in the amount of NADP in the freshly fertilized egg plays an important part in preparing the zygote for the many synthetic reactions that will soon follow.

We can also see from Figure 5–12 that there is a rise in oxygen consumption about 35 to 45 seconds after the sperm binds to the egg. At about the same time, another enzyme becomes activated. This enzyme, glucose-6-phosphate dehydrogenase, is very important in initiating the metabolic reactions involved in sugar metabolism. Some of the synthetic reactions that follow fertilization are apparently begun by the activation of the relevant enzymes.

With modern technology, it is relatively easy to obtain evidence regarding the sequence of events in the fertilization process. For example, to show that protein synthesis increases about 350 seconds after sperm binding, one simply incubates eggs with radioactive amino acid at various times after sperm binding. New protein synthesis is measured by examining the protein with a scintillation counter and measuring the amount of radioactive amino acid incorporated into protein. Many such experiments have been carried out in different systems by different investigators. The studies described in this chapter represent only a tiny sampling of the volumes of data in this area. They are intended only to introduce the reader to the findings that have been made in this field.

Protein synthesis in sea urchins increases rapidly after fertilization, as shown in Figure 5–13. Protein synthesis is measured by incubating eggs with radioactive amino acid (in this case methionine), precipitating the protein from sea urchin embryo homogenates with acid, and measuring the amount of radioactive amino acid incorporated in the precipitate. There is a rapid increase in protein synthesis during the first two hours after fertilization. By

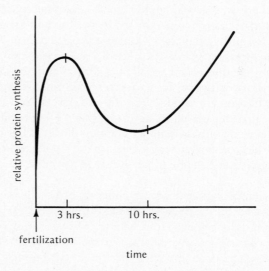

Figure 5–13. Protein synthesis in the sea urchin embryo. Based on experiments by H. Ursprung and K. D. Smith, *Brookhaven Symp. Biol.* 18 (1965):1–13, Figure 1.

four hours, protein synthesis decreases, but increases again at about ten hours. These data suggest that, upon fertilization, a large variety of proteins are required for the numerous reactions and syntheses that occur in the new embryo. Of immediate importance are the spindle proteins and chromosomal proteins needed for chromosome replication and cell division. New membrane proteins are also needed for the synthesis of new membranes as cells divide.

The sudden stimulation of protein synthesis upon fertilization is not well understood. Factors possibly involved in this activation could include (1) The activation of inactive ribosomes; (2) The appearance of new ions, enzymes, or cofactors (cofactors are small molecules that are needed to activate an enzyme or reaction); (3) The activation of messenger RNA templates, or transfer RNA, or activating enzymes. The ribosomes of unfertilized eggs can act as the seats of protein synthesis if synthetic messenger RNA is added to these ribosomes in an *in vitro* protein-synthesizing system. Recent results by Danilchik and Hille, however, show that ribosomes obtained from unfertilized eggs of the sea urchin *Strongylocentrotus purpuratus* became fully active in protein synthesis only after a characteristic, reproducible delay of up to 15 minutes. Ribosomes obtained from blastula embryo polyribosomes, on the other hand, are active in synthesizing protein without delay. This suggests that ribosomes from unfertilized eggs are bound to some sort of inhibiting molecule, such as a protein that blocks ribosome function. Intracellular physiological changes, such as the change in pH that occurs at fertilization, may separate the inhibitory component from the ribosomes, thereby activating them.

It may be that the messenger RNA in unfertilized eggs is inactive, or that the message is in some way masked, possibly by a protein coat. Upon fertilization, there is a rapid increase (maximum at 3 minutes after fertilization) in the activity of enzymes (proteases) that catalyze the hydrolysis (breaking down) of proteins (Figure 5–14). We can speculate that these enzymes may digest away some of the protein masking messenger RNA molecules, in this way activating the messages.

Protein synthesis may be activated by other mechanisms. Addition of polyadenylic acid residues to messenger RNA appears to be important in activating the message so that it can function in protein synthesis. At fertilization, adenylation of messenger RNA occurs in the cytoplasm. This may activate the messages. Steinhardt, Winkler, Grainger, and Minning have shown that both increased intracellular pH and increased intracellular calcium ion concentration play roles in the activation of protein synthesis. Other work with a different sea urchin species, however, suggests that pH may not be a key factor in such activation. Clearly, the total picture of what activates protein synthesis at fertilization still is not well understood.

The experiments described in Figure 5–13 give the picture for total cell protein synthesis. Although the rate of total cell protein synthesis may

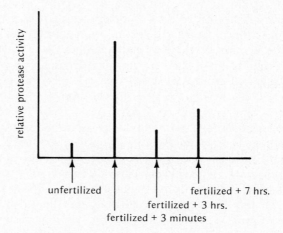

Figure 5-14. Protease activity in the sea urchin egg. Based on experiments by Lundblad and Lundblad.

increase at a given time, this does not mean that synthesis of all of the proteins in the cell is increasing. Figure 5-15, for instance, shows that the rate of synthesis of two specific proteins can be quite different at different times following fertilization in the sea urchin. Chapter 14 will describe in detail how specific protein synthesis is determined. The results graphed in Figure 5-15 were derived by labeling sea urchin embryos with radioactive amino acids during the first hour after fertilization and during the eighth hour after fertilization. The embryos were then homogenized and placed on chromatographic columns that separate specific proteins according to molecular size

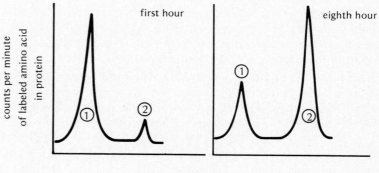

Figure 5-15. Synthesis of individual proteins after fertilization in the sea urchin. Each peak represents a specific protein. The results indicate that more of protein 1 than protein 2 is synthesized during the first hour. The opposite is true during the eighth hour. (Based on results of Ellis, *J. Exp. Zool.* 163:1-22, Figures 3, 4, 6.)

and charge. As can be seen, more of protein 1 than protein 2 is synthesized during the first hour after fertilization. During the eighth hour, however, the opposite is true. Different amounts of the same protein are thus synthesized at specific times after fertilization. This finding is consistent with the contention that developing embryos require specific amounts of different proteins at given times during early development.

HYPOTHETICAL RELATIONS AMONG THE EVENTS IN FERTILIZATION

Figure 5–16 outlines a tentative, largely hypothetical scheme proposed by Epel to interlink the events occurring during fertilization. This scheme, though it is based upon considerable experimental evidence, will surely be altered over the next few years as additional experiments are performed. A

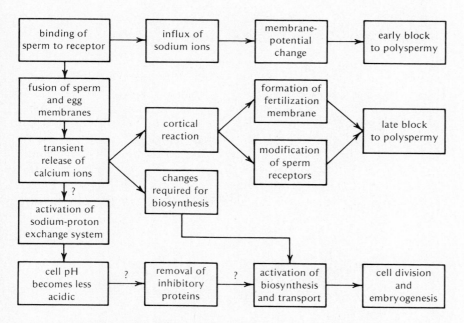

Figure 5–16. Flow chart of hypothetical relations among early calcium-dependent changes upon fertilization and late pH-dependent changes. The mechanism by which calcium turns on the sodium-proton exchange system to reduce the acidity of the egg cytoplasm (an increase in pH) is not yet understood. How this pH shift initiates protein and DNA synthesis is also not known, but the mechanism may involve removing inhibitory proteins from functional proteins such as those of ribosomes or from structural proteins such as actin or tubulin subunits. In the latter case, the removal of an inhibitory protein might enable the subunits to polymerize into filaments, changing the cell structure in ways that could initiate a variety of biochemical events. From D. Epel, Scientific American 237 (1977): 129. Copyright © 1977 by Scientific American, Inc. All rights reserved. Used with permission.

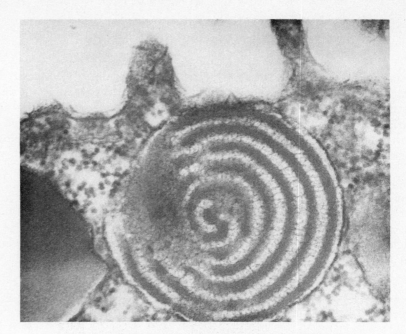

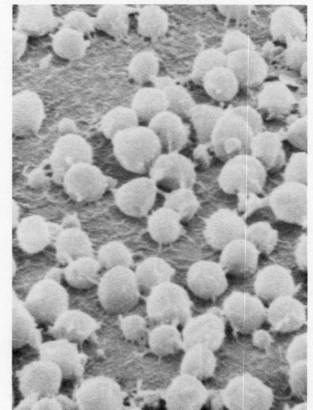

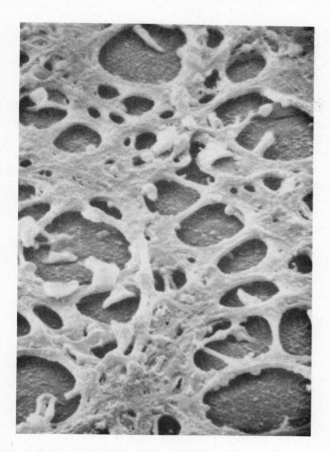

Figure 5–17. Calcium-initiated discharge of isolated sea urchin cortical granules.
(Left) Top sectioned cortical granule (transmission electron micrograph). (Below, left)
Intact cortical granules on inside-out plasma membrane (scanning electron micrograph).
(Above, right) Bottom after addition of calcium, cortical granules fuse (scanning
electron micrograph). Courtesy of Victor Vacquier.

few pieces to the puzzle of what triggers each event have been put in place.
For example, the release of calcium ions in the egg cell following sperm-egg
contact apparently triggers the cortical reaction and other early changes.
Vacquier performed an elegant experiment to support this notion (Figure
5–17). He isolated sea urchin egg plasma membranes with their associated
cortical granules. When calcium was added, the cortical granules released
their contents. Thus the release of free calcium ions in the egg appears to
trigger the cortical reaction directly.

In vertebrates, secondary oocytes are arrested in their development,
with the chromosomes on the equatorial plate of the second meiotic meta-

phase. Only after the egg is fertilized and the second polar body has been extruded is meiosis complete. A substance called *cytostatic factor* may be responsible for arresting cells at metaphase. The release of free calcium may inactivate the cytostatic factor, releasing the block at the second meiotic metaphase.

What is the significance of the reduction of acidity of the egg cytoplasm, another important event shown in Figures 5-12 and 5-16? Experiments by Nishioka, Epel, Shen, and Steinhardt suggest that the pH need be raised only for ten minutes for development to begin. Since many of the metabolic changes occur afterwards, the pH increase probably initiates a cellular change that allows activation to occur. Such a change may involve the dissociation of an inhibitor from enzymes or ribosomes, enabling synthetic reactions to proceed.

Synthetic reactions necessary for development to proceed are controlled by enzymes. Enzymes are often activated by specific ions. We have already seen that the release or influx of calcium and sodium ions occurs early in fertilization. These changes may activate certain synthetic reactions by activating the relevant enzymes.

MOLECULAR BIOLOGY OF SEA URCHIN FERTILIZATION

Some interesting (and rather unexpected) results have recently been obtained from studies of the molecular biology of sea urchin fertilization. Some of these results were obtained using a fairly new technique called two-dimensional gel analysis. One-dimensional separation methods separate proteins according to only one property, such as molecular weight. Clearly, different proteins that happen to have the same molecular weight will not be resolved by such a method. Two-dimensional gel analysis separates proteins according to two properties, such as molecular weight and electric charge, thus resolving somewhat similar proteins.

Infante and Heilmann showed by two-dimensional gel analysis that the several hundred translation products of messenger RNA extracted from unfertilized sea urchin eggs were indistinguishable from the translation products of messenger RNA extracted from early embryos. Brandhorst showed by the same method that almost all of the 400 species of proteins synthesized within an hour of fertilization (from maternal mRNAs) were also translated at a low rate in unfertilized eggs. From these data, one can conclude that fertilization in the sea urchin speeds up protein synthesis dramatically—by a factor of 100 several hours after fertilization—but that it does not substantially alter the set of mRNA sequences being translated.

This increased rate of protein synthesis results, in part, from an increase in the assembly of polyribosomes from maternal components. By the 16-cell stage, about 30% of the egg ribosomes are in polysomes and 10% or less of the mRNA is of embryonic rather than maternal origin. Polysome content in the embryos continues to increase until about 60% of the egg ribosomes are in polysome configurations. The rate of protein synthesis increases similarly from about 120 picograms per hour at the time of the first cleavage to at least 500 picograms per hour by the blastula stage.

The increased rate of protein synthesis after fertilization is not only reflected in increased assembly of polyribosomes. Fertilization also increases the translational elongation rate of newly forming polypeptide chains by a factor of about 2.5 as compared to the rate in the polysomes of an unfertilized egg. The increase in intracellular pH (from about 6.84 to 7.27) that occurs about 60 seconds after fertilization may trigger the increased translational elongation rate. Winkler, Steinhardt, and colleagues directly demonstrated that the elongation rate at pH 6.9 is lower than that at pH 7.4 by incubating *in vitro* components of the protein synthesizing machinery from unfertilized sea urchin eggs. The pH increase and other ionic changes that occur soon after fertilization may also activate the maternal mRNA contained in ribonucleo-protein structures in unfertilized eggs so that the mRNA can be used in translation.

SUMMARY

In this section we have examined from a causal viewpoint the sequence of events that occurs during fertilization. Which event triggers the other events? As yet, we cannot fully answer this question. We have seen, how-ever, that releasing specific ions such as calcium and altering the cellular pH appear to be important triggers of other events.

MOLECULAR ASPECTS OF SPERM-EGG RECOGNITION

Now that we have surveyed fertilization from the ultrastructural and physio-logical viewpoints, we shall briefly consider an important question: How does a sperm cell recognize an egg cell? Again we center our attention on the sea urchin.

In the ocean, sea urchins release up to 400 million eggs per female and 100 billion sperm per male during their three- to eight-month breeding season. How does a sea urchin sperm know that it has reached a sea urchin egg of the same species? In certain organisms, such as the fern and moss, male gametes are attracted to female gametes by specific chemicals given off by the female gametes. In the fern, L-malic acid seems to be the chemoattractant; in moss, sucrose appears to serve a similar function. In the water mold, male gametes are attracted to a specific organic compound, L-sirenin, released by the female gametes. The male cells respond to levels of L-sirenin as low as 0.5×10^{-10} M in solution! In most (but not all) animals, however, female gametes do not chemically attract male gametes over long distances; indeed, most of the movement of animal sperm appears to be random. The key aspect of sperm-egg recognition appears to be the specific sticking of sperm to egg. What is the nature of the molecules involved in this important process? To answer this question we must consider two processes: the attachment of sperm to the outer egg coats, and the recognition of the underlying egg plasma membrane by the sperm.

The jelly coat of sea urchin eggs contains a substance *fertilizin* (named by Lillie). This substance is an acid mucopolysaccharide (a protein containing polysaccharide and sulfate ions). It appears to bind sperm. If fertilizin is acid-extracted from sea urchin egg jelly coats and added to free sperm, it will dump them. This suggests that fertilizin has several receptor sites for sperm. Fertilizin (or a molecule associated with it) also appears to activate sperm by increasing their respiration and motility and by triggering the acrosomal reaction.

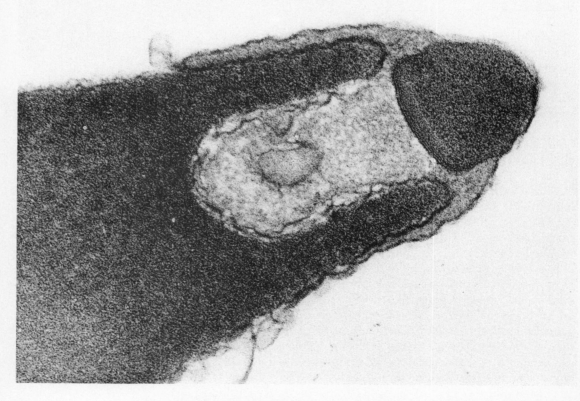

Figure 5-18. Head of sea urchin sperm, showing the acrosomal granule at the tip.
Courtesy of Victor Vacquier.

As described earlier in this chapter, Vacquier and colleagues at the University of California at Davis isolated bindin from the acrosomal granules of sea urchin sperm (Figures 5–18 and 5–19). Bindin agglutinates eggs only from the same species of sea urchin. Eggs whose surfaces have been treated with carbohydrate-destroying agents are not agglutinated by bindin. This suggests that bindin recognizes specific carbohydrate portions of egg-surface glycoproteins (Figures 5–20 and 5–21). Vacquier has also isolated a molecule (apparently a glycoprotein) from the egg surface. This molecule has a species-specific affinity for bindin and may be the egg-surface receptor site for bindin.

Many questions still remain unanswered. Is the bindin receptor site on the egg surface fertilizin? What is the relationship between bindin and a postulated molecule on the sperm cell surface, *antifertilizin*, that binds to

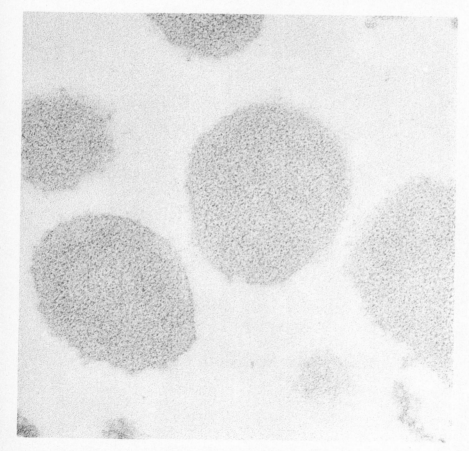

Figure 5–19. Isolated acrosomal granules from sea urchin sperm. Courtesy of Victor Vacquier.

reciprocal motion   rotary motion

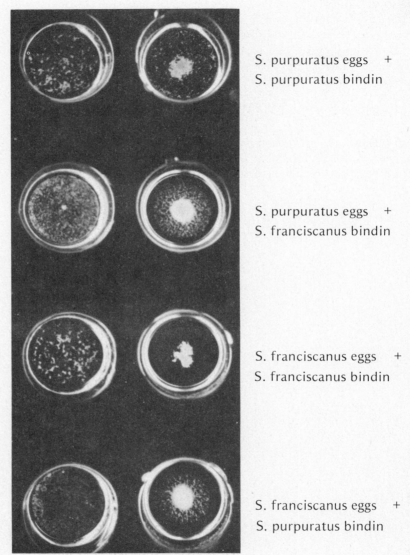

S. purpuratus eggs   +
S. purpuratus bindin

S. purpuratus eggs   +
S. franciscanus bindin

S. franciscanus eggs   +
S. franciscanus bindin

S. franciscanus eggs   +
S. purpuratus bindin

Figure 5-20. Specificity of sea urchin sperm bindin in agglutinating sea urchin eggs. The figure shows that the best agglutination occurs with bindin isolated from sperm of the same species. Courtesy of Victor Vacquier.

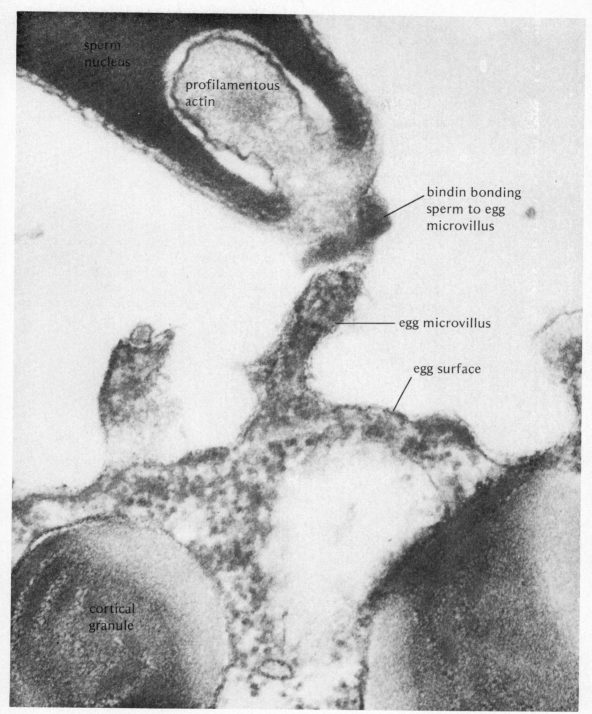

Figure 5-21. Sea urchin sperm bound to egg microvillus by bindin. Courtesy of Victor Vacquier.

fertilizin? What are the functional groups of bindin and their receptor sites that cause the interaction of these molecules? The preliminary evidence cited above suggests that we are dealing with the interaction of protein (bindin) with sugar portions of the egg surface receptor sites (Figure 5–22). Protein-carbohydrate interaction appears to play an important role in many cell-cell recognition phenomena besides sperm-egg interactions.

SUMMARY

The fertilization process is important to us all. It marks the beginning of the new organism, and provides for mixing the genetic information from the mother and father to form a new, unique being. This chapter gave a glimpse of the complexities involved in fertilization. We examined how the sperm penetrates the outer egg coats and how it fuses with the egg plasma membrane. We observed the egg cortical reaction and how it can provide a block to polyspermy. At the molecular level, we noted some of the many biochemi-

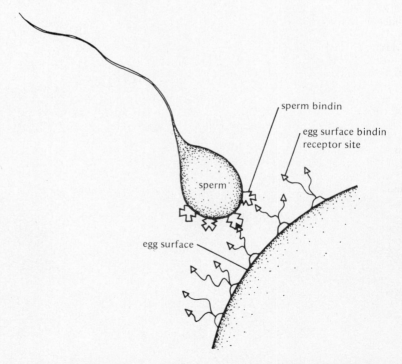

sperm bindin

egg surface bindin
receptor site

sperm

egg surface

Figure 5–22. Model of the molecular basis of sperm-egg recognition. The lock and key scheme may involve a protein (bindin) on the sperm interacting with sugar groups on the egg surface. The size relationships of these hypothetical molecules are greatly exaggerated.

cal reactions that become activated upon fertilization and described how events such as release of ions and change in cellular pH trigger some of the other events. Finally, we looked at the molecular basis of sperm–egg recognition and examined molecules such as fertilizin and bindin that have been implicated in mediating gamete contact. The exploration of fertilization, however, has barely begun. It is likely that exciting new work during the next several years will provide answers to many of the outstanding questions about the fertilization reaction. Perhaps the question posed at the beginning of the chapter will be fully answered during the next few years: "How can a tiny sperm that fuses with only 0.0002 percent of the egg surface trigger the multitude of changes that occur in the new zygote?"

READINGS AND REFERENCES

Carroll, E. J., and Epel, D. 1975. Isolation and biological activity of the proteases released by sea urchin eggs following fertilization. *Develop. Biol.* 44: 22–32.

Colwin, L. H., and Colwin, A. L. 1964. Role of gamete membranes in fertilization, 22nd Symposium of The Society for the Study of Developmental Growth. Chicago: Academic Press.

Danilchik, M. V., and Hille, M. B. 1981. Sea urchin egg and embryo ribosomes: Differences in translational activity in a cell-free system. *Develop. Biol.* 84: 291–98.

Davidson, E. H., Hough-Evans, B. R., and Britten, R. J. 1982. Molecular biology of the sea urchin embryo. *Science* 217: 17–26.

Epel, D. 1977. The program of fertilization, *Scientific American* 237: 129–39. This review is unusually stimulating because it describes possible causal relationships among the events of fertilization and provides evidence for these relationships.

Masui, M., and Clarke, H. J. 1977. Oocyte maturation. *Int. Rev. Cytol.* 57: 186–282.

Mazia, D. 1937. The release of calcium in arbacea eggs upon fertilization. *J. Cell Comp. Physiol.* 10: 291–304.

Metz, C. B., and Monroy, A. 1967. *Fertilization*. Chicago: Academic Press.

Singer, S. V., and Nicolson, G. L. 1972. The fluid mosaic model of the structure of the cell membrane. *Science* 175: 720–31.

Steinhardt, R. A., and Epel, D. 1974. Activation of sea urchin eggs by calcium ionophore. *Proc. Natl. Acad. Sci. U.S.* 71: 1915.

Tegner, M. J., and Epel, D. 1973. Sea urchin sperm-egg interactions studied with scanning electron microscopy. *Science* 179: 685–88.

Tyler, A. 1955. Gametogenesis, fertilization and parthenogenesis. In B. H. Willier, P. Weiss, and V. Hamburger (eds.), *Analysis of Development*. Philadelphia: Saunders.

Ursprung, H., and Schabtach, E. 1965. Fertilization in the tunicate: Loss of the paternal mitochondrion prior to sperm entry. *J. Exp. Zool.* 159: 379–84.

Vacquier, V. D., and Moy, G. W. 1977. Isolation of bindin: The protein responsible for adhesion of sperm to sea urchin eggs. *Proc. Natl. Acad. Sci. U.S.* 74: 2456–60.

Winkler, M. M., Steinhardt, R. A., Grainger, J. L., and Minning, L. 1980. Dual ionic controls for the activation of protein synthesis at fertilization. *Nature* 287: 558–60.

6
CLEAVAGE

Fertilization has occurred. The zygote has become "activated." Many synthetic reactions have begun. What next? When we think of an embryo, we think of a multicellular organism. At this point, all that exists is a single cell, the fertilized egg. The division of the fertilized egg, called *cleavage*, transforms the single cell into many cells, called *blastomeres* in early embryos. Cleavage is visually dramatic and can easily be observed in many embryos in the laboratory. Such observations lead to some interesting conclusions. One is that the pattern of cleavage is not the same in all organisms. In this chapter, we shall briefly examine some of the different ways in which embryos cleave. In addition, we shall look at some of the mechanisms and factors that influence cleavage patterns and the cleavage process itself. By the end of this chapter, our embryos will be multicellular, ready to twist and turn to form into something that begins to resemble a "real" organism.

We shall first examine various patterns of cleavage and describe some of the factors that appear to influence these patterns. Representative embryos will be selected that illustrate the concepts of interest.

TOTAL CLEAVAGE

Total, or *holoblastic*, cleavage is a cleavage in which the divisions pass through the entire fertilized egg. This type of cleavage occurs in embryos with a small or moderate amount of yolk, such as the sea urchin, *Amphioxus*, the frog, and most mammalian embryos. Dense accumulations of yolk retard cleavage.

FROG EMBRYO AND THE INFLUENCE OF YOLK ON CLEAVAGE

The influence of yolk on cleavage is nicely shown in the frog egg, which is *moderately telolecithal*. That is, yolk is substantially more concentrated in the *vegetal hemisphere* of the egg than in the *animal hemisphere*. The animal hemisphere is the region of the egg where the nucleus resides; the vegetal hemisphere is the other half. When eggs are free to rotate, the animal hemisphere is usually up; the more dense, yolky vegetal region is down. A low power lens reveals that the first cleavage division in the frog begins in the pigmented animal hemisphere and slowly moves down vertically (*meridonal* cleavage) through the more yolky vegetal hemisphere. The yolk retards the movement of the cleavage furrow through the vegetal region. Before the first division is completed, the second division often begins in the animal region, also vertically, at right angles to the first (Figure 6–1). Yolk thus retards the movement of the division plane into the vegetal region. The first two cleavages are total, however, and eventually pass completely through the yolky vegetal area. The third cleavage in the frog embryo is horizontal (*equatorial*). This cleavage does not separate the embryo exactly in half. Instead, it is displaced toward the animal region so that the upper (animal) cells are smaller than the lower, more yolky vegetal cells. The cleavage

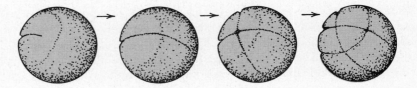

Figure 6-1. Cleavage in the frog.

pattern of the frog is thus total and unequal. Continued division results in many small animal cells and fewer, but larger, vegetal cells (Figure 6–1).

Yolk is obviously important in frog cleavage. The nucleus lies at the top of the dense yolk. Since the mitotic spindle forms near the nucleus, the cleavage furrows start at the animal pole region and move toward the vegetal area, retarded by the dense yolk. The third cleavage is also influenced by the yolk because of its position above the equator. Since the mitotic spindles form in conjunction with the nuclei, they too are displaced above the equator. In this way, the yolk directly causes the third cleavage to occur closer to the animal pole. Smaller animal cells (*micromeres*) now separate from the larger vegetal cells (*macromeres*). When they do so, the nuclei divide and the vegetal nuclei sit atop the dense yolk of the vegetal cells. The smaller, less yolky animal cells continue to cleave more rapidly than the yolky vegetal cells. In this way, the frog embryo takes form with yolk determining the size of the cleavage blastomeres.

In vertebrates and in certain other organisms (cephalopods, nematodes, and tunicates), the blastomeres become arranged across a plane of bilateral symmetry at some time during cleavage. In other words, a right and a left side of the embryo become established that are mirror images of each other. When this occurs, the pattern is called *bilateral cleavage*.

AMPHIOXUS
The primitive chordate *Amphioxus* possesses a small amount of evenly distributed yolk (an *oligolecithal* or *isolecithal* egg). The first three cleavages are much like those in the frog embryo. Because of the smaller amount of yolk, however, the third cleavage separates the animal and vegetal regions more equally. Thus, the animal and vegetal cells formed are nearly equal in size, with the vegetal blastomeres only slightly larger.

SEA CUCUMBER
Many other embryos, such as the sea cucumber, have only a small amount of evenly distributed yolk (an oligolecithal egg). Such cleavages are total and equal (see Figure 6–2). In the sea cucumber the first cleavage is vertical, and the second cleavage is also vertical but at right angles to the first. The third cleavage is horizontal. The third cleavage in the sea cucumber, however, separates the embryo into upper and lower cells that are about equal in size.

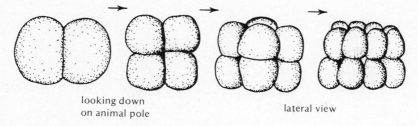

looking down
on animal pole   lateral view

Figure 6–2. Radial cleavage in the sea cucumber.

The upper cells lie exactly over the lower cells. Such an arrangement is called radial cleavage because the pattern of cells is radially symmetrical around the polar axis of the egg. These patterns are influenced, in part, by yolk distribution.

UNEQUAL CLEAVAGE IN THE SEA URCHIN EMBRYO

The sea urchin, which exhibits a radial cleavage pattern and has an oligolecithal egg, cleaves rather routinely until the fourth cleavage. The first cleavage is vertical (see Figure 6–3). The second cleavage is vertical at right angles to the first, and the third is equatorial. Up to now, the cleavage pattern in the sea urchin has been similar to that in *Amphioxus* and the sea cucumber. The fourth cleavage in the sea urchin, however, provides a surprise. The four animal cells now divide vertically and equally, forming eight middle-sized cells called mesomeres. The four vegetal cells, however, cleave horizontally and unequally, producing four larger cells (macromeres) and four tiny cells (micromeres) at the extreme tip of the vegetal region (the vegetal pole) as shown in Figures 6–4 and 6–5. The 16-cell stage, therefore, consists of eight animal mesomeres, four large vegetal macromeres, and four tiny vegetal micromeres.

The mechanisms that cause this unequal fourth cleavage are not well understood. Ikeda found that during cleavage cycles in the sea urchin embryo, the quantity of certain proteins that contained sulfhydryl groups varies in a cyclic way. A variety of experiments suggest that these so-called sulfhydryl cycles play a part in controlling cleavage. For example, if the sulfhydryl cycles are blocked with ether and the embryos are then returned to ether-free sea water, the fourth cleavage is not unequal and micromeres are not formed. A cleavage can also be blocked with ultraviolet light or with the metabolic inhibitor, 2, 4-dinitrophenol. These treatments do not inhibit the embryo's sulfhydryl cycle, but cause embryos to lose a cleavage. When the third cleavage occurs in these treated embryos, it is unequal, producing micromeres at the vegetal pole. Since the sulfhydryl cycles are not blocked at the time of the third cleavage, the fourth sulfhydryl cycle occurs. Thus, the fourth sulfhydryl cycle somehow causes an unequal cleavage. Future work will undoubtedly shed more light upon this problem.

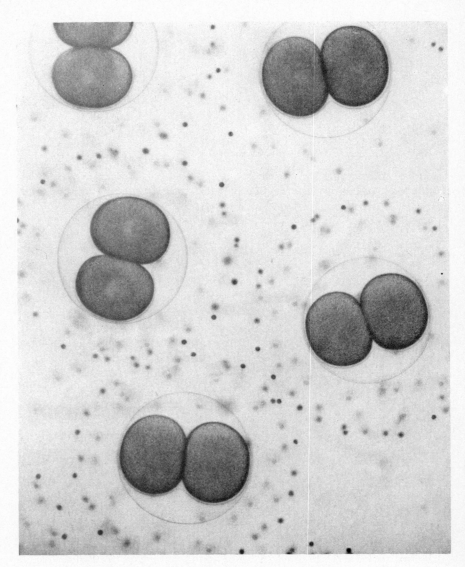

Figure 6-3. First cleavage in the sea urchin. Courtesy of Victor Vacquier.

The fourth cleavage in the sea urchin embryo uncovers yet another interesting problem. The eight medium-sized mesomeres give rise to most of the ectoderm, the outer layer of the embryo. The four large macromeres give rise to some ectoderm and to all of the endoderm, the innermost layer of the embryo. The four small micromeres become the primary mesenchyme cells that, in turn, form the skeletal elements of the embryo. The different sizes of these cell types makes it possible to determine their fates by watching what

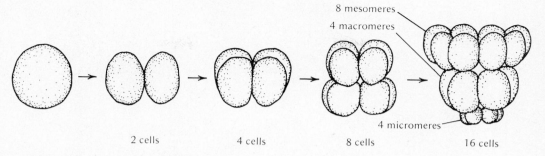

8 mesomeres

4 macromeres

4 micromeres

2 cells   4 cells   8 cells   16 cells

Figure 6-4. Cleavage in the sea urchin.

each cell type forms. In some species, some of the cell types also contain pigment granules that aid the observer in following the development of the cells.

At the 16-cell stage, the mesomeres, macromeres, and micromeres can be separated from one another and then recombined in specific ways. These experiments show that combinations of mesomeres and micromeres can form a normal embryo. Mesomeres plus macromeres can form a nearly normal embryo. Mesomeres, micromeres, or macromeres alone do not develop into a normal embryo. These experiments by Hörstadius indicate that in order to form a normal embryo, one must combine cells from the animal half of the embryo with cells from the vegetal half of the embryo. These results suggest that two basic kinds of cells exist in the embryo. One is most concentrated in the animal half of the embryo, the other in the vegetal region. If an embryo is to develop normally, both animal "stuff" and vegetal "stuff" must be present.

What is the nature of the animal and vegetal "stuff"? This is the $100,000 question. Before we produce the illusion that we really understand what is going on here, let's flatly say that nobody knows what the "stuff" is. Many interesting experiments have been done, however, that shed light upon this problem. It is unlikely that the "stuff" is a cytoplasmic material that can easily be displaced by centrifugation. This conclusion is based upon experiments in which the egg cytoplasm is rearranged by centrifugation. The blastomeres then contain materials different from those that they would have under normal conditions, but a normal embryo still develops. Many other experiments have been done with similar inconclusive results.

A variety of investigations suggest that important differences among mesomeres, macromeres, and micromeres may reside in the surface region of the cells. This region includes the cell membrane and the cortex, the region just below the cell membrane. Experiments using the large squid egg suggest that the cell surface contains key information that can control development. In these experiments (by Arnold), an ultraviolet light microbeam was used to irradiate small regions of the egg surface. These treatments prevented the

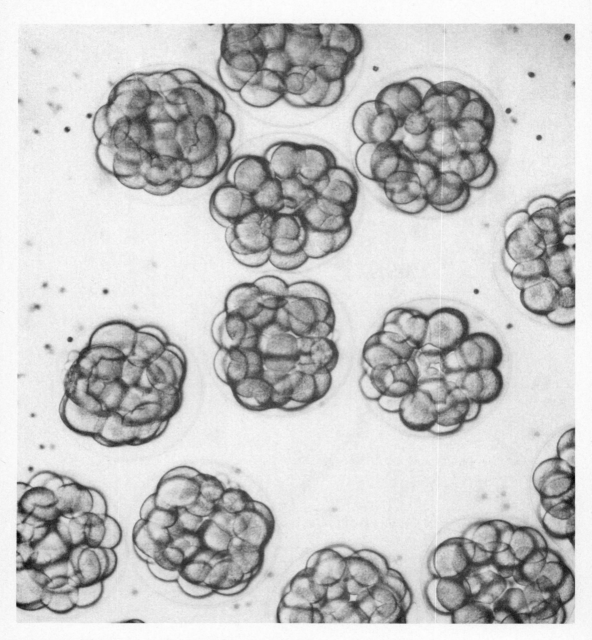

Figure 6-5. The 16- to 32-cell stage in the sea urchin. Courtesy of Victor Vacquier.

formation of specific organs or tissues that normally would form at each irradiated site.

But what about the sea urchin embryo? Are the surfaces of micromeres, mesomeres, and macromeres actually different? Roberson, Neri, and Oppenheimer showed that there are differences between the surface properties of micromeres and those of macromeres and mesomeres. These workers used a lectin (a carbohydrate-binding protein—see Chapter 12) called concanavalin A to study the surfaces of micromeres, mesomeres, and macromeres. Concanavalin A binds to certain receptor sites on the cell surface that contain sugar. This binding can be visualized by coupling a fluorescent dye to the concanavalin A. One can then use a fluorescence microscope to observe the binding of this material to the cell surface as a fluorescent glow in the region of binding. Using this technique Roberson et al. showed that the fluorescence moved to one pole only on the micromere cell surface. No movement was observed on the surface of the macromeres and mesomeres (Figure 6–6). These results suggest that certain sites on the micromere cell surface can move and accumulate at one end of the cell. On the macromeres and mesomeres, however, the sites are more rigidly embedded in the cell surface and do not move. What does all this mean? Clearly, the surfaces of micromeres are different from those of macromeres and mesomeres. Whether or not these surface differences are important in helping the cell types complete their specific lines of development remains to be determined. It is clear, however, that micromeres do become migratory mesenchyme cells that play an important role in the embryonic development of the sea urchin (Chapter 7). Perhaps their "mobile" surfaces facilitate their migration.

SPIRAL CLEAVAGE AND SPECIAL CYTOPLASM

Many embryos, including those of annelids, mollusks, some flatworms, and nemerteans, exhibit a spiral cleavage pattern. In this type of cleavage, the mitotic spindle is oriented obliquely. Therefore, the daughter cells do not lie directly above or below one another, but are shifted so that each upper cell lies over the junction of two lower cells (see Figure 6–7).

Two important developmental concepts can be learned from studying

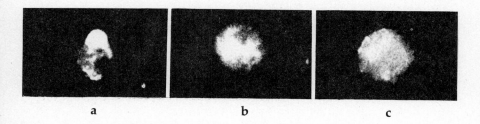

a   b   c

Figure 6–6. Cell surface differences of sea urchin micromere (a), mesomere (b), and macromere (c). Only the micromere cell surface exhibits mobile fluorescent concanavalin A receptor sites. This is shown by the accumulated fluorescence at one tip of the micromere. Other experiments have shown that fluorescence moves from the cell body to the tip. From M. Roberson, A. Neri, and S. B. Oppenheimer, *Science* 189 (1975): 639.

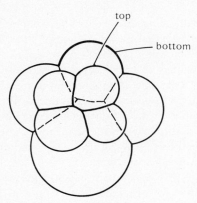

top

bottom

Figure 6-7. Diagram of spiral cleavage in the mollusc *Unio.* Each top cell is directly above the junction of two lower cells. After Lillie, in Kellicott *General Embryology,* 1914.

spirally cleaving embryos. These will be explained more fully in other sections of the text. It is useful, however, to consider them here in the context of cleavage.

The first concept deals with the direction of cleavage. What causes spiral cleavages to be left-handed or right-handed? In other words, when a cleavage occurs, the top cell will form to the left or right of the bottom cells. What orients the spindle so that left- or right-handed cleavages will result? A partial answer to this question comes from a study of the snail *Limnaea*. The shells and organs of these snails are coiled in either a left- or a right-handed manner. Such coiling results from early cleavages that were left-handed (counterclockwise) or right-handed (clockwise), as shown in Figure 6-8. A single gene appears to determine the orientation of the cleavage. Right-handedness is dominant over left-handedness. One would expect, therefore, that if a homozygous left-handed coiled (and cleaving) female were mated with a homozygous right-handed coiled (and cleaving) male, all of the offspring would possess right-handed cleavage and coiling. However, just the opposite occurs. All first-generation offspring exhibit left-handed cleavage and coiling. The control of cleavage pattern thus resides in the genes of the mother's body. If the mother has recessive genes for left-handed cleavages, her body produces some substance that gets into the cytoplasm of the oocytes, causing the fertilized egg to cleave in the left-handed manner. If the mother, however, has both right and left genes, her offspring will cleave in a right-handed spiral because right is dominant, and the maternal right-handed factor supersedes the genetic characteristics of the offspring. Thus, the mother forms a cytoplasmic substance for the oocytes that later controls the cleavage pattern. The genes of the fertilized egg do not seem to affect the pattern. The genes of the mother, in this case, appear to be of major importance in producing the substance that controls the direction of cleavage

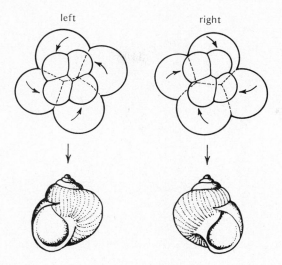

Figure 6-8. Diagram of left-handed (counterclockwise) and right-handed (clockwise) spiral cleavage and shell coiling in the snail. After Conklin, in T. H. Morgan, *Experimental Embryology*, Columbia University Press, 1927.

spindle orientation. Thus, cytoplasmic substances can play an important role in controlling a significant developmental event.

What is the second important concept that can be learned from studying spirally cleaving embryos? Not only are cytoplasmic substances important in controlling the pattern of cleavage, they are also important in controlling the fate of the blastomeres. This is nicely shown in the spirally cleaving embryo of the mollusc *Dentalium*. The egg cytoplasm of *Dentalium* is divided into three distinct regions: a clear region at the animal pole, a broad equatorial region of granular cytoplasm, and a second clear region of cytoplasm at the vegetal pole (see Figure 6-9). Just before the first cleavage, the clear vegetal cytoplasm extrudes a *polar lobe*. As the first division finishes, the clear vegetal cytoplasm withdraws into the embryo, but into only one of the two daughter cells, the CD cell. The other cell (the AB cell) has only two of the original cytoplasmic regions, whereas the CD cell has all three. As the second cleavage finishes, the clear vegetal cytoplasm is segregated again, this time in only one of the four blastomeres (the D cell). The A, B, and C cells do not possess the clear vegetal material. Thus, certain cells receive certain cytoplasms during cleavage. The blastomeres can be separated at the two-cell and four-cell stage. Only blastomeres containing the clear vegetal cytoplasm (CD cell or D cell) give rise to a normal (though smaller) larva. The other cells give rise to larvae without a middle germ layer (mesoderm layer). This suggests that the clear vegetal cytoplasm is important for the formation of mesoderm.

This conclusion is supported by the following experiment. If the polar

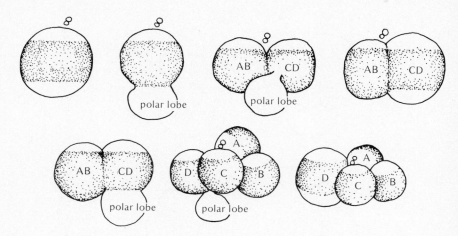

Figure 6-9. Cleavage of the mollusc *Dentalium*. After E. B. Wilson, *J. Exp. Zool.* 1 (1904): 1-72.

lobe, which contains only the special cytoplasm and no nucleus, is nipped off before the end of the first or second cleavage, the remaining embryo forms a larva without mesoderm. Thus it appears that the cytoplasm (or surface region), not the nucleus, makes blastomere D able to give rise to mesoderm.

Note that the special, segregated cytoplasms may contain genetic information possibly in the form of messenger RNA. This genetic information might be the key component of the special cytoplasm in some cytoplasmic segregations. A vegetal body has been observed in the polar lobe of the fresh water snail, *Bithynia tentaculata*. This body contains a substance having the staining characteristics of RNA. Such bodies, however, have not been found in many other lobe-forming embryos. It remains to be determined whether or not the material in the body has important morphogenetic functions.

The marine snail *Ilyanassa* also forms polar lobes during development. Using this organism, Clement found evidence that the so-called morphogenetic determinants were present in the surface membrane (or in the cortical cytoplasm bound to the surface membrane). *Ilyanassa* eggs were inverted and centrifuged, driving the vegetal cytoplasmic components into the animal hemisphere. The eggs then fragmented in the centrifuge. The animal halves contained vegetal cytoplasm; the vegetal halves lacked most of the vegetal cytoplasmic components, but contained animal cytoplasm. The vegetal halves still formed polar lobes and produced complete larvae. The animal halves did not form polar lobes and formed larvae lacking lobe-dependent structures. Such abnormal larvae are typically formed when the polar lobe has been removed from the embryo. These experiments suggest that (at least for *Ilyanassa*) the special cytoplasm responsible for normal, complete development is strongly bound to the cell membrane of the vegetal region (Figure 6-10).

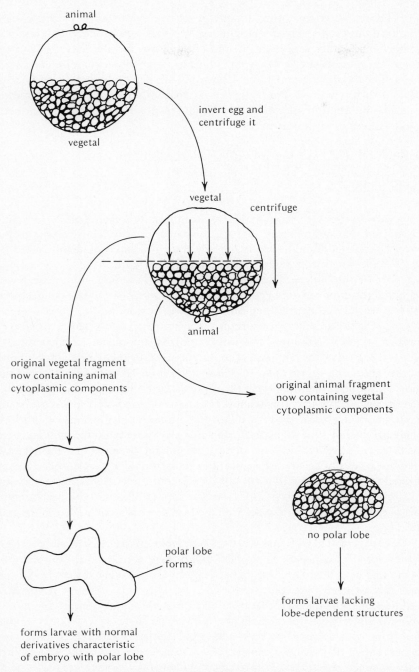

animal

invert egg and
centrifuge it

vegetal

vegetal   centrifuge

animal

original vegetal fragment
now containing animal
cytoplasmic components

original animal fragment
now containing vegetal
cytoplasmic components

polar lobe
forms

no polar lobe

forms larvae with normal
derivatives characteristic
of embryo with polar lobe

forms larvae lacking
lobe-dependent structures

Figure 6-10. Experiment suggesting that polar lobe determinants in the Ilyanassa egg are tightly bound to the cell membrane in the vegetal region. Based on work of A. C. Clement, *Devel. Biol.* 17 (1968): 165–68.

There are many other examples of the importance of cytoplasmic materials in embryonic development. We shall see in Chapter 11 that all embryos probably segregate certain cytoplasms during cleavage, and that these cytoplasms play a key role in determining the developmental fate of cells. The example given above illustrates this principle well because of the visible differences in the cytoplasmic regions. In many other embryos, cytoplasmic regions are not visibly distinct so that the cytoplasmic influence on development is not apparent.

ROLE OF CALCIUM IN CONTROLLING CYTOPLASMIC SEGREGATION IN THE TUNICATE

Several investigators, including Jaffe, Robinson, Nuccitelli and Jeffery, have suggested that cytoplasmic segregation in early embryos is related to gradients of free calcium. Recently, Jeffery at the Marine Biological Laboratory in Woods Hole, Massachusetts, completed an elegant set of experiments in support of this notion. *Boltenia villosa* eggs (an ascidian or tunicate) were selected for these studies because they exhibit an easily observed segregation of pigmented cytoplasmic regions after fertilization.

Cytoplasmic segregation begins when cortical cytoplasm, composed of orange pigment granules and associated mitochondria, streams into the vegetal region of the egg, and accumulates as an orange crescent around the point where the sperm entered. Clear cytoplasm from the germinal vesicle flows into the vegetal region in the wake of the orange cytoplasm. This clear cytoplasm temporarily forms another crescent immediately inside the orange crescent, and eventually returns to the animal hemisphere along with the male pronucleus. A cytoplasmic lobe also protrudes from the vegetal region of the egg. The first cleavage divides the orange crescent between the two daughter cells. The orange crescent cytoplasm becomes specifically partitioned into cells that will form tail muscle and mesenchyme in the larvae.

Calcium ionophore A23187, a substance that liberates calcium bound to intracellular membranes, also produced orange crescents, even in unfertilized eggs.

Next, small glass fibers were soaked in ionophore A23187 and placed on cover slips. This produced an A23187 gradient, the highest concentration occurring near the fiber. Unfertilized eggs were placed near the fibers so that one side of each egg contacted a fiber or was very close to it. In 20–30 minutes, most of the eggs that became activated formed orange crescents on the side nearest the fiber (see Figure 6–11). The clear crescent and cytoplasmic lobe also formed on that side. Eggs positioned between two fibers generally formed a single orange crescent centered between the points where they touched the fibers. A few eggs formed small orange crescents at the points of contact.

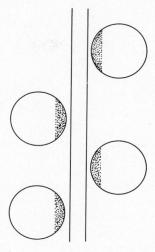

Figure 6-11. Formation of orange crescent on side of eggs adjacent to fiber coated with calcium ionophore A23187. Based upon results of W. R. Jeffery (1982).

The simplest interpretation of these results is that local elevations in calcium concentration activate cortical microfilaments and cause cytoplasmic polarizations in the ascidian egg. These changes in intracellular calcium may be important in organizing the distribution of the cytoplasmic determinants that control the pattern of early embryonic development.

INCOMPLETE CLEAVAGE

In some organisms, the egg yolk is so concentrated in the vegetal region that all of the nonyolky cytoplasm floats on the yolk as a little cap. Such *strongly telolecithal* eggs are produced by sharks and rays, bony fish, reptiles, and birds. We already know that a moderate amount of yolk retards cleavage. What happens when almost the entire egg is dense yolk? Here, the cleavages divide only the nonyolky cytoplasmic cap, and do not affect the dense yolk at all. This is called incomplete cleavage (see Figures 6-12 and 6-13).

Insects and many other arthropods have eggs with the yolk concentrated in the interior of the egg (*centrolecithal eggs*). The nonyolky cytoplasm is located in a thin surface layer around the yolk and in a small patch in the center of the egg. The central patch contains the nucleus (Figure 6-14). In these eggs, the nucleus divides before any cytoplasm divides. Surrounded by small amounts of central cytoplasm, the divided nuclei move outward toward the egg surface. The surface now consists of many nuclei in an uncleaved layer of cytoplasm. The cytoplasm then divides into compartments, each still connected to the central yolk. In time the cells separate from

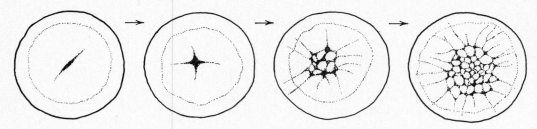

Figure 6-12. Incomplete cleavage of chick egg. Surface view, looking down onto blastodisc at animal pole. (Modified from J. T. Patterson, *J. Morph.* 21 (1910): 101–34.)

the yolk and use the yolk for nourishment. This type of incomplete cleavage is called *superficial cleavage*.

Thus we see that in eggs with a small or moderate amount of yolk, cleavage is usually total. In eggs with highly concentrated regions of yolk, cleavage occurs only in the nonyolky cytoplasm. Yolk, therefore, is an important factor in controlling the cleavage pattern, along with other factors such as the cycles of certain proteins and other, genetically controlled cytoplasmic factors.

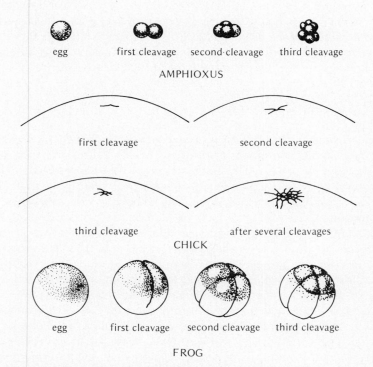

Figure 6-13. Comparison of *Amphioxus*, chick, and frog cleavage.

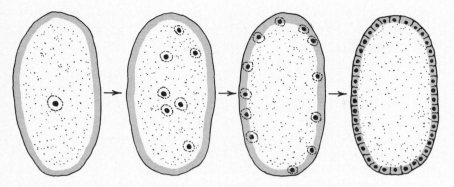

Figure 6–14. **Superficial cleavage in a centrolecithal egg (insect egg).**

MECHANICS OF CLEAVAGE

Cleavage involves the division of the cell cytoplasm and the nucleus. What are the mechanics of these processes? A ring of microfilaments 30–70 Å in diameter, probably composed of the protein actin, can be observed just below the cell surface of many eggs. These filaments are composed of globular subunits that can polymerize and depolymerize in response to such cytoplasmic factors as ionic strength (see Chapter 15). These filaments form a contractile ring that separates the cytoplasm much as one pulls purse strings to decrease the size of the purse opening. A drug, cytochalasin B, disrupts cytoplasmic microfilaments. It also interferes with the cytoplasmic division of cultured mammalian cells. If the drug is removed, cell division resumes as the microfilaments reappear. Such experiments suggest that microfilaments are involved in cytoplasmic division. It should be stressed, however, that cytochalasin B also affects other cell processes, such as protein synthesis, sugar uptake and respiration, and so on. Thus, studies involving this drug must be interpreted with caution. Cytoplasmic division may indeed be controlled by microfilaments, but cytochalasin B studies cannot be cited as definitive evidence for such a conclusion.

There is evidence that the spindle—or more correctly, the asters— initiate cytoplasmic division by producing a diffusible factor that acts on the contractile ring. This was shown by experiments in which asters were removed or separated. Asters must be present in the area where the cleavage furrow occurs. Asters, therefore, may set up conditions that are required for the organization of a constriction mechanism.

The mitotic spindle apparatus involved with nuclear division is composed mainly of the proteins, tubulin A and B (see Chapter 15). This protein is present in the egg and blastomeres (in subunit form) even when the mitotic spindle is not present. At the proper time for spindle formation, these subunits probably are triggered to come together (polymerize), forming the visible mitotic apparatus. In Chapter 15, we shall discuss in more detail how cellular organelles polymerize from preformed subunits.

What are some of the metabolic events that occur during cleavage? Large amounts of DNA are synthesized to provide each blastomere with a set of chromosomes. Many proteins are also required for cleavage. Some of these are stored in the oocyte and need not be synthesized; others are synthesized during cleavage. If protein synthesis is inhibited with the drug puromycin, cleavage stops completely. This suggests that protein synthesis is essential for cleavage. Inhibitor studies, however, should also be approached with caution. Some inhibitors may interfere with processes other than those known to be affected by them.

The synthesis of large amounts of RNA, however, is not needed during early cleavage. RNA synthesis can be blocked with the drug actinomycin D. If embryos are treated in this way, protein synthesis and cleavage still occur. This experiment suggests that most of the RNA required to synthesize proteins, ribosomes, and so on is stored in the oocyte. Again, this drug study should be interpreted with caution. For instance, a tiny amount of RNA may be synthesized even in the presence of inhibitor, and this amount may be sufficient to drive the remaining reactions.

Activated frog eggs can be enucleated—that is, the nucleus with its chromosomal DNA is removed. The enucleated eggs nonetheless cleave and develop to the blastula stage. These results provide firm evidence that continued RNA synthesis is not absolutely required for cleavage in the frog embryo. In some other embryos, however, this is not necessarily the case.

Let us move on to the next stage of development. Repeated cleavages have now produced a ball or cap of cells. These embryos still do not resemble a real being that we can recognize. In the next chapters, we shall see how the embryo turns into something that begins to resemble an organism. Let us introduce these events by briefly looking at how the cleaved cell mass readies itself for the events to follow.

BLASTULA

Many embryos become transformed from a solid ball of cells (often termed a *morula*) into a hollow ball called a *blastula*. This transformation prepares the embryo for the various cellular rearrangements that shape the embryo in the next stage of development.

The forces involved in causing the solid ball of cells to become hollow in the sea urchin and similar embryos include osmotic pressure exerted from the influx of water into the ball. A buildup of macro-molecules occurs in the center of the ball. This causes an influx of water and the pressure may push the cells outward. Other forces involved may include cell-cell adhesiveness, surface tension, and the fact that the cells are attached to an outer elastic

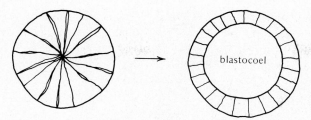

Figure 6-15. Possible mechanism of blastocoel formation. Dividing cells push outward, forming a central cavity. After Gustafson and Wolpert, *Biol. Rev.* 42 (1967): 442–98.

layer called the hyaline layer. If the cells are attached to each other as they divide, they will push the elastic hyaline layer outward to form a central cavity (Figure 6–15).

Blastulas vary in appearance in different species. Many are hollow balls of cells with the cavity (the *blastocoel*) located in the center. Such blastulas are often found in organisms such as the oligolecithal eggs, sea urchin and *Amphioxus* (Figures 6–16 and 6–17). In embryos with a more uneven yolk distribution, such as the frog, the blastocoel is displaced toward the animal pole (Figure 6–16).

A blastula-like stage also occurs in very yolky eggs where the embryo proper develops as a cap atop an uncleaved mass of yolk, as in birds, bony fish, and reptiles. In these embryos, however, a cavity does not form in the yolk. Instead, it forms between the cap or disc of cells (*blastodisc*) and the

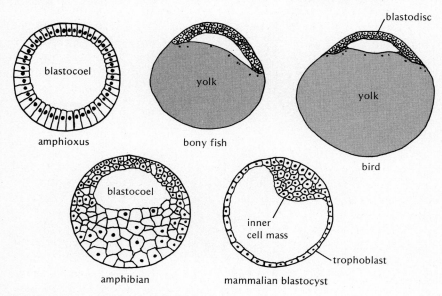

Figure 6-16. Comparison of blastulas.

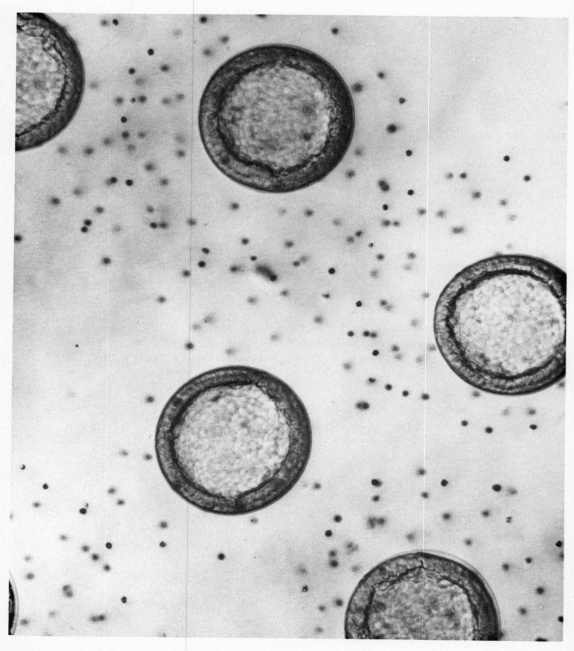

Figure 6-17. Sea urchin blastulas. Courtesy of Victor Vacquier.

uncleaved yolk. Some researchers do not consider such a cavity a true blastocoel because it lies outside (below) the embryo proper. Other cavities also appear within the blastodisc (within the embryo proper), as described in Chapter 7.

No cavity appears at all in insects that have centrolecithal eggs and superficial cleavage. The cells form at the periphery of the embryo; the center remains filled with yolk. This can be considered a kind of blastula, if one imagines that the central cavity is filled with yolk.

In most mammals, cleavage forms a solid ball of cells (morula). The outer layer of this ball will form membranes that surround the embryo proper; the embryo itself forms from cells in the inner part of the morula (the *inner cell mass*). The outer layer (trophoblast) lifts away from one side of the inner cell mass, forming a blastula-like stage called the *blastocyst* (Figure 6–16). During the blastocyst stage, the human embryo implants itself in the wall of the uterus, where it remains and is nourished until birth. The mammalian embryo resembles that of the chick, but the mammalian blastocyst is not filled with yolk. Instead, nourishment is supplied by the mother. Since mammals evolved from yolky-egged ancestors, it is reasonable to assume that at one time the blastocyst cavity was filled with yolk. When a system for nourishing the embryo by the mother developed, the yolk was probably lost. The early stages of the embryo and blastocyst, however, did not need to change. In this way we can explain the similarities between mammalian development and that of the reptile and bird.

We now have a ball or cap of cells. In the next chapter we shall see how these rather nondescript embryos undergo a remarkable process that ends in an embryo that begins to resemble what we think of as "real" organisms. We shall see how embryos become transformed by twists, turns, and cellular rearrangement.

READINGS AND REFERENCES

Clement, A. C. 1968. Development of the vegetal half of the *Ilyanassa* egg after removal of most of the yolk by centrifugal force compared with the development of animal halves of similar visible composition. *Devel. Biol.* 17: 165–186.

Dohmen, M. R., and Verdonk, N. H. 1974. The structure of a morphogenetic cytoplasm, present in the polar lobe of *Bithynia tentaculata* (Gastropoa, Prosobranchia). *J. Embryol. Exptl. Morphol.* 31: 423–33.

Gustafson, T., and Wolpert, L. 1967. Cellular movement and contact in sea urchin morphogenesis. *Biol. Rev.* 42: 442–98.

Jeffery, W. R. 1982. Calcium ionophore polarizes ooplasmic segregation in ascidian eggs. *Science* 216: 545–47.

Morgan, T. H. 1927. *Experimental Embryology*. New York: Columbia University Press.

Neri, A., Roberson, M., Connolly, D. T., and Oppenheimer, S. B. 1975. Quantitative evaluation of concanavalin A receptor site distributions on the surfaces of specific populations of embryonic cells. *Nature* 258: 342–44.

Patterson, J. T. 1910. Studies on the early development of the hen's egg. I. History of the early cleavage and accessory cleavage. *J. Morph.* 21: 101–34.

Rappaport, R. 1974. Cleavage. In J. Lash and J. R. Whittaker, eds., *Concepts of Development*. Sunderland, MA: Sinauer.

Roberson, M., Neri, A., and Oppenheimer, S. B. 1975. Distribution of concanavalin A receptor sites on specific populations of embryonic cells. *Science* 189: 639–40.

Schroeder, T. E. 1975. Dynamics of the contractile ring. In S. Inoue and R. E. Stephens, eds., *Molecules and Cell Movement*. New York: Raven Press, p. 304.

Wilson, E. B. 1904. Experimental studies on germinal localization. I. The germ regions in the egg of *Dentalium*. II. Experiments on the cleavage-mosaic in *Patella* and *Dentalium*. *J. Exp. Zool.* 1: 1–72.

Wolpert, L. 1960. The mechanics and mechanism of cleavage. *Intl. Rev. Cytol.* 10: 164.

7

GASTRULATION

We have reached the point in embryonic development at which the embryo is a ball or cap of cells. It does not even slightly resemble the adult organism. How does this nondescript group of cells change into a layered embryo that resembles a familiar organism? We shall examine the important process of gastrulation in which the hollow ball of cells, the blastula, changes into a layered embryo, the gastrula, by cell movements and selective adhesion. Let us first, however, consider briefly the topic of fate maps. These are projections of how specific areas of early embryos develop at later stages. They will help us understand the movements that occur during gastrulation.

FATE MAPS

Fate maps are constructed by marking specific areas on early embryos and watching these marks to see where they reside in later stages. Embryos are marked by pressing agar impregnated with vital stains (such as Nile blue sulfate or neutral red) against specific areas of early embryos, a technique first developed by Vogt. By observing the location of these marks in the older embryos, Vogt could construct fate maps that showed the fate of each embryonic region. Fate maps of *Amphioxus*, amphibian, and bird embryos are given in Figure 7–1. The term *prospective* (or *presumptive*) is used in the fate maps and relevant text to denote what a specific region will become as the embryo grows older.

AMPHIOXUS AND AMPHIBIAN FATE MAPS

The fate maps of *Amphioxus* and the amphibian are similar (see Figure 7–1). The animal region of each embryo gives rise to *epidermis* and to the *neural* plate, which becomes the nervous system and associated structures. The prospective epidermis and prospective neural plate will make up the outer layer of the gastrula, the *ectoderm*. The middle or marginal zone of each embryo gives rise to the middle region of the gastrula, the *mesoderm*, which is composed of prospective *notochord* and prospective non-notochordal mesoderm. The notochord is a rod that develops below the neural tube and supports the developing embryo. It usually disappears in the adult. The nonnotochordal mesoderm develops into many structures, including limbs, heart, muscles, kidneys, and gonads. The vegetal region of the *Amphioxus* and amphibian embryos gives rise to the *endoderm*, the inner embryonic layer. This layer forms the gut and the derivatives of the gut tube. The structures that develop from all of these regions will be described in detail in Chapters 9 and 10.

Recent work by Keller and colleagues suggests that Vogt's amphibian fate map may need reevaluation for amphibians such as *Xenopus* (South African clawed toad, a frog). The prospective notochord and other mesoderm are in a different position in a fate map projected upon a late blastula. In Vogt's map, these structures are on the surface of the embryo. In *Xenopus*,

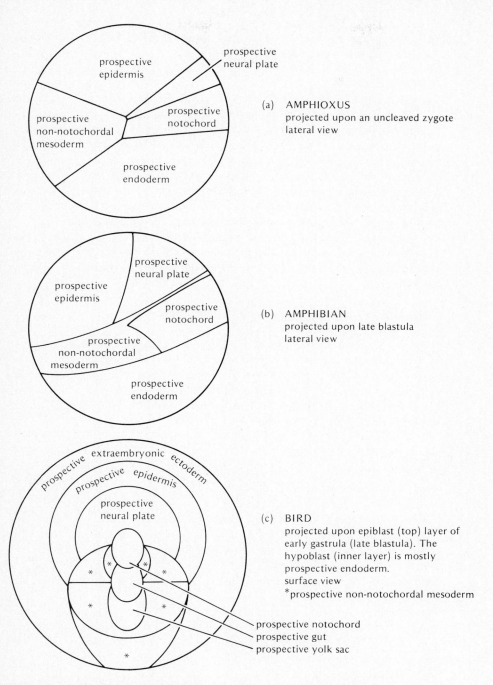

(a) AMPHIOXUS
projected upon an uncleaved zygote
lateral view

(b) AMPHIBIAN
projected upon late blastula
lateral view

(c) BIRD
projected upon epiblast (top) layer of
early gastrula (late blastula). The
hypoblast (inner layer) is mostly
prospective endoderm.
surface view
*prospective non-notochordal mesoderm

Figure 7–1. Diagrammatic fate maps of major areas in three embryos.

they appear to be located in deeper layers of the blastula and during gastrulation, move even deeper.

BIRD FATE MAP

The bird fate map shown in Figure 7–1 is different from the other two: the bird blastula is not a hollow ball, but a layered cap sitting atop the yolk. The fate map of the top (*epiblast*) layer is shown in Figure 7–1. This layer gives rise to the *extraembryonic* ectoderm (the membranes that lie outside the embryo), the epidermis; the neural plate; part of the notochord and all of the nonnotochordal mesoderm; the yolk sac; and the gut endoderm. The lower (*hypoblast*) region gives rise to the endoderm and to the rest of the notochord. Thus, epiblast and hypoblast are not at all equivalent to ectoderm and endoderm. Details of the fate of these regions are given in Chapters 8, 9, and 10.

Fate maps describe the fate of specific regions of the early embryo. We shall now describe how these regions get to their respective destinations. Gastrulation is the important process that facilitates these key cellular rearrangements.

GASTRULATION IN THE SEA URCHIN

In previous chapters, we noted that the sea urchin embryo is ideal for a variety of studies in developmental biology. This is true also in studies of the gastrulation process. The embryos are relatively transparent, allowing the observer to analyze the components of the process. We shall thus begin with the sea urchin, then move on to organisms that achieve gastrulation by somewhat different mechanisms.

SEQUENCE OF EVENTS

Figure 7–2 shows the process of gastrulation in the sea urchin, as revealed by time-lapse cinematography. The sea urchin egg (an isolecithal egg) contains a rather even distribution of a relatively small amount of yolk. Therefore, no excessive quantity of yolk can interfere with the buckling in, or invagination, of the blastula. Gastrulation begins with changes in the shape of certain cells in the vegetal plate. Some of the cells lose adhesion with their neighbors and are forced into the blastocoel (see Figure 7–3). These are the primary mesenchyme cells. Partly as a result of the loss of vegetal plate cells, the vegetal area begins to indent. This indentation forms a cavity or tube called the *archenteron*, or primitive gut. The opening of this cavity (at the vegetal end) is the *blastopore*. The tip of the archenteron advances toward the animal pole. At the tip, certain cells extend projections called *filopodia*. These cells are the *secondary mesenchyme cells*, which become the mesoderm. The tips of the filopodia appear to "feel" the inner surface of the gastrula and finally stick to the animal end. They then contract, pulling the archenteron tube toward the animal end (see Figure 7–4).

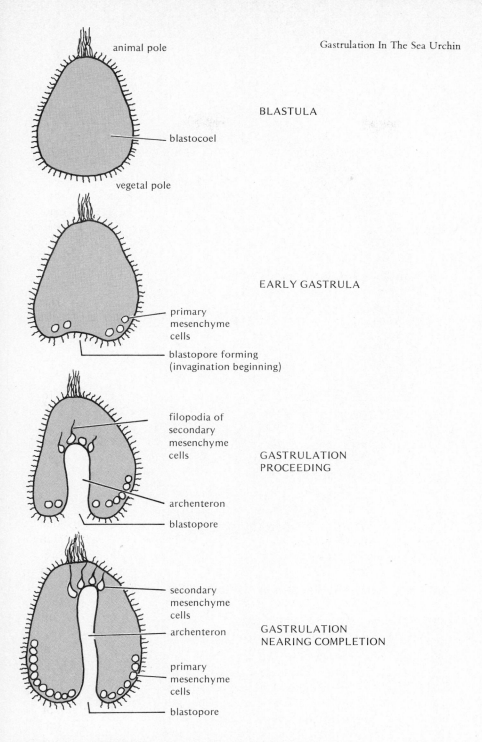

animal pole

BLASTULA

blastocoel

vegetal pole

EARLY GASTRULA

primary
mesenchyme
cells

blastopore forming
(invagination beginning)

filopodia of
secondary
mesenchyme
cells

GASTRULATION
PROCEEDING

archenteron

blastopore

secondary
mesenchyme
cells

archenteron

GASTRULATION
NEARING COMPLETION

primary
mesenchyme
cells

blastopore

Figure 7–2. Gastrulation in the sea urchin. After T. Gustafson, *Experimental Cell Research, 32* (New York: Academic Press).

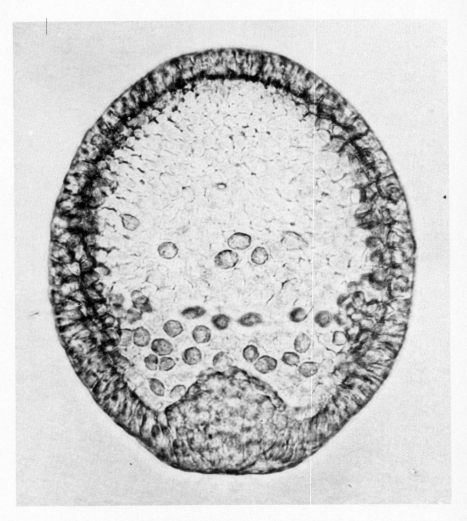

Figure 7-3. Early sea urchin gastrula. Courtesy of Victor Vacquier.

MECHANISMS OF GASTRULATION

What mechanisms are involved in the gastrulation of the sea urchin embryo? Obviously, cell movement or cell motility is one important mechanism, as seen from the activity of the mesenchyme cells. A second important mechanism is selective adhesiveness. The primary mesenchyme cells lost adhesiveness with their neighbors in the vegetal plate, while the secondary mesenchyme cells preferentially adhere by their filopodia to the animal end. Finally, the contractility of the filopodia plays a role in pulling the gut tube toward the animal pole.

Figures 7-4 b, c, d, e show that a network of fibers appears on the inner blastula surface facing the blastocoel during the gastrulation of the sea urchin blastula. Kawabe, Armstrong, and Pollock believe that this fibrillar matrix helps guide cells to their destinations. It may also stabilize the forming archenteron.

Gastrulation has transformed the sea urchin blastula, a hollow ball of cells, into a gastrula with a gut cavity. This gastrula begins to resemble the body of the adult sea urchin. We have seen that during gastrulation, mechanisms including cell motility, selective adhesiveness, and contractility appear to control the cellular rearrangements needed for the embryo to begin taking the form of the adult. We shall examine these mechanisms in detail in Chapter 12.

MOLECULAR BIOLOGY OF EARLY SEA URCHIN DEVELOPMENT

The sea urchin embryo has recently provided some unexpected insights into the activity of genes during cleavage and early development.

No extensive "switching on" of new genes occurs during the early development of the sea urchin embryo; all or most of the mRNA sequences found in the gastrula stage are represented in stored maternal RNA. Davidson and Britten (as well as others), showed that about 60% of the sequences found in maternal mRNA are still present in newly synthesized transcripts of the late embryo. There are exceptions, however. A specific rare maternal sequence appears to be loaded on polysomes in the 16-cell embryos, but thereafter disappears from the embryo cytoplasm. A limited number of new RNA species also appear during blastulation.

The structural complexity of the early embryo progressively increases during development. However, the molecular complexity of the embryo mRNA is highest at the beginning of development, and most maternal mRNA species are merely replaced by new transcripts of the same set of genes that were active in oogenesis. Perhaps most of the proteins required for embryonic morphogenesis do not assemble into biological structures until long after these proteins are synthesized. This would explain why large amounts of different mRNAs need not be transcribed after oogenesis.

How do the different cell types in the sea urchin embryo synthesize specific products needed for differentiation? Senger and Gross showed that, for example, the ratio of histone to nonhistone protein synthesis is higher in the tiny sea urchin micromeres than in the rest of the 16-cell embryo.

Differences of this sort may trigger the differential expression of other genes in different cell types. The basic distinction between the micromeres and other cell types, for example, may result from early differences in nuclear expression. These, in turn, may be caused by a cytoplasmic environ-

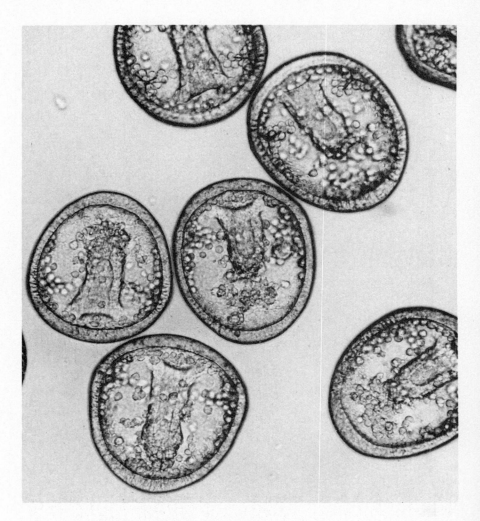

Figure 7-4. (a) Sea urchin gastrula, gut forming. Courtesy of Victor Vacquier.
(b) Sagittal section of the sea urchin embryo in the late blastula stage. The thickened vegetal plate and primary mesenchyme cells are shown. The inner boundaries of the cells lining the blastocoel cavity are clearly visible, unobscured by matrix. Courtesy of T. Kawabe and E. G. Pollock.
(c) Transverse section of the sea urchin embryo in the midgastrula stage, looking toward the prospective anterior end. The fibrillar matrix is visible on the inner blastocoel surface. Courtesy of T. Kawabe and E. G. Pollock.
(d) Sagittal section of the embryo in the midgastrula stage. The archenteron has extended toward the site of oral contact. Some secondary mesenchyme and primary mesenchyme cells are visible. A dense fibrillar matrix covers the inner blastocoelar walls of the embryo. The hyaline layer adheres tightly to the surface of the embryo except over

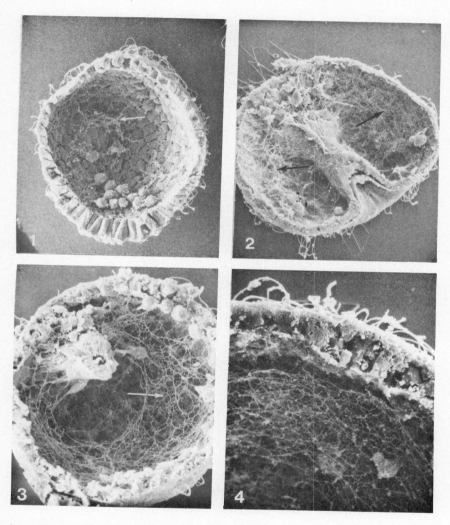

the surface of the archenteron, where it has separated. Cilia are visible on the outer surface of the embryo. Courtesy of T. Kawabe and E. G. Pollock.
(e) Higher magnification of the fibrillar matrix. Courtesy of T. Kawabe and E. G. Pollock.

ment in the micromeres that is different from that to which the nuclei of the other cell types are exposed. Recall that the micromeres essentially bud off from the adjacent macromeres, so that micromeres contain very little cytoplasm.

It is appropriate to end this section with some speculations on the overall scheme of early sea urchin development suggested by Davidson, Hough-Evans, Britten, and others. During oogenesis, the transcription and translation of most macromolecules needed for cell division and morphogenesis occur. Fertilization triggers biosynthetic activities in the oocyte and also spatial reorganizations of the cytoplasm. During cleavage, blastulation, and gastrulation, diverse domains are established. The nuclei of cells in these different domains are induced to function differentially. Note, for example, micromeres *versus* macromeres and mesomeres. Maternal mRNA and proteins are replaced by embryonic mRNA and proteins, plus other specific products from genes differentially activated in cells of a given lineage. These products are needed for specific cellular differentiation. Embryogenesis is terminated when a constant ratio of nucleus to cytoplasm is obtained, when all transcripts on polysomes are of embryonic rather than maternal origin, when specific patterns of gene expression characteristic of the major cell lineages are established, and when the morphogenetic program whose execution began in the oocyte nucleus has been carried to completion (Figure 7–5). It is clear from our brief examination that a great deal must still be learned before we can fully understand the mechanisms that control embryonic development. We shall examine the molecular changes that occur during development in more detail in Chapters 12 through 15.

GASTRULATION IN AMPHIOXUS

Like the sea urchin egg, the *Amphioxus* egg is isolecithal. No excessive amount of yolk hinders invagination. Let us examine gastrulation in *Amphioxus* and compare it with that in the sea urchin.

Gastrulation in *Amphioxus* also occurs by invagination, but not with the aid of secondary mesenchyme cells. Instead, the vegetal area first flattens, then bends inward or invaginates (see Figure 7–6). The embryo begins to resemble a punched-in ball. The outer portion of the embryo, the ectoderm, consists of prospective epidermis and prospective neural system. The inner part of the cup, where your fist would strike as the ball is punched in, consists mainly of endoderm, the prospective gut and gut derivatives. The mesoderm (prospective notochord and nonnotochordal mesoderm) extends from the rim into the cup.

Like the sea urchin, the *Amphioxus* gastrula now possesses a primitive gut, or archenteron. The opening of the archenteron to the outside is the blastopore. After examining gastrulation in the amphibian and bird embryo, we shall return to *Amphioxus* to investigate the next phases of development. These are the separation of the mesoderm and endoderm and the development of the nerve tube (see Figure 7– 7).

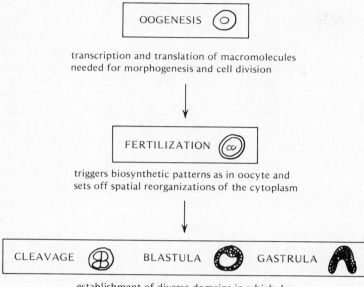

establishment of diverse domains in which, by
some unknown process, the nuclei of different
embryonic cell lineages are induced to function
differentially; replacement of maternal RNAs
and proteins; also activation of sets of
cell-lineage-specific genes (products of such
genes may be needed for some specific
cellular differentiations)

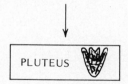

termination of embryogenesis when a constant
ratio of nucleus to cytoplasm is attained, when
the transcripts on polysomes are all of embryonic
rather than maternal origin, when specific patterns
of gene expression characteristic of the major cell
lineages are established, and when the morphogenetic
program that began to be read out in the oocyte
nucleus has been carried to completion

Figure 7–5. Simplified hypothetical scheme explaining some features of sea urchin embryogenesis. Based upon ideas of E. H. Davidson, B. R. Hough-Evans, and R. J. Britten, *Science* 217 (1982): 17–26.

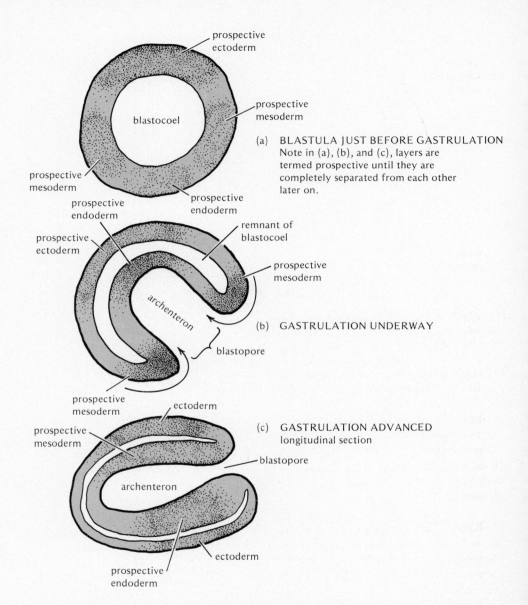

prospective
ectoderm

prospective
mesoderm

blastocoel

(a) BLASTULA JUST BEFORE GASTRULATION
Note in (a), (b), and (c), layers are
termed prospective until they are
completely separated from each other
later on.

prospective
mesoderm

prospective
endoderm

prospective
endoderm

prospective
endoderm

prospective
ectoderm

remnant of
blastocoel

prospective
mesoderm

archenteron

(b) GASTRULATION UNDERWAY

blastopore

prospective
mesoderm

ectoderm

(c) GASTRULATION ADVANCED
longitudinal section

prospective
mesoderm

blastopore

archenteron

ectoderm

prospective
endoderm

Figure 7-6. Gastrulation in *Amphioxus*. After Conklin, *J. Morph.*, 54 (1932): 69–118.

GASTRULATION IN AMPHIBIANS

We have seen that both the sea urchin and *Amphioxus* embryos possess only a small amount of yolk, and that that amount is rather evenly distributed. The amphibian egg (and embryo), on the other hand, is moderately telolecithal;

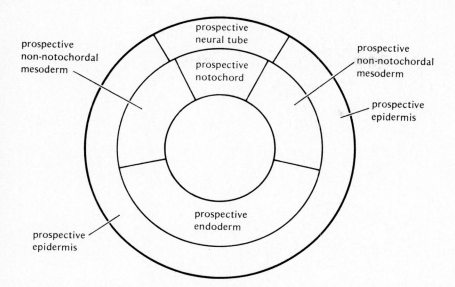

Figure 7-7. Cross section through the *Amphioxus* gastrula. Note that the mesoderm and endoderm are not yet separated. Thus, the inner layer is called the *mesendoderm*. The outer layer is the ectoderm. Modified from Conklin, *J. Morph.*, 54 (1932): 69–118.

the vegetal region contains much more yolk than the animal region. Cleavage in the amphibian embryo is unequal—the yolk retards cleavage in the vegetal region. Gastrulation is influenced by yolk as well. In the amphibian, unlike in *Amphioxus* and the sea urchin, invagination or "punching in" of the embryo does not occur. Instead, cells move from the exterior to the interior of the embryo by active migration through a groove that forms at the surface of the embryo. The vegetal region of the amphibian is too thick and laden with yolk to allow the type of invagination found in the sea urchin and *Amphioxus*.

SEQUENCE OF EVENTS: SURFACE VIEW
Figure 7-8 shows surface views of gastrulation in the amphibian. A slit, the blastopore, forms just below the equator. Cells from the surface of the embryo move inside through the blastopore. At first, this migration occurs only in a small region below the equator between the animal and vegetal hemispheres. The area just dorsal to this cleft is termed the dorsal lip of the blastopore. This forms at the site of the grey crescent described in Chapter 6. Cells migrate over this lip, through the blastopore, and into the interior of the embryo.

As can be seen in Figure 7-8, the blastopore lengthens and becomes crescent-shaped, then semicircular, and finally a full circle. This results from the inward movement of cells on the surface of the embryo. The first cells to move in are from the dorsal area of the embryo. As the blastopore becomes

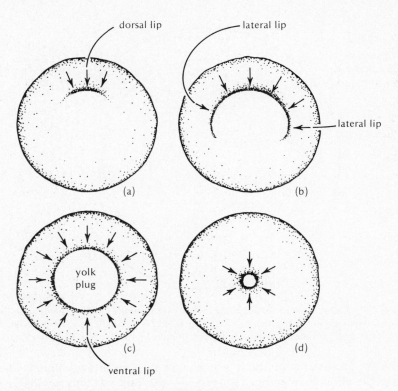

Figure 7–8. Surface views of amphibian gastrulation.

crescent shaped, cells from lateral regions of the embryo move in. Finally, ventral cells move in, completing the circular blastopore. The lateral lips and ventral lip of the blastopore are simply the regions (lateral and ventral, respectively) over which cells migrate into the embryo through the blastopore. When the blastopore is complete (forming a full circle) the center of the circle is filled with yolky endoderm cells. This plug of yolky endoderm is called the *yolk plug* (Figure 7–9).

MECHANISMS OF GASTRULATION

What causes surface cells of the amphibian embryo to move inward? Cells begin to change shape, first at the dorsal lip region. Then cellular movement begins. Cells expand and contract, reminiscent of the changes in the secondary mesenchyme cells of the sea urchin embryo. This expansion and contraction appears to play an important role in the inward movement of the active cells, as well as cells attached to the active cells. The forces involved in amphibian gastrulation are not well understood. They do, however, act even if the dorsal lip is removed and transplanted to a different part of the embryo. In this case, the dorsal lip cells will migrate inward just as though they were in the proper position in the embryo.

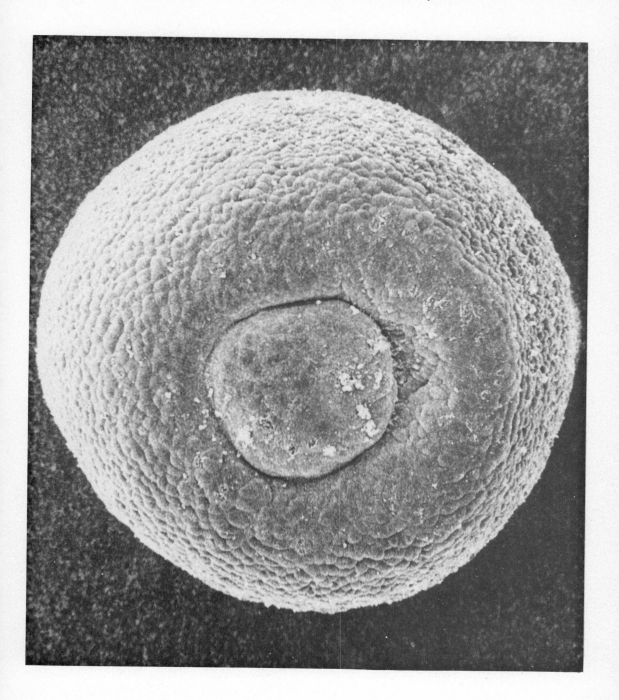

Figure 7-9. Frog yolk plug (late gastrula). Courtesy of Peter Armstrong.

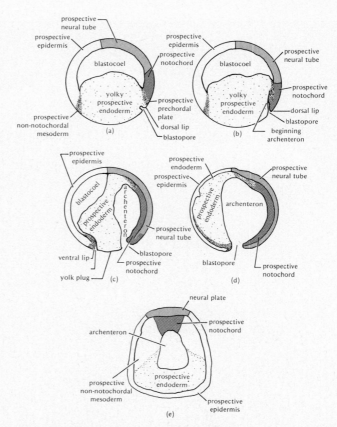

Figure 7-10. Gastrulation in the amphibian embryo. (a) to (d) Sagittal sections. (e) Cross section. (f) Frog yolk plug, sagittal section. After W. Vogt, *Roux Arch. 120* (1929): 385–706.

SEQUENCE OF EVENTS: CROSS-SECTION

Let us now slice the amphibian embryo in half and examine what occurs inside during gastrulation (see Figure 7–10). The notochord forms from cells that move from the dorsal surface of the embryo, over the dorsal lip region, through the blastopore, and stop inside the embryo below the prospective neural tube. The notochord forms the roof of the archenteron. The prospective epidermis and the prospective nervous system (which together comprise the ectoderm) expand over the entire surface of the embryo. The blastopore becomes circular, the notochordal mesoderm enters dorsally and the non-notochordal mesoderm enters laterally and ventrally. As the rim of the blastopore contracts, the yolk plug endoderm is pulled inside and disappears from the surface of the embryo.

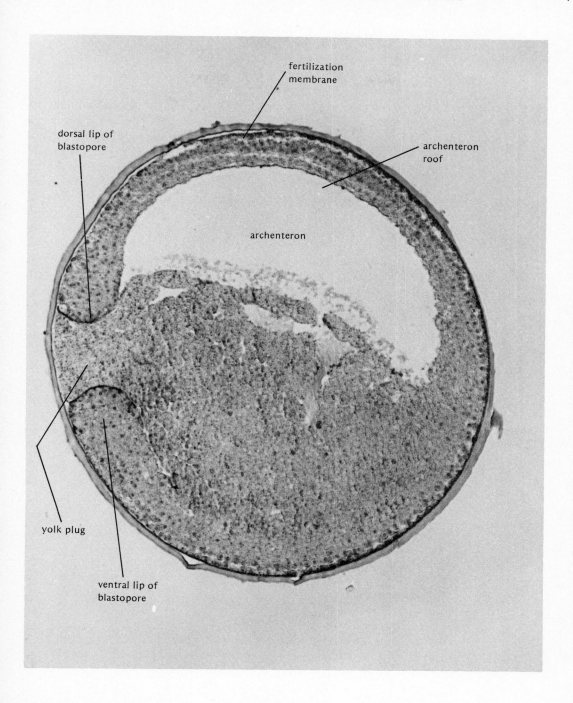

(f)

GASTRULATION IN THE BIRD EMBRYO

We have examined gastrulation in the sea urchin, *Amphioxus*, and the amphibian. In embryos consisting of cells with relatively small amounts of yolk (sea urchin and *Amphioxus*), gastrulation occurs by in-pocketing or invagination. In the amphibian embryo, the large, yolky vegetal cells prevent invagination; instead cells migrate into the embryo through a narrow slit. As discussed in Chapter 6, the bird embryo is quite different from those of the sea urchin, *Amphioxus*, and the amphibian. Cleavage occurs only in the little cytoplasmic cap that sits atop the large amount of yolk in the bird egg. At the blastula stage, the cytoplasmic cap (blastodisc) has cleaved to form a multi-layered embryo, the blastoderm. The blastoderm sits above a *subgerminal space* that separates most of the embryo from the yolk below it.

Gastrulation begins in the bird embryo when the blastoderm separates into two layers, the top *epiblast* layer and the bottom *hypoblast* layer (see Figure 7–11). The space between these layers is the *cleft space*.

The second step is the movement of lateral cells toward the center. This produces a thickening, the *primitive streak*, at one end of the blastodisc.

An indentation called the *primitive groove* forms down the midline of the primitive streak. The primitive groove functions in the same way as the blastopore of the other embryos. The primitive streak elongates, and cells move from the epiblast surface through the primitive groove into the cleft space. Now recall the bird fate map (Figure 7–1). Some of the incoming epiblast cells form endoderm. These cells enter the hypoblast. The other incoming epiblast cells form the middle mesoderm layer. In the bird, therefore, gastrulation occurs only in the nonyolky cell cap. The yolk remains uncleaved and uninvolved in the gastrulation process.

GENESIS OF EARLY EMBRYONIC STRUCTURES

How is a flat embryo, such as the bird gastrula, transformed into a three-dimensional entity? How are the structures that surround developing embryos formed? Let us begin with the bird gastrula and examine how this embryo takes shape.

BODY FOLDS OF THE BIRD EMBRYO

The bird embryo is transformed into a three-dimensional organism by a series of folds that occur in the embryo proper. Recall that the bird embryo sits atop the yolk as a layered sheet of cells. The embryo lifts itself off of the yolk by folds that occur anteriorly, posteriorly, and laterally. These folds are shown in Figure 7–12. Similar foldings occur in both reptilian and mammalian embryos.

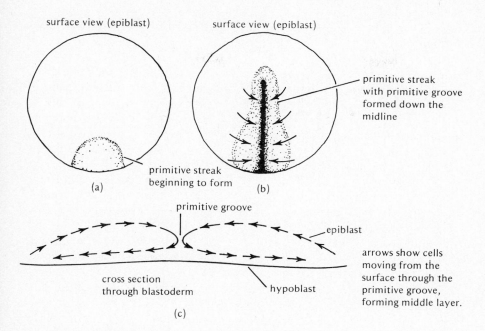

surface view (epiblast)

surface view (epiblast)

primitive streak
with primitive groove
formed down the
midline

primitive streak
beginning to form

(a)

(b)

primitive groove

epiblast

arrows show cells
moving from the
surface through the
primitive groove,
forming middle layer.

cross section
through blastoderm

hypoblast

(c)

Figure 7–11. Bird gastrulation.

The head fold occurs at the anterior end of the embryo, the tail fold at the posterior end, and the lateral body folds at the sides of the embryo. As a result, the embryo becomes a three-dimensional cylinder above the yolk. The midgut region of the embryo remains open to the yolk. Neural folds also occur, to be described in Chapter 8.

EXTRAEMBRYONIC FOLDS OF THE BIRD EMBRYO

The bird embryo now begins to resemble a real three-dimensional organism. The body folds have transformed a flat sheet into a cylinder that sits atop the yolk. Another series of folds occurs in the cells surrounding the embryo proper. These extraembryonic folds form membranes that protect the embryo and store its wastes (see Figure 7–13).

The anterior extraembryonic *somatopleure* is the combination of the ectoderm and the mesoderm in contact with it anterior to the head fold of the embryo proper. This membrane folds over the head fold of the embryo. Similarly, the posterior extraembryonic somatopleure folds over the tail fold of the embryo. Laterally, the extraembryonic somatopleure folds over its sides. These folds are called the head, tail and lateral folds of the amnion, respectively. They cover the embryo anteriorly, posteriorly, and laterally and fuse together to form a double "helmet" that covers the embryo. The outer membrane of the double helmet is called the *chorion*; the inner mem-

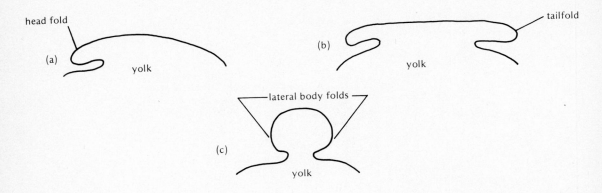

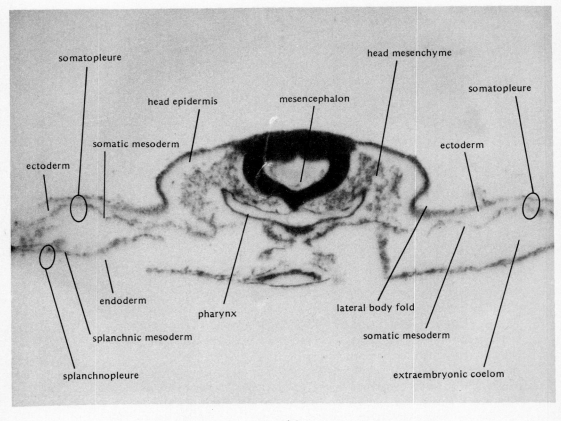

(d)

Figure 7-12. Body folds of the chick embryo. (a) and (b) are longitudinal sections. (c) is a cross section. (d) is a cross section through a 33 hr. chick embryo showing lateral body folds.

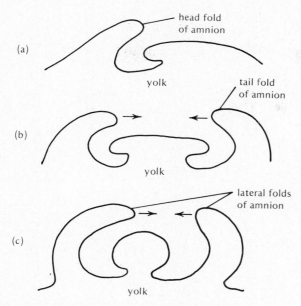

(a)

head fold
of amnion

yolk

tail fold
of amnion

(b)

yolk

lateral folds
of amnion

(c)

yolk

Figure 7-13. Extraembryonic folds of the chick embryo.

brane, the *amnion*. Each consists of somatopleure. The space between the amnion and chorion is called the *extraembryonic coelom*; the space between the amnion and the embryo proper is called the *amniotic cavity*. The amniotic cavity contains amniotic fluid, which protects the embryo from drying out and cushions it against mechanical injury.

Another important sac, the *allantois*, develops as an outpocketing of the hindgut (see figure 7-14). It consists of *splanchnopleure*, the combination of endoderm and splanchnic mesoderm (mesoderm that touches the endoderm). The allantois extends into the extraembryonic coelom and serves as an embryonic excretory organ in which uric acid is deposited. The combination of the allantois and the chorion is called the chorioallantoic membrane. This membrane becomes full of blood vessels and functions as a respiratory organ until the chick hatches. Oxygen from the outside passes through the shell into the chorioallantoic vessels; carbon dioxide passes out of the vessels and passes through the shell to the outside.

MAMMALS

Gastrulation in mammals, including humans, resembles that of the bird. However, most mammalian eggs contain little or no yolk (except in the egg-laying monotremes). If yolk quantity and distribution influence gastrulation, why is human gastrulation similar to that of the bird? Yolk probably

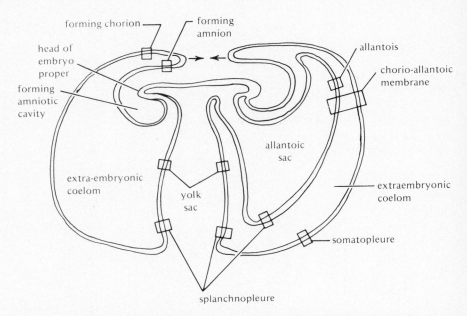

Figure 7–14. Diagram of the early chick embryo with its forming extraembryonic membranes. Longitudinal view.

disappeared from most mammalian eggs when development inside the mother came about. Perhaps the pattern of gastrulation was not significantly altered with the loss of yolk and still resembled that of its yolky-egged ancestors.

The mammalian blastodisc consists of an epiblast and hypoblast layer like that of the bird. A primitive streak forms. Cells from the primitive streak migrate to form a middle layer between the epiblast and the hypoblast.

Obviously, it is much more difficult to study embryonic development in mammals than in sea urchins, frogs, and birds. Mammal embryos must be in the mother to develop normally; whereas sea urchin, frog, and bird embryos develop nicely outside of the mother under easily controlled conditions. Human eggs and early embryos are occasionally available for study after medical operations, or by treating volunteers with hormones to induce oocyte development. This sort of experiment, however, is controversial. Mammalian eggs can be removed from the mother and fertilized in a test tube. Early development can be observed outside of the mother, but the embryo soon deteriorates unless it is implanted in the uterus of a receptive female. A human egg was removed from the ovary of a female with blocked oviducts. The egg was fertilized outside the mother with the husband's sperm and implanted in the mother's uterus. The egg developed normally, and a normal baby girl was born. This has also been done in other mammals, such as mice. At this point, however, detailed information is not available about some aspects of embryonic development in many mammals. This can be attributed to the complex conditions needed to study mammalian embryos and to the lack of sufficient quantities of such embryos for study.

SUMMARY

Let us now review some of the important mechanisms that are in gastrulation. In the sea urchin, invagination occurs as a result of cell motility, selective adhesiveness, and contractility. The active movement of the primary mesenchyme cells and the contraction of the filopodia of the secondary mesenchyme cells that specifically adhere to the animal end cause the vegetal plate of the sea urchin embryo to invaginate. Yolk quantity and distrubution also affect the gastrulation process. In the sea urchin and in *Amphioxus*, the relatively small amount of yolk permits invagination to occur. As the quantity of yolk increases, the pattern of gastrulation changes. In the amphibian, cells migrate into the embryo through a narrow slit. Invagination does not occur. In the bird embryo, the blastodisc sits atop the yolk. The

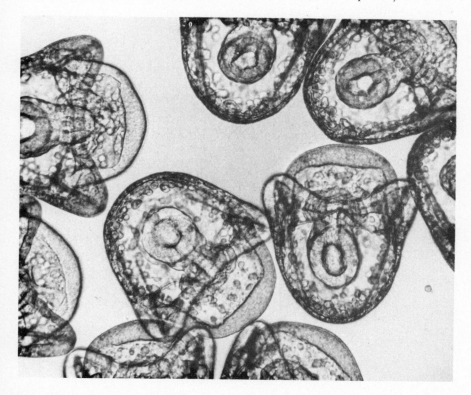

Figure 7–15. Sea urchin larvae (early pluteus). This stage follows gastrulation. Basic body plan is nearly complete. Courtesy of Victor Vacquier.

entire gastrulation process, like the entire cleavage process, occurs in the nonyolky cap.

Although many different patterns of gastrulation occur in various organisms, the end result is the same. Gastrulation rearranges the cells of a nondescript ball or cap to give them form and structure. In Chapter 12, we shall investigate the molecular mechanisms (such as selective cell adhesiveness) that control the cell rearrangements during gastrulation and other stages.

In the next chapter, we shall focus on how the embryo achieves a three-layered state, and how the nervous system forms. No attempt will be made to review these processes in all—or even most—types of organisms. Instead, certain systems will be described that provide basic examples of the relevant mechanisms.

We began our study of embryology with sperm and egg cells. The embryo has developed until it is more than just a blob of cells. It still does not resemble the adult organism, but the foundations of the basic body plan are almost complete (see Figure 7–15). In the next chapter, the basic plan will be fully established.

READINGS AND REFERENCES

Davidson, E. H., Hough-Evans, B. R., and Britten, R. J. 1982. Molecular biology of the sea urchin embryo. *Science 217*: 17–26.

Guidice, G. 1973. *Developmental Biology of the Sea Urchin Embryo*. Chicago: Academic Press.

Gustafson, T., and Wolpert, L. 1967. Cellular movements and contact in sea urchin morphogenesis. *Biol. Rev.* 42: 442–98.

Holtfreter, J. 1943. A study in the mechanics of gastrulation I. *J. Exp. Zool.* 84: 261–318.

Holtfreter, J. 1944. A study in the mechanics of gastrulation II. *J. Exp. Zool.* 95: 171–212.

Keller, R. E. 1976. Vital dye mapping of the gastrula and neurula of *Xenopus laevis*. *Develop. Biol.* 51: 118–37.

Keller, R. E. 1980. Cellular basis of epiboly. An SEM study of deep cell rearrangement during gastrulation in *Xenopus laevis*. *J. Embryol. Exptl. Morph.* 60: 201–34.

Kirschner, M. W., and Gerhart, J. C. 1981. Spatial and temporal changes in the amphibian egg. *BioScience* 31: 381–88.

Moore, A. R. 1941. On the mechanics of gastrulation in *Dendraster excentricus*. *J. Exp. Zool.* 87: 101–11.

Nicolet, G. 1971. Avian gastrulation. *Adv. Morphogen.* 9: 231–62.

Spratt, N. J., and Haas, H. 1960. Integrative mechanisms in the development of the chick embryo I. *J. Exp. Zool.* 145: 97–137.

Trinkaus, J. P. 1969. *Cells Into Organs: The Forces that Shape the Embryo*. Englewood Cliffs, NJ: Prentice Hall.

Vogt, W. 1929. Gastrulation and mesoderm formation in urodeles and anurans. *Roux Arch.* 120: 385–706.

8

NEURULATION AND GERM LAYER FORMATION

In the last chapter we examined gastrulation in several representative systems. We did not, however, reach the point at which the embryo is composed of three distinct layers, ectoderm, mesoderm, and endoderm. How does the three-layered state come about in various embryos? How does the neural tube form in the embryo? What are the mechanisms of these events? The first two questions can be fully answered. The question about mechanism is more difficult to answer because these mechanisms are, as yet, not well understood. We shall, however, learn a great deal about the forces involved in these events. Chapters 11 and 12 deal with these mechanisms in a more thorough manner.

THE THREE-LAYERED STATE

AMPHIOXUS

In the last chapter we left *Amphioxus* when it was composed of two layers, an outer ectodermal layer and an inner mesendoderm layer. As can be seen in Figure 8-1, the mesoderm separates from the endoderm as a result of the formation of pouches. The mesoderm separates into three regions. The two lateral mesodermal pouches migrate between the ectoderm and the endoderm and fuse ventrally. Thus, the middle germ layer, or mesoderm, forms when the mesendoderm (prospective mesoderm plus prospective endoderm) forms pouches and the lateral mesodermal pouches migrate between the endoderm and ectoderm. The middle mesodermal region forms the notochord. A space called the body cavity, or *coelom*, forms within the mesodermal pouches at the anterior region of the embryo by the separation of solid mesoderm in the more posterior regions of the embryo.

In summary, the mesendoderm layer of the primitive chordate *Amphioxus* reached the inside of the embryo during gastrulation by invagination (Chapter 7). The mesoderm separated from the endoderm by pouching of the mesendoderm layer. The body cavity or coelom developed by pouching in the anterior portion of the embryo, and by the separation of solid mesoderm in the posterior portion.

AMPHIBIAN

Figure 8-2 shows that the amphibian embryo is also in a two-layered state before the mesoderm forms. As in *Amphioxus*, the amphibian gastrula consists of an outer ectoderm and inner mesendoderm layer. In the amphibian, however, mesoderm does not form by pouching of the mesendoderm; instead, the mesendoderm separates into mesoderm and endoderm. The mesoderm migrates between the ectoderm and endoderm, fusing ventrally.

The amphibian coelom does not form by pouching, either. Rather, the mesoderm splits, forming somatic mesoderm (in contact with ectoderm) and splanchnic mesoderm (in contact with endoderm). The space between these regions is the coelom. The mesoderm splits off four major subdivisions:

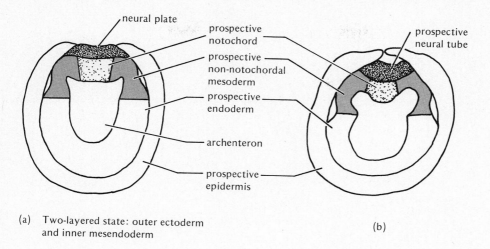

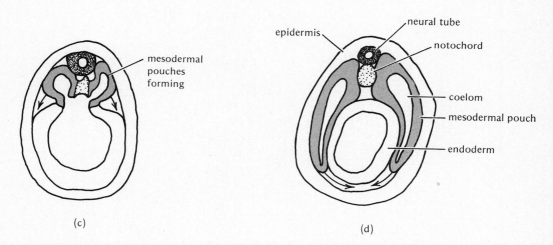

Figure 8-1. Formation of the three-layered state in *Amphioxus*. Mesoderm and coelom form by pouching. Arrows show the movement of mesodermal pouches between epidermis and endoderm. The three-layered state is nearly complete. Modified and redrawn from Hatohek in Korschelt, *Vergleichende Entwicklungsgeschichte Der Tiere* (Jena: G. Fischer, 1936).

notochord, *somite, intermediate mesoderm*, and *hypomere* (somatic and splanchnic mesoderm). We shall define these regions and their derivatives in the next chapters.

BIRD
The three-layered state develops similarly in birds, reptiles, and mammals. Most of the present discussion also applies to the latter two groups. In the last chapter, the bird embryo was at a stage in which the prospective mesoderm

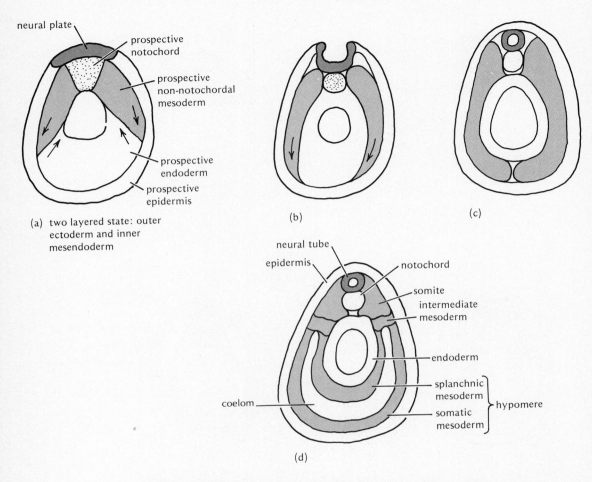

Figure 8-2. Formation of the three-layered state in the amphibian. Mesoderm and coelom form by splitting.

and some prospective endoderm from the epiblast (top layer) were migrating into a central cleft space between the epiblast and the hypoblast (bottom layer). The prospective endoderm cells originating from the epiblast enter the hypoblast. The prospective mesoderm cells remain between the epiblast and the hypoblast. Note that the endoderm of the gut tube originates in the epiblast, so that the epiblast is not exactly equivalent to the amphibian ectoderm. Also, some hypoblast cells end up in the notochord, so that hypoblast is not equivalent to endoderm. The end points of the various migrations were found by marking regions of the embryo with radioactive tracers and carbon particles and observing the final location of marked cells.

After the mesoderm moves from the epiblast through the primitive groove, it occupies the cleft space between the epiblast and the hypoblast.

The presumptive notochordal mesoderm accumulates in an anterior-to-posterior direction. The notochord begins to form at the anterior end of the embryo. Additional cells move in from the epiblast, elongating the notochord posteriorly. The epiblast cells directly above the prospective notochord are prospective neural plate.

The bulk of the non-notochordal mesoderm has also moved into the space between the epiblast and the hypoblast. As described in the next chapter, the parts of the mesoderm adjoining the neural tube become somite and intermediate mesoderm. The remainder splits, as in the amphibian, into outer somatic and inner splanchnic mesoderm (Figure 8–3). The space between the two is the coelom.

In summary, we have seen how the three-layered state forms in *Amphioxus*, the amphibian, and the bird. The mechanisms of coelom formation differ, but the end is the same. In *Amphioxus* the coelom forms by the

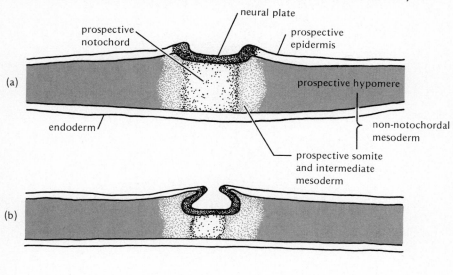

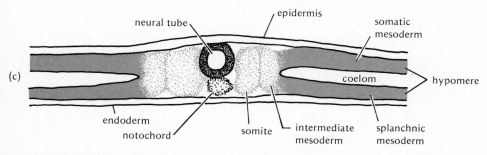

Figure 8–3. (a) to (c) Mesodermal separation in chick embryo. (d) (overleaf) Cross section through 50-hr old chick embryo showing mesodermal separation. Photo part d by Richard L. C. Chao.

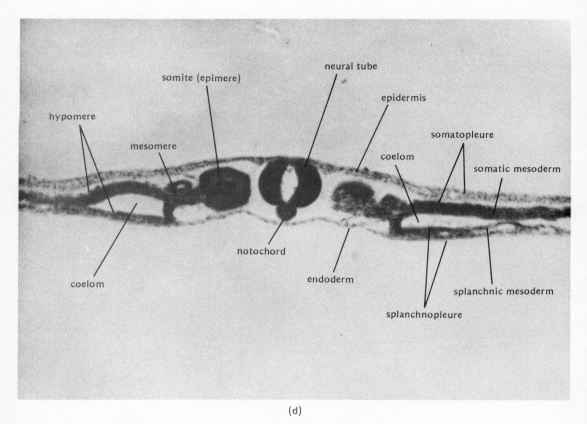

(d)

Figure 8-3. (*continued*)

pouching and splitting of the mesoderm; whereas in the amphibian and bird, the mesodermal tissue separates to form the coelom. Chapter 12, "Mechanisms of Morphogenesis," deals with some of the factors involved in the development of form in the embryo. For example, changes in the adhesiveness of tissue cells may cause the tissue to separate into two parts. We shall now turn to neurulation, the formation of the nerve tube.

NEURULATION

AMPHIOXUS

Neurulation in *Amphioxus* occurs in three basic steps: (1) The neural plate (prospective neural tube) separates from the prospective epidermis; (2) The neural plate folds; (3) The neural tube forms and the overlying epidermis fuses. (See Figure 8-4.) It seems plausible that, for example, the prospective

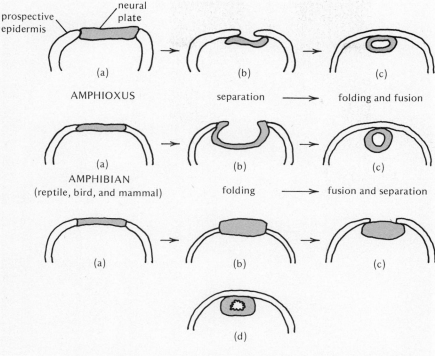

Figure 8-4. Neurulation in *Amphioxus*, Amphibian, and bony fish.

neural tube separates from the prospective epidermis as the result of adhesive changes in the cells of the ectoderm. More will be said about mechanisms of neurulation in the following sections.

AMPHIBIAN, REPTILE, BIRD, AND MAMMAL

The neural tube forms similarly in the amphibian, reptile, bird, and mammal. The sequence of events is slightly different from that in *Amphioxus*. It can be summarized as follows (see Figure 8-4): The neural plate folds, the crests of the neural plate fuse, and the neural tube separates from the epidermis and drops below the surface. Meanwhile, the overlying epidermal regions fuse to form an intact outer epidermal layer. Neural crest cells (Chapter 9) also separate from the ectoderm and drop below the surface. Thus in the amphibian, reptile, bird, and mammal, the folding of the neural plate is the first step in neurulation, followed by the fusion of the neural folds and of the overlying epidermis (Figures 8-5 and 8-6).

Figure 8-5. Amphibian neural plate stage. Courtesy of Peter Armstrong.

MECHANISMS OF NEURULATION

In Chapter 12, we shall examine the mechanisms of morphogenesis in some detail. Mention only some of the forces that influence the neurulation process. The formation of the neural tube in the systems discussed involves cell-surface adhesive changes in the ectoderm. In addition, the contraction of peripheral microfilaments oriented parallel to the short axis of neural plate cells causes some folding to occur (Figure 8-7). Baker and Schroeder have found bundles of microfilaments just below the surface of folding neural plate cells in certain amphibian embryos (*Hyla* and *Xenopus*). Microtubules also play a role in neurulation. Waddington and Perry found many microtubules in the neural plate cells of certain amphibian embryos (*Triturus alpestris*). These workers suggested that neurulation occurs because of cell elongation caused by microtubules (Figure 8-7). The role of microtubules and microfilaments in neural folding will be examined in Chapter 12. Thus, factors such as changes in cell adhesiveness, the contraction of microfilaments, and the

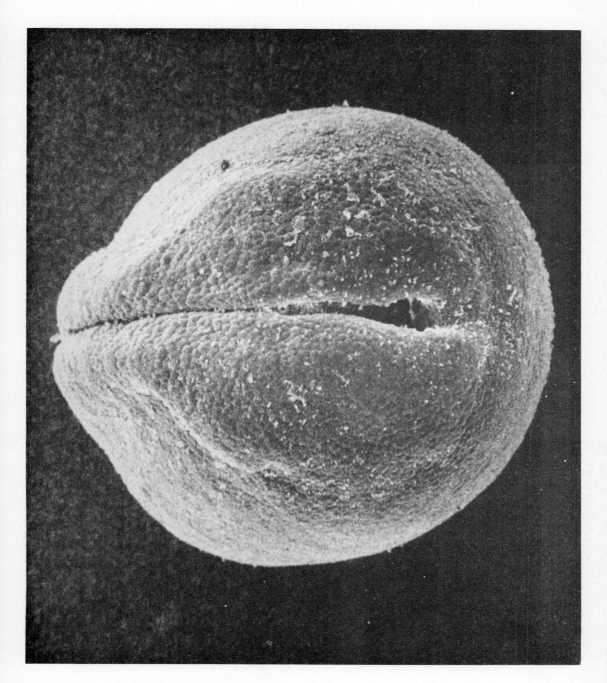

Figure 8-6. Amphibian embryo, neural folds closing. Courtesy of Peter Armstrong.

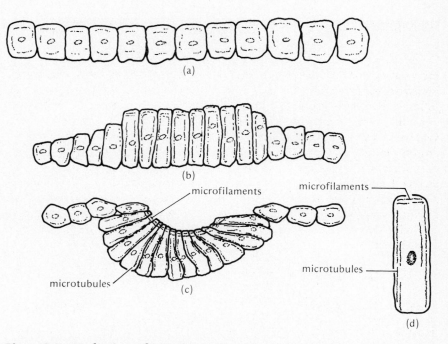

Figure 8-7. Mechanism of neurulation. Two factors that may be involved in neural folding in newt embryos are the elongation of cytoplasmic microtubules parallel to the long axis of the cells and the contraction of a surface layer of cytoplasmic microfilaments parallel to the short axis of the cells. After Burnside, *Devel. Biol.* 26 (1971): 434.

elongation of microtubules are important in the neurulation process. We shall conclude our discussion of neurulation by looking at this process in one other system—the bony fish. The neural tube forms in this embryo by a mechanism completely different from those described earlier.

BONY FISH (TELEOST) NEURULATION

As seen in Figure 8-4, the neural plate in the bony fish thickens, but it does not fold. This thickened neural plate separates from the rest of the ectoderm, sinks below the surface and the epidermis fuses over it. The prospective neural tube, however, is still a solid rod. How does it become a hollow tube? The neural tube forms in the bony fish by cavitation. Cells in the solid rod die. The hollow nerve tube in the bony fish, therefore, forms by differential cell death in the center of the prospective neural tube. Cell death plays other important roles in morphogenesis, as will be seen in Chapter 12.

SUMMARY

In this chapter we saw that, although the mechanisms differ slightly, the three-layered state and body cavity formed in all of the embryos examined. Likewise, the nerve tube forms by various mechanisms, including separation followed by folding, folding followed by separation, and selective cell death in a solid core. The basic body plan of chordates and vertebrates has been achieved. The embryos we have examined now contain three layers and possess neural tubes.

What structures are derived from each germ layer, and what are the mechanisms of their formation? We shall begin answering this question in the next chapter, with an overview of the primary germ layer derivatives. In the following two chapters, we shall deal in more detail with organogenesis and with the mechanisms controlling the development of embryonic form and structures.

READINGS AND REFERENCES

Baker, P. C., and Schroeder, T. E. 1967. Cytoplasmic filaments and morphogenetic movement in the amphibian neural tube. *Devel. Bio.* 15: 432–50.

Balinsky, B. I. 1975. *An Introduction to Embryology*, 4th ed. Philadelphia: W. B. Saunders.

Burnside, B. 1973. Microtubules and microfilaments in amphibian neurulation. *Am. Zool.* 13: 989–1006.

Karfunkel, P. 1972. The activity of microtubules and microfilaments in neurulation in the chick. *J. Exp. Zool.* 181: 289–302.

Spemann, H. 1938. *Embryonic Development and Induction.* New York: Hafner Publishing.

Torrey, T. W. 1967. *Morphogenesis of the Vertebrates*, 2nd ed. New York: John Wiley.

Waddington, C. H., and Perry, M. M. 1969. A note on the mechanism of cell deformation in the neural folds of the amphibian. *Exp. Cell Res.* 41: 691–93.

Weiss, P. 1955. Nervous system (Neurogenesis). In B. H. Willier, P. Weiss, and V. Hamburger, eds., *Analysis of Development*, p. 346. Philadelphia: W. B. Saunders.

9

EARLY HUMAN EMBRYO AND PRIMARY GERM LAYER DERIVATIVES

SECTION 1   THE EARLY HUMAN EMBRYO

So far, we have examined early development in several groups of organisms, but have made only casual references to development in mammals. We shall survey early mammalian development here because of our special interest in the development of humans and their relatives in the Class Mammalia.

GAMETES

Mature human spermatozoa are formed from spermatogonia in the seminiferous tubules of the testis. The primordial germ cells that form the spermatogonia apparently originate in the yolk-sac endoderm and migrate to the gonad. Mature spermatozoa are the final product of a growth phase, two meiotic divisions, and a differentiation phase of the spermatids into spermatozoa. The steps leading to the formation of mature sperm are:

spermatogonia
(growth)

primary spermatocytes
(meiosis 1)

secondary spermatocytes
(meiosis 2)

spermatids
(differentiation)

mature spermatozoa

Human sperm are about 65 μ in length. There are about 100 million human sperm in a milliliter of semen.

Human eggs are formed in the ovary from oogonia. The primordial germ cells that form human oogonia also probably originate in the yolk-sac endoderm (the inner lining of the yolk sac) and migrate to the gonad. In the human female, the outer region of the gonad (cortex) develops into the ovary. Only the primordial sex cells that entered the outer region of the gonad develop into oogonia. In the human male, only the primordial sex cells that enter the inner (medulla) region of the primitive gonad develop into sperm. Thus, it is the inner region of the primitive gonad that develops into the testis in males (see Chapter 10).

The development of human eggs from oogonia, as noted in Chapter 4, involves the following sequence of events:

oogonia
(growth)

primary oocytes
(meiosis 1)

secondary oocytes
(meiosis 2)

mature eggs

Recall that one oogonium gives rise to only one mature egg; whereas, one spermatogonium gives rise to four mature spermatozoa. This occurs because the meiotic cytoplasmic divisions are unequal in the female, yielding large cells (secondary oocyte and egg) and tiny cells (polar bodies). In this way the stores of molecules and other components required for embryonic development are not divided wastefully among four cells. Instead, one large egg with a full supply of stored material is formed. Since mammals usually obtain nourishment from the placenta, their eggs need not be as large as those of some other organisms. Sperm, on the other hand, develop from equal cytoplasmic divisions, so that one spermatogonium yields four sperm cells. Unlike eggs, sperm do not accumulate large stores of material.

The human egg is about 120–150 μm in diameter. Human polar bodies, which are not fertilizable, are at most 10 μm in diameter. In the human female, as many as two million primary oocytes are present at birth. Only 300–400, however, ever reach maturity. At puberty and thereafter until menopause, one oocyte (usually) matures during each menstrual cycle. Thus, about once a month, one egg matures in the human female.

The maturation of human oocytes in their expanding chambers (follicles) is controlled by the combined presence of hormones from the anterior pituitary gland (*follicle-stimulating hormone* and *luteinizing hormone*) and a hormone, *estrogen*, secreted from the follicles themselves. The pituitary hormones, in turn, are secreted in response to secretions from the hypothalamus. *Ovulation*, the release of the oocyte with a surrounding layer of follicle cells (corona radiata) from the follicle chamber (and ovary), is controlled by a burst of luteinizing hormone secretion that may dissolve the surface layer of the follicle (Figures 9–1 and 9–2).

During maturation, the follicle expands. The follicle cells secrete fluid into the expanding follicle and also nourish the developing oocyte (see Chapter 4). Just before ovulation, the follicle greatly enlarges and fills with liquid (liquor folliculi). This mature follicle is called a *Graafian follicle*. After ovulation, the collapsed Graafian follicle (now empty) is transformed into a structure called the *corpus luteum*. The corpus luteum secretes estrogen and an additional hormone, *progesterone*. These hormones act on the hypothalamus, which signals the anterior pituitary to begin synthesizing follicle-stimulating hormone and luteinizing hormone.

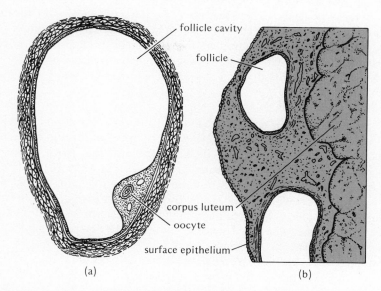

Figure 9–1. (a) Human Graafian follicle. (b) Section of human ovary. From H. Tuchmann-Duplessis et al., *Illustrated Human Embryology*, vol. I (New York: Springer-Verlag, 1972; Paris: Masson, Editeur, 1972).

Estrogen and progesterone from the corpus luteum also prepare the uterus for the implantation of the ovulated oocyte (if fertilization occurs). The corpus luteum hormones cause the uterine lining (endometrium) to thicken and enrich its blood and gland supply. Note that the human oocyte is ovulated as a secondary oocyte at the second meiotic metaphase (where the

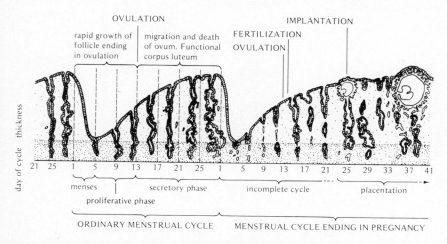

Figure 9–2. Changes in human uterine lining during the menstrual cycle. From B. M. Patten, *Human Embryology* (New York: McGraw-Hill, 1974).

chromosomes are ready to separate into two daughter cells). Fertilization stimulates the completion of the second meiotic division. If fertilization does not occur, the oocyte does not implant into the uterine lining, and the corpus luteum, by some unknown means, stops secreting hormones. This causes the rich uterine lining to slough off, a process called menstruation.

HUMAN GAMETE ANOMALIES

Before we discuss fertilization, implantation, and early embryonic development in human beings, we shall describe briefly some of the anomalies that can occur in human gametes. For instance, some of the sperm in normal semen (the fluid containing the sperm) are double (Figure 9–3). These may result from a failure of cell separation during spermatogenesis. In the fetal ovary, unusual oocytes are also occasionally seen. These include oocytes with two nuclei, or double oocytes in the same follicle (Figure 9–4). An embryo formed from grossly abnormal gametes is probably defective and would spontaneously abort early in development.

Anomalies at the chromosomal level can also occur during meiosis. Sometimes one cell gets both chromosomes of a pair, while the other gets neither (Figure 9–5). Any embryo resulting from gametes with such abnormal chromosome numbers is likely to have severe problems. Chromosomal abnormalities can occur in either autosomes or sex chromosomes. Thus, a gamete may end up with either an X or Y chromosome (normal), or with no sex chromosome, or with an XY or XX set in a single gamete. When fertilization involves an abnormal gamete and a normal gamete, a variety of abnormalities can result. These include the XXY *karyotype* (chromosome pattern) called Klinefelter's syndrome and the XYY syndrome. About one in 500 male births is a Klinefelter's baby. The genitalia are male, but reduced in size. This results in lowered androgen levels, causing reduced body hair and

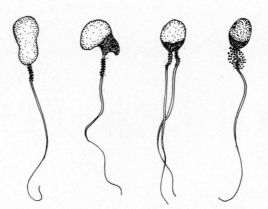

Figure 9–3. Some abnormal human sperm cells.

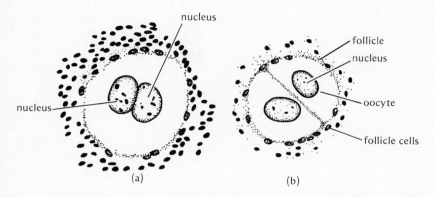

Figure 9-4. Some abnormal human egg cells. (a) Oocyte with two nuclei. (b) Two oocytes sharing the same follicle.

reduced sexual activity. Typically, Klinefelter's individuals are taller than average, are sterile, and exhibit some degree of mental retardation. XXXY and XXXXY individuals also exhibit the basic characteristics of the XXY individual. With additional X chromosomes, the genitalia become more ambiguous.

The XYY karyotype often results in taller individuals. Originally, XYY males were thought to exhibit aggressive and antisocial behavior. Note, however, that many individuals have these syndromes, yet possess normal behavior. It may be that only a minority of them (those diagnosed in prisons or mental institutions) have social problems.

The XO karyotype develops into a sterile female. Only about 1 in 1000 births, however, is an XO individual. This condition is called Turner's syndrome, and the low birth rate is paired with a higher rate of spontaneous XO abortions. Those XO individuals who survive have hormonal deficiencies, short stature, reduced sexual development, and are sterile.

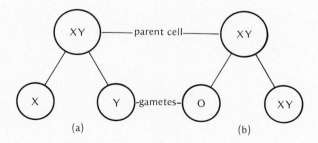

Figure 9-5. Distribution of sex chromosomes. (a) Normal distribution. The haploid parent cell, when divided, results in two sperm cells, one carrying the X (female) chromosome and one carrying the Y (male) chromosome. (b) Abnormal distribution (nondisjunction) produces one gamete with both sex chromosomes, and one with neither.

Some XXX females exhibit a decrease in mental abilities, but they are sometimes fertile and generally exhibit normal characteristics even though they have an inceased "dose" of X.

The more common autosomal (nonsex) chromosomal anomalies include Trisomy 21 (Down's syndrome or Mongolism). Persons with Down's syndrome possess three, not two of chromosome number 21 (Figure 9-6). Down's individuals are mentally retarded and have a flat facial profile, unusual external ears, and slow bone growth. Males with Down's syndrome are not fertile. At age forty-five, the risk of a woman giving birth to a Down's baby is 1 in 60, while it is only 1 in 1500 for women at age twenty. Trisomy-18 individuals possess three 18 chromosomes instead of two. About 1 in 3000 babies are born with this anomaly. Only about 10 percent survive to the first year. Such children are very weak. Trisomy 13 results from three instead of two 18 chromosomes. This anomaly occurs in 1 of 5000 births. Fewer than 20 percent survive to the first year. These individuals have improperly developed forebrains, poorly developed optic and olfactory nerves, and severe facial abnormalities.

These anomalies occur as a result of the nondisjunction of chromosomes during meiosis, as indicated earlier. One of the resulting gametes has an extra

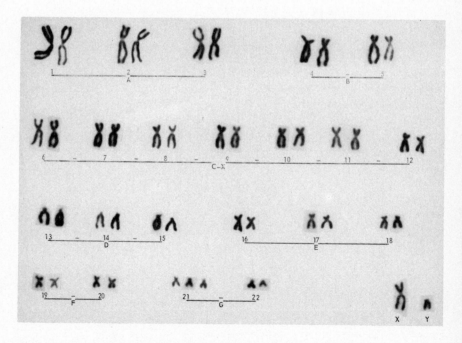

Figure 9-6. Karyotype of male with Down's syndrome (Trisomy 21). Note that there are three chromosomes in group 21 rather than the normal two.

chromosome; the other lacks a chromosome. Nondisjunction, however, can also occur after fertilization or during cleavage—in short, during mitotic division in which chromatids pass to daughter cells. The earlier it occurs, the more cells are involved and the more widespread the effects. In other cases, chromosomes lag behind on the spindle at anaphase in mitosis, resulting in one daughter cell that receives no chromosome, while the other is normal. In addition, radiation and a variety of mutagenic chemicals can interfere with chromosome replication and cause chromosome damage with physiological consequences. The anomalies discussed (and a host of others) occur because certain genes are missing, altered, or present in the wrong dose. The manner by which genes lead to normal differentiation of cells is discussed in Chapters 13, 14, and 15.

FERTILIZATION, IMPLANTATION, AND EARLY DEVELOPMENT

In humans, fertilization usually occurs inside the distal third of the Fallopian tube, the end nearest the ovary. The mature oocyte is ovulated from the Graafian follicle and picked up by the expanded end of the Fallopian tube surrounding the ovary. Fertilization occurs if sperm are present in the distal third of the Fallopian tube. Human sperm take about ten hours to reach this part of the Fallopian tube, so that fertilization usually occurs hours after coitus. If the egg is not fertilized within 24 hours after ovulation, it dies.

Upon fertilization, the second polar body is extruded and the nuclei of the sperm and egg fuse, restoring the normal diploid chromosome number. Sex is determined at fertilization by the sperm. A sperm containing an X chromosome forms an XX zygote that is genetically female; one containing a Y chromosome yields an XY zygote that is genetically male.

After fertilization, the embryo moves through the Fallopian tube to the uterus, developing in transit. The 2-cell stage occurs at about 30 hours; the 4-cell and 8-cell stages occur at 40–50 hours and 60 hours, respectively. When the embryo approaches the entrance to the uterus, it is in the 12–16 cell stage, the morula. This occurs on the fourth day. The morula emerges from its enclosing coat, the zona pellucida, on the fifth day. The free embryo, now in the 32–64 cell stage, is called a blastula or blastocyst. It possesses a cavity filled with fluid, called the blastocyst cavity. The free blastocyst has reached the uterus and usually implants into the uterine wall during the sixth or seventh day (Figures 9–7 through 9–18). In abnormal situations, the blastocyst may become implanted in the ovary, in the Fallopian tube, or in the wall of the body cavity or intestine, leading to ectopic pregnancy.

The outer layer of the blastocyst is called the *trophoblast*. In one region of the blastocyst, a cluster of cells forms below the trophoblast. This cluster is called the inner cell mass. The inner cell mass forms the embryo proper; the trophoblast forms the extraembryonic membranes (the membranes that

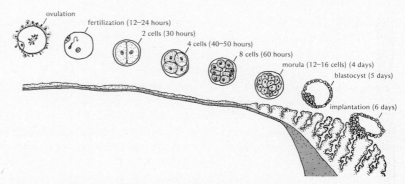

Figure 9-7. Stages in human embryonic development, from ovulation to implantation in the uterus.

surround the embryo). The trophoblast implants itself by invading the uterine tissue. According to some researchers, this process resembles the invasion of tumors into surrounding tissue. The blastocyst cavity, however, is below the embryo proper, not in its middle, so it is not exactly like the blastocoel of other embryos.

A space forms above the inner cell mass. This becomes the amniotic cavity that protects the embryo. During the second week, the epiblast and hypoblast of the blastocyst differentiate in the inner cell mass. As in the chick, the upper layer is the epiblast and the lower layer is the hypoblast. Meanwhile, the embryo continues to bury itself in the uterine lining. By the end of the second week, the primitive streak appears on the surface of the embryonic disc. This marks the start of gastrulation. A primitive groove forms down the middle of the primitive streak. Cells from the top layer (epiblast) move into the primitive groove to form the third embryonic layer, the mesoderm. Figures 9-19 to 9-22 illustrate some of the events occurring during early human development.

Human gastrulation is similar to that of birds. A major difference between the two, however, is the massive amount of yolk in the bird but none in the higher mammals. Thus, although the mammalian blastodisc sits atop a space filled only with fluid, gastrulation occurs much as it does in birds and reptiles. This strongly suggests that mammalian ancestors once had large yolky eggs, but the yolk disappeared as development inside the mother evolved.

PLACENTA

The placenta brings the blood of the mother into close contact with that of the fetus, transferring nutrients and oxygen to the developing fetus and removing fetal wastes. After three months of gestation, the placenta is well defined and continues to develop as the fetus matures. The human placenta forms from both maternal and fetal tissues.

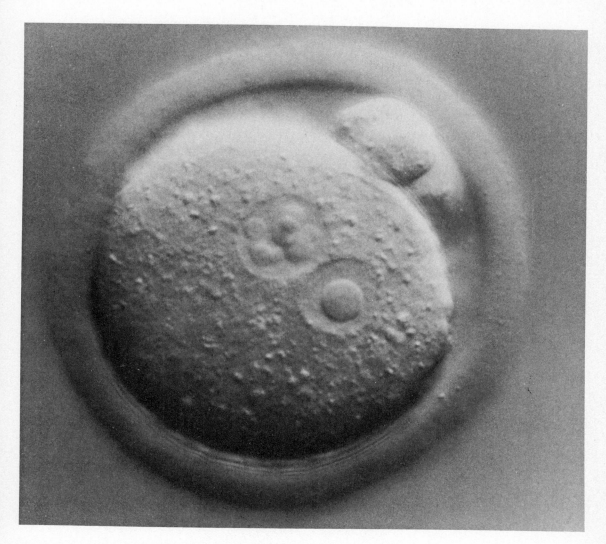

Figure 9–8. Mouse zygote, magnified by Nomarski optics. Courtesy of Dr. P. Calarco.

Before the placenta develops, the human embryo absorbs nutrients from uterine fluid through its surface epithelium (surface layer). After the embryo implants itself in the uterine wall, the blastocyst trophoblast layer grows and spreads out, sending finger-like projections (villi) deep into the uterine lining. Many spaces develop inside the trophoblast. The invading trophoblast reaches blood capillaries of the uterine lining and breaks down the walls of these vessels. Maternal blood therefore flows into the trophoblast spaces to

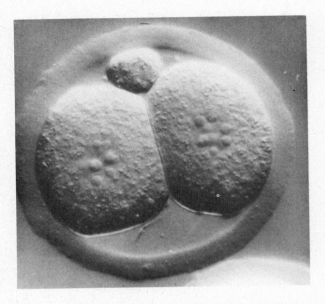

Figure 9-9. Nomarski micrograph of 2-cell mouse embryo. Courtesy of Dr. P. Calarco.

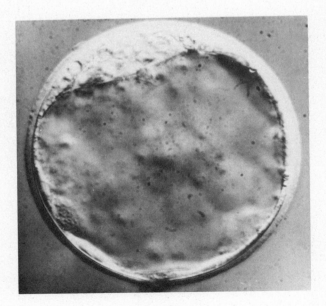

Figure 9-10. Mouse blastocyst, magnified by Nomarski optics. Note that this blastocyst is at the same stage as that shown in Figure 9-16. Courtesy of Dr. P. Calarco.

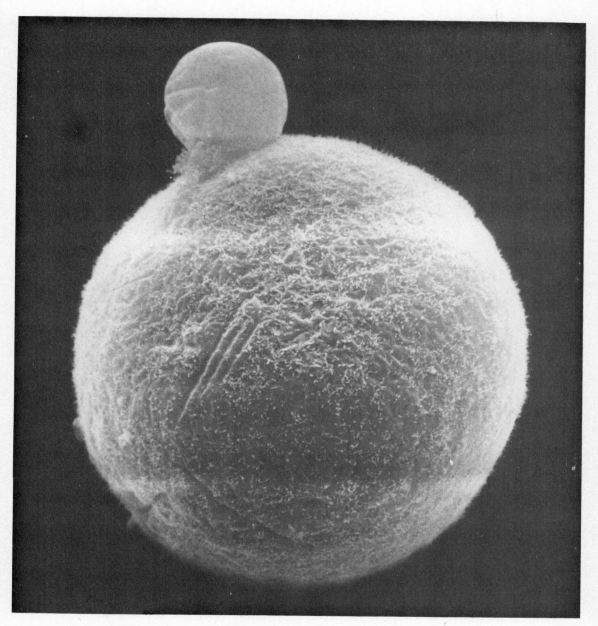

Figure 9–11. Scanning electron micrograph of fertilized mouse zygote. Early developmental stages are similar in most mammals. Courtesy of Dr. P. Calarco.

Figure 9-12. Scanning electron micrograph of 2-cell mouse embryo. Courtesy of Dr. P. Calarco.

nourish the fetus. The placenta is a combination of the embryonic trophoblast villi and the uterine wall. The villi become bathed in maternal blood as a result of the breakdown of the uterine blood vessels. Thus, gas exchange and nutrient diffusion is facilitated between mother and fetus. The fetus can also

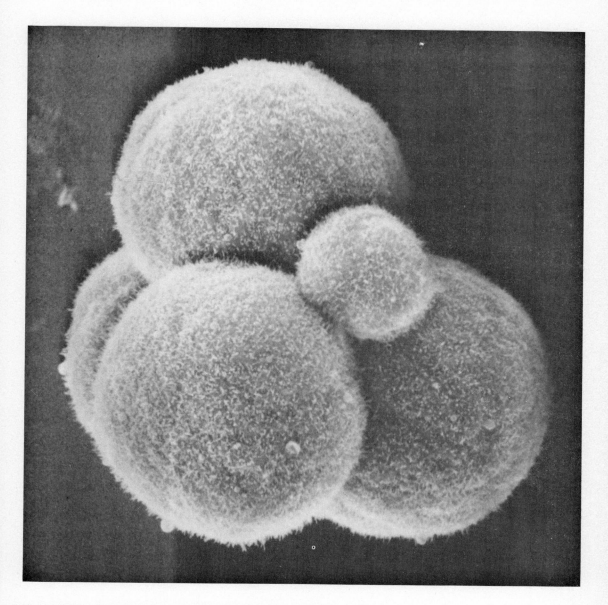

Figure 9-13. Scanning electron micrograph of 4-cell mouse embryo. Courtesy of Dr. P. Calarco.

acquire immunity to certain diseases as the result of maternal antibodies passed from the mother's blood to the fetus via the placenta.

Many drugs and chemicals (including alcohol) are passed to the fetus through the placenta. Embryonic organs are in their most sensitive states

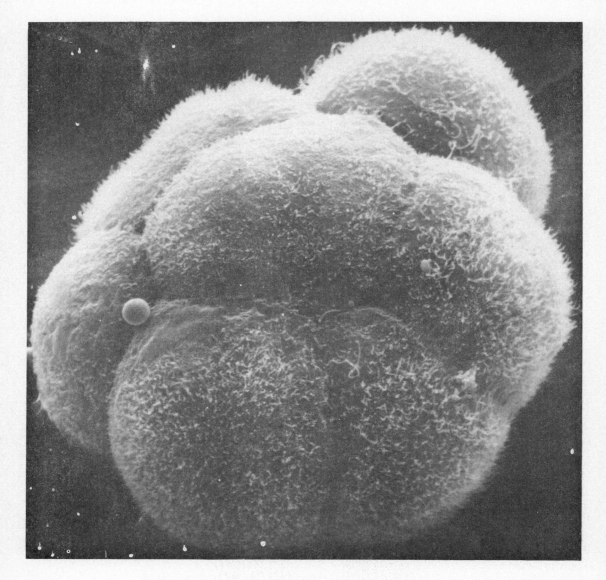

Figure 9-14. Scanning electron micrograph of 8-cell mouse embryo. Courtesy of Dr. P. Calarco.

during early embryonic development and are likely to be damaged by drugs, chemicals, alcohol, and pathogens. Mothers who take drugs, smoke, or drink alcohol are exposing their developing children to agents known to damage embryonic tissues. The sedative thalidomide is one example of a drug that causes extensive damage to the fetal limbs, heart, and gut when taken by

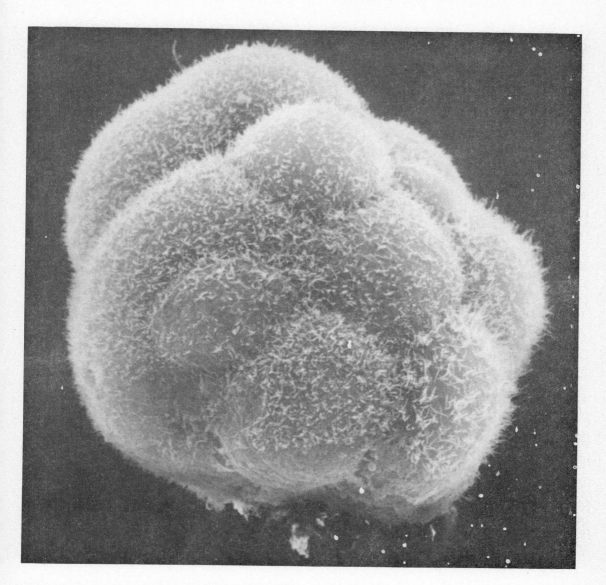

Figure 9–15. Scanning electron micrograph of mouse morula (12–16 cells). Courtesy of Dr. P. Calarco.

mothers during the first two months of pregnancy, when these organs are developing.

HUMAN BIRTH DEFECTS

About two or three percent of human births result in malformed babies. Many kinds of malformations can occur, and many mechanisms cause them.

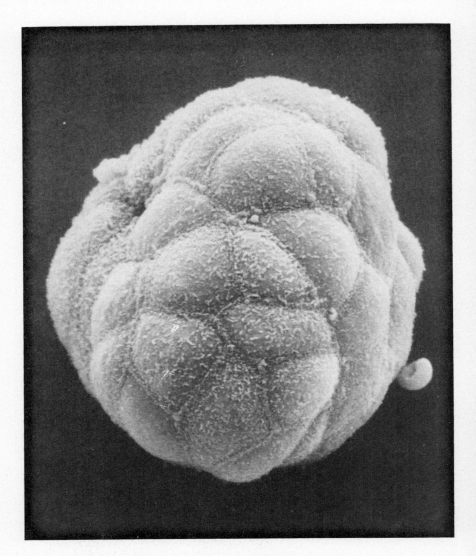

Figure 9-16. Scanning electron micrograph of mouse blastocyst. Courtesy of Dr. P. Calarco.

Ten percent of the women who took the drug thalidomide gave birth to babies with severe limb anomalies (which are otherwise quite rare). Some babies are born with a defect in the closure of the abdominal wall. In these cases, the viscera protrude externally. Other babies are born with an open neural tube. Some twins are born unseparated. Some babies are born with a single median eye. Some others are born with holes between their heart chambers, gill clefts, blindness, or deafness.

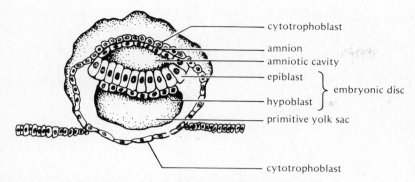

cytotrophoblast
amnion
amniotic cavity
epiblast
} embryonic disc
hypoblast
primitive yolk sac

cytotrophoblast

Figure 9-17. Human blastocyst implanted in the uterine wall.

Developmental abnormalities are caused by such factors as infective agents (such as viruses), X rays, certain drugs and food additives, improper nutrition, mutations and chromosomal abnormalities arising from biological error, and possibly others.

The factors that cause defects do so by a variety of mechanisms. These include improper gene dose arising from abnormal chromosome number (for instance, Klinefelter's syndrome and Trisomy 21); effects on inducers or on the responding tissues; failed fusion of bilateral structures; persistence of embryonic conditions; incomplete development of an organ or tissue; and inhibited cell proliferation. It is obvious that factors can affect almost any aspect of embryonic development. Exactly how these agents cause defects is not well understood, however.

A key concept in understanding how these agents and factors cause developmental defects is that of the sensitive period. Certain structures are

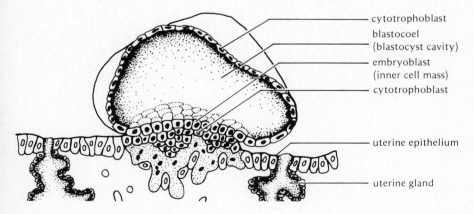

cytotrophoblast
blastocoel
(blastocyst cavity)
embryoblast
(inner cell mass)
cytotrophoblast

uterine epithelium

uterine gland

Figure 9-18. Human blastocyst on the eighth day. Blastocyst has already embedded itself in the uterine wall. This figure represents a later stage than Figure 9-17.

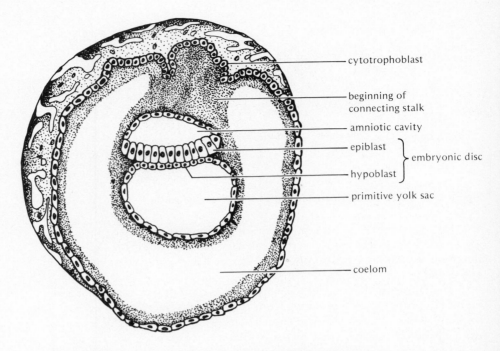

cytotrophoblast

beginning of
connecting stalk

amniotic cavity

epiblast

hypoblast

} embryonic disc

primitive yolk sac

coelom

Figure 9–19. Human blastocyst at the end of the second week.

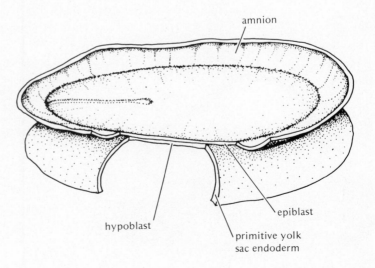

amnion

hypoblast

primitive yolk
sac endoderm

epiblast

Figure 9–20. Human blastodisc at the end of the second week.

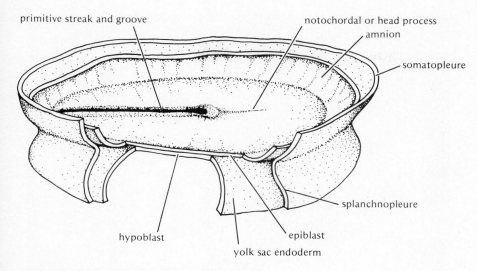

Figure 9–21. **Human blastodisc at 15–16 days.**

sensitive to specific agents at given times and only at given times. The German measles virus, for instance, causes infant blindness only if the mother has this disease between about the fifth and eighth weeks of her pregnancy because the virus must be present when the lens cup closes. Once it closes, the virus can no longer enter the lens and destroy it. Similar sensitive periods are known for the effects of German measles virus on other structures. Ear destruction occurs only during the seventh to twelfth weeks of pregnancy. Heart damage occurs from the fourth to the ninth weeks.

Some drugs, such as those that inhibit RNA synthesis, cause defects only if they are present when the embryonic cells are synthesizing RNA. For example, drugs such as actinomycin D that inhibit RNA synthesis also inhibit development because new RNA is synthesized during organogenesis. At this time RNA synthesis inhibitors have major effects on development. Once organs are formed and little new RNA is being synthesized, such inhibitors do not cause major defects. Thus, an agent causes defects only if that agent is present at a time when developmental processes are occurring that are

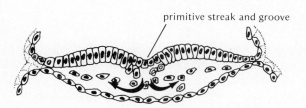

Figure 9–22. **Movement of cells through the primitive groove.**

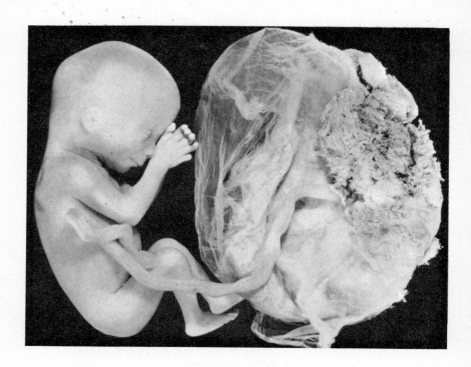

Figure 9-23. Four-month-old human fetus, with placenta. From H. Tuchmann-Duplessis et al., *Illustrated Human Embryology*, vol. I (New York: Springer-Verlag, 1972; Paris: Masson, Editeur, 1972).

sensitive to that specific agent. We should also keep in mind that some agents produce severe defects in some species, but not in others. Thalidomide, for example, causes severe limb defects in humans and rabbits, but not in rats at comparable doses. Thus, to conclude that a given drug is safe for humans because it does not harm a given animal species is, at best, risky. Drugs must be tested on several different mammalian species. Even then, their safety with respect to humans is seldom known for sure.

EXPERIMENTS WITH MAMMAL EMBRYOS: ARE BLASTOMERE FATES FIXED?

Beatrice Mintz and others have performed a set of exciting experiments showing that mammalian blastomeres can form tissues other than those they normally form.

Separate cleaving mouse embryos (or cells from such embryos) were aggregated together and implanted into the uterus of foster mother mice. These embryos developed and were born as normal mice. Specific cells were marked with radioactive labels or with genetic markers such as specific enzyme characteristics. By combining parts of embryos and determining the fate of labeled cells, it was found that each cell of the 4-cell embryo could form trophoblast or inner cell mass. In an experiment with rabbit embryos, it was also found that if seven of the eight cells of the 8-cell stage were destroyed, the remaining cell could form an entire embryo and develop into a fully formed adult. In yet another experiment, one or several labeled cells from the inner cell mass of a mouse blastocyst were injected into another mouse blastocyst. A large portion of the developing host embryo tissues contained descendants of the labeled cell. These results suggest that the fates of cells in the early mouse embryo are not fixed and that a great deal of damage could be done to early mammalian embryos without affecting their normal development. This should help the embryo survive traumas occurring early in development, before implantation.

By the blastocyst stage, however, although the inner cell mass cell fates may not be fixed with respect to the tissue they can form in the embryo proper, the fates of many blastomeres appear to be fixed, at least with respect to their ability to form embryo proper *versus* trophoblast. The position of a cell in the early embryo seems to be important in fixing the fate of that cell. Hillman and Graham found that fifteen whole mouse morulas could be aggregated around a central labeled morula. A giant composite embryo formed, and the central morula formed inner cell mass only and not trophoblast. If the surrounded morula is inside such a composite for more than eight hours, it loses its ability to form trophoblast. By the blastocyst stage, trophoblast cells of one embryo cannot form part of the embryo proper if injected into another blastocyst.

These and other experiments suggest that the position of cells in the early embryo fixes the fate of those cells. Early in development, the fates of cells in forming trophoblast or embryo proper are not fixed. By the blastocyst stage, however, cells become limited to forming either trophoblast or embryo proper as a result of factors that depend on their position in the embryo. These questions will be discussed in greater detail in Chapters 10 and 11. We should keep in mind, however, that although a cell is committed to forming trophoblast or embryo proper by the blastocyst stage, the cells within the inner cell mass are not yet committed to forming specific parts of the embryo proper.

The unfixed status of the cells (or cell nuclei) within the inner cell mass of early mammalian embryos has been suggested by totally different experiments. Illmensee and Hoppe removed inner cell mass cells from blastocyst-stage embryos of a gray mouse. They removed the nuclei from these cells with a micropipette and placed one of these nuclei in each of many fertilized eggs from a black mouse. (Egg and sperm nuclei from the fertilized eggs were previously destroyed.) The eggs were cultured in artificial culture medium until they developed into blastocyst stage embryos, then inserted into the uterus of a pregnant white mouse. The white mouse gave birth to gray mice genetically identical to the embryo that donated the nuclei (Figure 9-24).

Several mice were produced in this way with cell nuclei from other embryos. The transplanted nuclei control the development of the enucleated fertilized eggs into which they were inserted. The results indicate that the nuclei of single inner cell mass cells from blastocyst-stage mouse embryos can regulate the development of the entire organism.

Similar experiments (described in Chapter 13) have been done with frog embryos. In some of these experiments, donor nuclei were taken from more differentiated tissues. In some cases, such nuclei could regulate the develop-

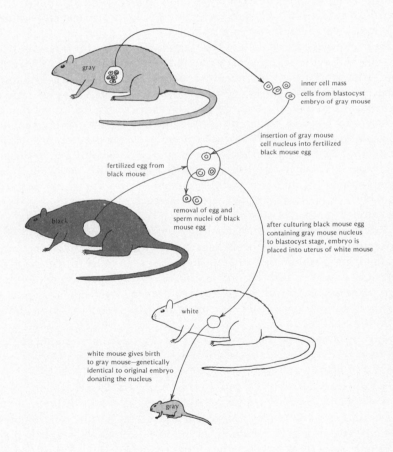

Figure 9-24. Nuclear transplantation in mouse. Based on experiments by Illmansee and Hoppe.

ment of a normal frog. This is probably also possible in mammals, but the techniques needed to deal with tiny, mature mammalian tissue cells are not yet perfected. Illmensee and Hoppe's results suggest that, if mice can develop from eggs with transplanted nuclei, so can human beings. These matters will undoubtedly receive the careful scrutiny they deserve. Such experiments will probably never be done with human cells. Many of the experiments would surely result in deformed embryos and infants, and would be considered morally wrong in any case.

SUMMARY

In summary, the embryos of higher mammals develop in the uterus and are nourished through the placenta. Thus, the egg does not need a great deal of stored yolk. The early development of mammals such as humans is similar to that of reptiles and birds. The major difference is that the eggs and embryos of higher mammals contain little yolk. Yolk probably existed in the ancestral eggs, because the development of mammals uses many of the mechanisms that apparently evolved with the yolky eggs of reptiles and birds.

The early development of the human embryo can be studied in culture. Steptoe and his group removed human eggs from the ovary, fertilized them outside the body, cultured them outside of the mother for several days, and then implanted the blastocysts into the mother's uterus. These embryos developed to term, resulting in the birth of normal babies. Successful culture of early human embryos outside the mother will lead to a better understanding of early human development. Intimate knowledge of human development after the blastocyst stage, however, will probably lag behind our knowledge of development in related mammals such as the mouse. As our understanding of artificial culture improves, mammalian embryos will be developed in culture for longer and longer periods, allowing careful study and continuous monitoring at all times. Mouse embryos can now develop in culture for nearly half of their normal gestation period (to about $8^{1}/_{2}$ days). Society will probably not allow similar long-term experiments with human embryos; it would be inhumane to subject them to conditions that could cause pain, defects, or death. Our understanding of human development, however, will still progress by experiments with mouse and rabbit embryos, for example. Such understanding will lead to medical advances in the area of birth defects, genetic diseases, and other maladies of human infants.

SECTION 2   PRIMARY GERM LAYER DERIVATIVES

In this brief section, we shall outline the major structures derived from the ectoderm, mesoderm, and endoderm of vertebrates. Some differences from the general plan exist in specific organisms; these will be dealt with in later chapters. In the next chapter (on organogenesis) we shall examine in detail the development of many of these derivatives. We look at exceptions to the rule and discuss the mechanisms of organ formation. This outline serves the useful purpose of providing an overview of the primary germ layer derivatives as an introduction to the more detailed study of organogenesis.

ECTODERMAL DERIVATIVES

The outer embryonic layer, the ectoderm, is composed of three regions: the prospective neural tube, the prospective neural crest, and the prospective epidermis. The major derivatives from each of these regions are described below.

NEURAL TUBE DERIVATIVES

As described in Chapter 8, the neural plate forms the neural tube as a result of interactions with the archenteron roof that are discussed further in the chapter on induction. The neural tube similarly differentiates into (a) the brain, (b) the posterior pituitary gland, (c) the optic vesicles (giving rise to the retina, as described in Chapter 10), (d) the spinal cord, and (e) the motor nerves that originate in the ventral portion of the neural tube and innervate muscles.

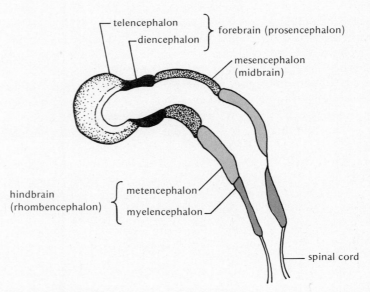

Figure 9-25. (a) Divisions of the vertebrate brain. (b) (facing page) 10 mm pig embryo, sagittal section, showing brain divisions. Photo by Richard L. C. Chao.

The brain differentiates into three major regions: the forebrain (prosencephalon), the midbrain (mesencephalon) and the hindbrain (rhombencephalon) (Figure 9–25). The anterior portion of the forebrain, the telencephalon, forms the cerebral hemispheres and the olfactory centers. The posterior portion of the forebrain, the diencephalon, forms the thirst center

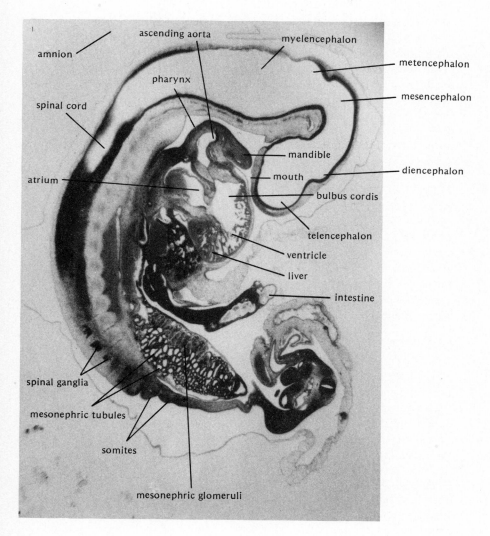

(b)

(thalamus), the hunger center (hypothalamus), the posterior pituitary, and the optic vesicles. The midbrain, or mesencephalon, forms the visual interpretation centers (the optic tecta). The anterior portion of the hindbrain, the metencephalon, gives rise to the cerebellum; the posterior portion of the hindbrain, the myelencephalon, forms the medulla (Figure 9-25).

NEURAL CREST DERIVATIVES

In Chapter 8 we noted that as the neural tube separates from the rest of the ectoderm and drops below the surface of the embryo, a narrow region at the top of the neural folds also moves into the embryo. This region, the neural crest, consists of cells that migrate to distant parts of the body. At their destinations, they form sensory nerves and *ganglia*, which receive impulses from sense organs; autonomic ganglia, which control involuntary activities; the adrenal medulla, the inner part of the adrenal gland; all of the pigment cells in the body with the exception of the pigmented retina cells, which are derived from the neural tube; the cartilages in the voice box and head; and some of the ectodermal muscles.

EPIDERMAL DERIVATIVES

After the neural tube and neural crest have dropped below the surface of the embryo (Chapter 8), the remaining surface ectoderm is the epidermis. Epidermal derivatives can be divided into two types, those derived from epidermal thickenings (placodes) and those derived from the rest of the epidermis. Epidermal placode derivatives include some of the head nerves, the lens of the eye, the olfactory structures, the inner ear, and the taste buds. The remainder of the epidermis forms the outer layer of the skin; the hair, horns, and nails; the linings of the mouth and anus; and the anterior pituitary.

MESODERMAL DERIVATIVES

The middle embryonic layer, the mesoderm, is located between the ectoderm and endoderm as a result of the gastrulation process. Figure 9-26 shows the major subdivisions of the mesoderm. These are the dorsal or upper division (on the back of the animal), called the epimere or *somite*; the mesomere or intermediate mesoderm; and the hypomere, the remainder of the mesoderm.

EPIMERE (SOMITE) AND MESOMERE (INTERMEDIATE MESODERM) DERIVATIVES

The epimere or somite is divided into three regions: the *myotome*, the *dermatome*, and the *sclerotome*. The outer region of the somite is called the dermatome and produces the dorsal portion of the dermis of the skin (dermis on the back of the animal). The inner part of the somite consists of the

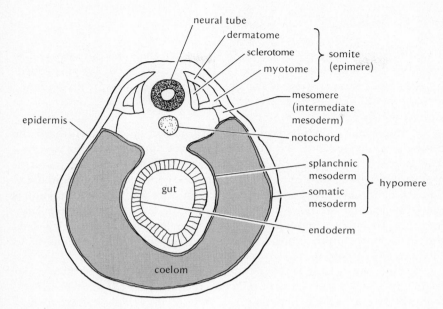

Figure 9–26. Mesoderm divisions.

myotome, which forms the back muscles, and the sclerotome, which gives rise to the vertebral column.

The mesomere, or intermediate mesoderm, gives rise to the kidneys, gonads, and associated structures.

HYPOMERE DERIVATIVES

The hypomere is composed of two parts, the somatic mesoderm and the splanchnic mesoderm. Somatic mesoderm is the hypomere material closely associated with the epidermis. It is the outer portion of the hypomere in the lateral (side) and ventral (belly) regions of the embryo. The lateral and ventral parts of the dermis, the limb buds, and the outer portion of the peritoneum (coelom lining) are derived from somatic mesoderm.

Splanchnic mesoderm is the hypomere material closely associated with the endoderm of the embryo. The inner part of the peritoneum (the lining of the body cavity), the smooth (gut) muscle, the heart and blood vessels, and the embryonic blood cells are derived from the splanchnic mesoderm.

Thus, the three regions of the mesoderm form a variety of structures. The development of many of them will be examined in Chapter 10. Having looked at the derivatives of the outer and middle embryonic layers, we now turn our attention to the inner germ layer, the endoderm.

ENDODERMAL DERIVATIVES

The inner embryonic layer, the endoderm, surrounds the gut cavity. The structures derived from the endoderm are either part of the gut tube proper or result from outpocketings of the tube. These relationships are diagrammed in Figure 9–27. Let us examine the major endodermal derivatives of the embryo.

The gut tube proper is divided into three major regions: the foregut, the midgut, and the hindgut. The anterior portion of the foregut is the pharynx. The middle region of the foregut forms the stomach, while the posterior region forms the duodenum. The small intestine consists of three regions: the duodenum, the jejunum, and the ileum. The anterior segment of the midgut is the jejunum, and the posterior region is the ileum and part of the large intestine. The hindgut is divided into the remainder of the large intestine and the cloaca. Some authors consider the entire large intestine to be midgut, but this is not an important distinction.

Let us briefly examine the endodermal derivatives of the gut tube proper (see Figure 9–27). Four pairs of pouches, the pharyngeal pouches, develop from the lateral walls of the pharynx. Parts of the middle ear are formed from pouch one. Tonsil tissue is derived from pouch two (and also from the pharyngeal walls). The thymus gland (involved in the formation of antibody-forming cells) and the parathyroid gland (involved in calcium and phosphorus metabolism) are derived mostly from pouches three and four. (Part of the thymus may be of ectodermal origin.) The pharynx itself forms the esophagus, which leads to the stomach. Ventrally, it forms the trachea, which ends

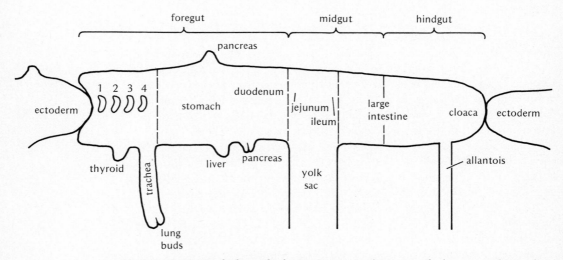

Figure 9–27. Endodermal derivatives. A diagram of the gut tube and its outpocketings.

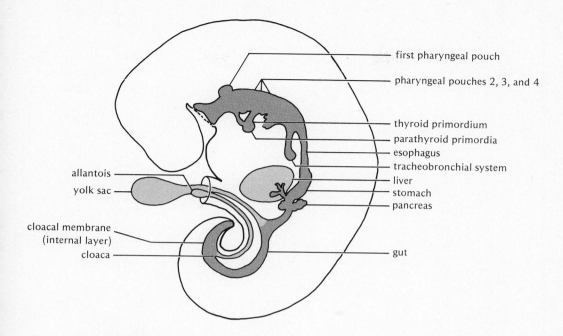

first pharyngeal pouch

pharyngeal pouches 2, 3, and 4

thyroid primordium
parathyroid primordia
esophagus
tracheobronchial system
liver
stomach
pancreas

allantois
yolk sac

cloacal membrane
(internal layer)
cloaca

gut

Figure 9–28. A diagram of a human embryo showing endodermal derivatives.
From H. Tuchmann-Duplessis et al., *Illustrated Human Embryology*, vol. I (New York:
Springer-Verlag, 1972; Paris: Masson, Editeur, 1972).

in the lung buds. The thyroid gland is an anterior outpocketing from the
ventral wall of the pharynx (Figure 9–28).

As we indicated, the rest of the foregut becomes the stomach and
duodenum. Outpocketings also arise from the posterior portion of the foregut. The pancreas arises from a dorsal outpocketing and a ventral outpocketing; the liver arises from the ventral wall, posterior to evagination of the
trachea. In bird and mammal embryos, where a yolk sac is present, the sac is
directly connected ventrally with the midgut. If an allantois is present, it is
derived from the hindgut and connected ventrally to it. The anterior region
of the hindgut forms part of the large intestine; the posterior region forms the
cloaca. The cloaca, in turn, gives rise to the urinary bladder, the urethra, and
the rectum.

This concludes our examination of the major derivatives of the endoderm. Note that there are three places in the body where the endoderm
directly contacts the ectoderm without any intervening mesoderm. These
are at the stomodeum (mouth opening), the cloaca, and the pharyngeal
pouches.

In this section we have outlined the major derivatives of the three germ
layers of the embryo and their relationships with one another. A more

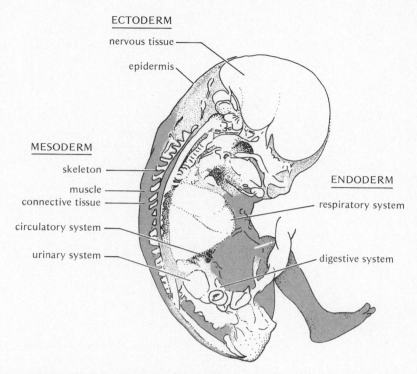

ECTODERM
nervous tissue
epidermis

MESODERM
skeleton
muscle
connective tissue
circulatory system
urinary system

ENDODERM
respiratory system
digestive system

Figure 9-29. (a) Major germ layer derivatives. From Tuchmann-Duplessis et al., *Illustrated Human Embryology*, vol. I (New York: Springer-Verlag, 1972; Paris: Masson, Editeur, 1972). **(b) 8 week old human embryo.**

detailed analysis of the development of these structures is found in Chapter 10. Figure 9–29 summarizes the major germ layer derivatives described in this section.

READINGS AND REFERENCES

EARLY HUMAN EMBRYO

Austin, C. R., and Short, R. V., eds. 1972. *Reproduction in Mammals*, book 2 : *Embryonic and Fetal Development*. Cambridge, England: Cambridge University Press.

Carr, D. H. 1969. Chromosomal abnormalities in clinical medicine. *Prog. Med. Genet.* 6: 1.

Corner, G. W. 1963. *Hormones in Human Reproduction*. New York: Atheneum.

Hillman, N., Sherman, M. I., and Graham, C. F. 1972. The effect of spatial arrangement on cell determination during mouse development. *J. Embryol. Exp. Morph.* 28: 263–78.

Hsu, Y. C. 1973. Differentiation *in vitro* of mouse embryos to the stage of early somite. *Develop. Biol.* 33: 403–11.

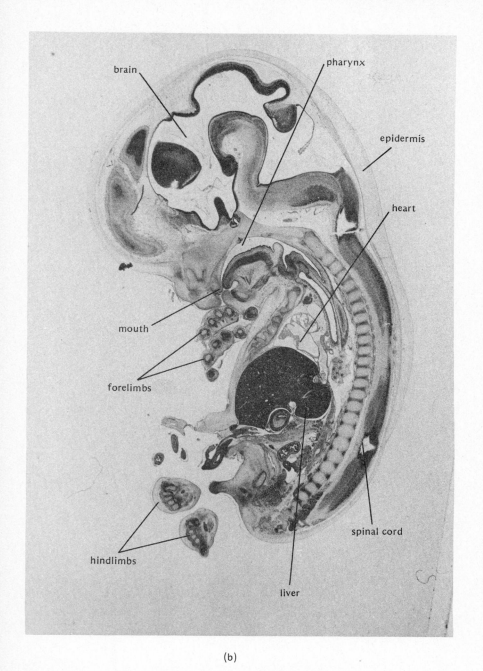

(b)

Mintz, B. 1971. Clonal basis of mammalian differentiation. *Symp. Soc. Exp. Biol.* 25: 345–69.

Moustafa, L. A., and Brinster, R. L. 1972. The fate of transplanted cells in mouse blastocysts *in vitro. J. Exp. Zool.* 181: 181–202.

Redding, A., and Hirschhorn, K. 1968. *Guide to Human Chromosomal Defects*. Birth Defects Original Article Series, 4. New York: National Foundation.

Rugh, R. 1968. *The Mouse: Its Reproduction and Development*. Minneapolis: Burgess.

Steptoe, P. C., and Edwards, R. G. 1976. Reimplantation of a human embryo with subsequent tubal pregnancy. *Lancet* i: 880.

Tuchmann-Duplessis, H., David, G., and Haegel, P. 1971. *Illustrated Human Embryology*. New York: Springer-Verlag.

PRIMARY GERM LAYER DERIVATIVES

Balinsky, B. I. 1975. *An Introduction to Embryology*, 4th ed. Philadelphia: W. B. Saunders.

De Haan, P. L., and Ursprung, H., eds. 1965. *Organogenesis*. New York: Holt, Rinehart & Winston.

Hamilton, W. J., Boyd, J. D., and Mossman, H. W. 1962. *Human Embryology*. Philadelphia: Blakiston Co.

Mossman, H. W. 1937. Comparative morphogenesis of the fetal membranes and accessory uterine structures. *Contrib. Embryol.* 26: 129–296.

Nelson, O. E. 1953. *Comparative Embryology of the Vertebrates*. New York: Blakiston Co.

Patten, B. M. 1946. *Human Embryology*. Philadelphia: Blakiston Co.

Patten, B. M. 1958. *Foundations of Embryology*. New York: McGraw-Hill.

Rugh, R. 1948. *Experimental Embryology*. Minneapolis: Burgess.

Witschi, E. 1956. *Development of Vertebrates*. Philadelphia: W. B. Saunders.

10
ORGANOGENESIS

SECTION 1   EYE DEVELOPMENT

We have examined the early development of embryos and outlined the derivatives of the ectoderm, mesoderm, and endoderm. We shall now zero in on some of these derivatives, stressing the development of organs and organ systems that convey concepts of general importance in developmental biology.

We shall examine organogenesis—the development of organs and organ systems—from both an anatomical and an experimental viewpoint. In this way, we shall see how the research in the field has expanded our knowledge of organ development. We shall first consider the vertebrate eye, then the neurological system, the heart, the limbs, the urogenital system, and the immune system.

THE VERTEBRATE EYE

The development of the eye has been a stimulating topic of investigation in many laboratories. Experiments on this topic often involve elegant operations on the eye-forming region. In this section, we shall look at some of these experiments to try to understand the mechanisms that control eye embryogenesis. We shall also describe some interesting experiments in the area of eye regeneration. First, however, let us look at the structural aspects of eye development to gain an understanding of the basic system.

EYE DEVELOPMENT: SUMMARY OF EVENTS

The major events in eye development can be briefly summarized as follows:

1. Contact between the roof of the archenteron (the notochord) and the eye cup rudiments on the neural plate (the prospective neural tube) causes them to develop into the *optic vesicles*.

2. The optic vesicles form as lateral evaginations (outgrowths) of the posterior portion of the forebrain (diencephalon).

3. The optic vesicles touch the prospective lens ectoderm. Contact with this area causes the optic vesicles to form the *eye cups*. The eye cups consist of an outer *pigmented layer* and an inner layer of *neural (sensory) retina*.

4. The lens-forming ectoderm (epidermis) has previously interacted with the underlying foregut endoderm and with portions of the prospective heart mesoderm. These interactions apparently maintain the lens-forming competence of the prospective lens ectoderm. The *lens* actually forms, however, only after the lens-forming ectoderm touches the tip of the optic vesicle. The lens-forming epidermis first thickens, then folds or undergoes cellular rearrangements to form the lens.

5. The newly formed lens and optic cup touch the overlying epidermis and mesenchyme. This induces the formation of the transparent protective covering of the eye, the *cornea*.

6. The *choroid coat* and *sclera*, the outer coats of the eye, develop from mesenchyme that accumulates around the eyeball.

7. The *iris* of the eye, the structure that regulates the size of the pupil, develops from the rim of the optic cup.

Some of these events are summarized in Figure 10-1. A drawing of a fully developed eye is shown in Figure 10-2.

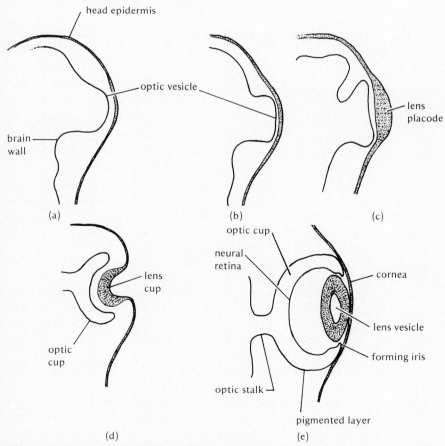

Figure 10-1. Steps in eye development. The retina and iris form from the brain wall, while the lens and part of the cornea form from the epidermis.

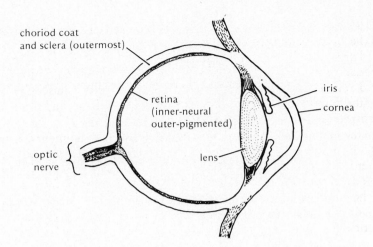

choriod coat
and sclera (outermost)

retina
(inner-neural
outer-pigmented)

iris

cornea

optic
nerve

lens

Figure 10–2. Drawing of sagittal section of human eyeball, showing the major parts.

MECHANISMS OF EYE DEVELOPMENT (OPTIC VESICLES)

No one knows exactly what causes the optic vesicles to form from the brain wall. We do know, however, that the archenteron roof induces the neural tube to differentiate into many components, including the optic vesicles. Specific genes in the prospective optic vesicle cells probably become activated and form the specific RNA messages that code for optic vesicle proteins. The actual process whereby the optic vesicles evaginate and form cups may involve cytoplasmic microtubules and microfilaments that change the shape of cells. Changes in adhesiveness may also be involved. These mechanisms are discussed in depth in Chapter 12. Recall that similar mechanisms were implicated in neural tube formation (Chapter 8) and in gastrulation (Chapter 7). The expansion of the optic vesicles appears to be aided by the accumulation of fluid in the brain ventricles, which are continuous with the optic vesicles. Coulombre and Coulombre inserted a glass tube through the wall of the eye to prevent fluid buildup. The eye then developed abnormally. Regional patterns of mitosis and cell growth probably play roles in optic vesicle development as well.

The optic vesicles seem to differentiate as a result of touching the surface ectoderm (prospective lens). If they do not touch, the prospective neural retina tends to form pigmented retina instead of neural retina. In addition, the optic cups normally form from the optic vesicles on the basis of contact with the developing lens ectoderm. Mesenchyme (embryonic connective tissue) that accumulates around the eyeball also appears to be important for normal eyeball development. If the mesenchyme is removed, the optic vesicles do not develop normally.

Let's sum up what we have learned so far. The optic vesicles evaginate from the brain wall. After contact with the prospective lens, the single-layered optic vesicles become pushed in, or cup shaped. The inner layer of the cup becomes the neural (sensory) retina; the outer layer becomes the pigmented retina. The optic vesicles develop as a result of induction by the archenteron roof, fluid buildup, contact with mesenchyme, and contact with the prospective lens epidermis.

In the neural plate and optic vesicle stages, the eye rudiment can be split in half. Each half can give rise to a complete eye. This implies that the eye rudiment can regulate itself. That is, the cells that remain after a part is removed are not fixed in their fate, but instead can change their fates so as to form the complete structure. This can also be demonstrated by placing ear vesicle or nasal placode in contact with the prospective pigmented retina of the optic cup. These can cause neural retina to develop from the pigmented layer.

The rods and cones, the light-sensitive receptor cells of the neural retina, develop in the outermost part of the retina—that is, in the layer closest to the choroid coat and farthest from the lens and pupil. To reach the rods and cones, light must first pass through a layer of ganglion cells and then a layer of bipolar neurons (Figure 10-3). The light excites the rods and cones. The nerve impulse is passed to the bipolar cells and thence to the ganglion cells, which transmit the message to the brain via the optic nerve. Optic nerve fibers from an eye generally enter the brain on the side opposite that eye. The optic chiasma is the place at which the nerve fibers from the two eyes cross. In mammals, the bundle of optic nerve fibers splits, and some enter the brain on the same side as the eye of origin. The neural retina differentiates first into the ganglion cells, followed by the bipolar neurons and finally by the rods and cones.

A very intriguing question now arises: What causes the nerve processes from the retina to enter the brain and connect to very specific regions there? In

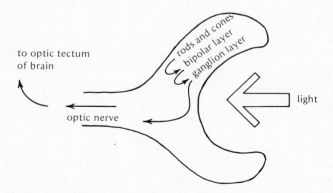

Figure 10-3. Layers of the neural retina. Arrows show impulse path.

other words, the nerve cells from the eye enter the optic tectum, the visual interpretation center of the mesencephalon of the brain. What causes these cells to attach to specific regions of that area? It is known that in chick embryos, for example, nerves from the dorsal part of the retina "home" to the ventral part of the optic tectum. Nerves from the ventral retina, in turn, "home" to the dorsal portion of the optic tectum. A fascinating approach to this problem has recently been taken by Barbera, Marchase, and Roth. The details of this work will be given in Chapter 12, but we shall summarize these experiments here in the context of eye development. These workers hypothesized that adhesive recognition helped govern where the nerve cell endings from the eye finally attach in the brain. That is, the cells stay where they "stick" the best. This hypothesis was tested by rotating labeled single cells from the dorsal *or* ventral retina, in turn, with pieces of dorsal *or* ventral optic tectum. The researchers measured the number of labeled dorsal or ventral retina cells that adhered to each tectal half. The results indicated that dorsal retina cells adhered best to ventral tectum and that ventral retina cells adhered best to dorsal tectum. The results suggested that adhesive recognition indeed plays an important role in controlling retino-tectal nerve hookups. Dorsal retina nerves hook up to ventral tectum and also stick best to ventral tectum as shown by Roth's group. Ventral retina cells, likewise, hook up to and stick best to dorsal tectum. At this point, however, we can suggest only that adhesiveness plays the key role in controlling the right "hookups." The future should provide new insights in this intriguing area.

The molecular nature of the selective adhesion of dorsal retina cells to ventral optic tectum and ventral retina cells to dorsal optic tectum is being investigated by Roth's group. Tentative preliminary evidence suggests that this adhesion involves specific enzymes on one cell surface binding to sugar chains on the adjacent cell surface. These experiments will be explored further in Chapter 12.

Before moving on to the parts of the eye derived from epidermis, we must note that the optic cup gives rise to structures other than the neural retina. The outer wall of the optic cup becomes the pigmented coat on the posterior side. On the anterior side, it thins out to form the iris and the basal part of the ciliary body that helps control lens shape. The pigmented coat cells give rise to cytoplasmic processes that interdigitate with the outer parts of the rods and cones, thus locking the neural retina and the pigmented coat together. The pigmented coat apparently helps nourish the neural retina by transporting nutrients between the blood vessels of the choroid coat and the cells of the neural retina.

MECHANISMS OF LENS DEVELOPMENT

Some of the mechanisms involved in lens development have been elucidated. We have already seen how contact with the lens-forming ectoderm plays a key role in the development of the optic cup. Let us begin our discussion of lens formation by examining the other part of the latter relationship. That is, when the optic vesicle touches the lens-forming epidermis, does this help induce lens formation? The answer is yes. The final induction or direct cause of the lens forming comes from contact or proximity with the optic vesicle (Figure 10–4). It is not well understood exactly how such contact causes the lens to form.

The ability of the prospective lens ectoderm to form lens is maintained by previous contact with the foregut endoderm and with portions of the prospective heart or head mesoderm. The final induction of the lens is caused by proximity to or contact with the optic vesicle. Molecules may pass from nearby tissue to the lens-forming ectoderm, causing the lens to form. Abnormal inducers such as ear vesicle and guinea pig thymus can also cause a lens to form in competent ectoderm. Environmental influences (such as temperature) also affect final lens induction. The nature of induction—at least what we know about it—is discussed in more detail in Chapter 11.

What is the nature of the inducer that appears to pass from the optic vesicle to the lens-forming ectoderm? This is not known, but there is evidence that substances pass from the optic vesicle to the lens-forming ectoderm. At the time of induction, one can observe a decrease in cytoplasmic basophilia (acidity) and a decrease in the number of ribosomes in the tip of the optic vesicle (prospective neural retina). At the same time, there is an increase in cytoplasmic basophilia and an increase in the number of ribosomes in the lens-forming ectoderm. These results, however, do not prove that something actually passes from the optic vesicle to the lens ectoderm. More direct evidence comes from experiments in which the optic vesicles were labeled with radioactive amino acid (C^{14} phenylalanine). This label began to appear in the lens-forming ectoderm. This result shows that something is passed between the optic vesicle and the lens-forming ectoderm, but not what is passed or what the nature of the hypothetical inducer is.

There have been some additional experiments that illustrate the relationship between the optic vesicle and the lens-forming ectoderm. If the optic vesicle is removed before it reaches the lens-forming epidermis, no lens usually forms. A thin layer of cellophane placed between the optic vesicle and the lens-forming ectoderm blocks lens formation. However, when a thin slice of agar (which may allow passage of molecules) was inserted between the optic vesicle and the lens-forming ectoderm, it did not block lens formation. Note, however, that in certain species and under certain conditions the lens can form, to some extent, without final induction by the optic vesicle. This may occur as the result of other factors that can influence the lens-forming ectoderm, such as interaction with the foregut endoderm or with the heart or head mesoderm, or environmental influences such as temperature. For example, low temperatures appear to favor lens induction by head mesoderm, making contact with the optic vesicles less necessary. However, most lenses that form without induction by the optic vesicles are

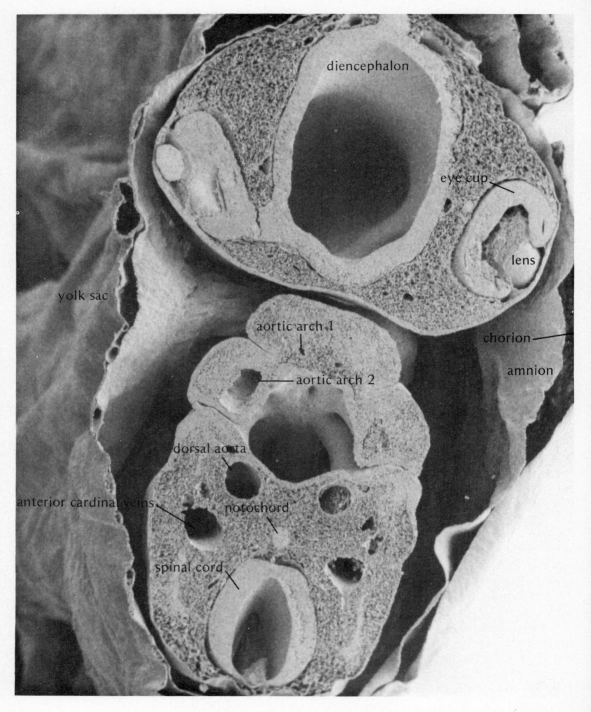

diencephalon

eye cup

lens

yolk sac

aortic arch 1

aortic arch 2

chorion

amnion

dorsal aorta

anterior cardinal veins

notochord

spinal cord

Figure 10-4. Chick embryo, 72 hours old, showing eyecups and lenses on either side of the large diencephalon cavity. Courtesy of Peter Armstrong.

not normal. Thus, we can conclude that although many factors can stimulate lens formation, contact or proximity with the optic vesicle normally provides the final induction for the proper development of the lens.

What actually causes the lens-forming ectoderm to fold and form the lens vesicle? In Chapters 8 and 12 we examine some of the factors that cause a flat sheet of cells to fold, forming a cup and then a vesicle. For instance, cytoplasmic microtubules and microfilaments change the shape of cells, so that cell sheets can fold. Also, the adhesiveness of cells with their neighbors may play a role in vesicle formation. Another way in which a sheet of cells can fold to form the lens vesicle may involve tight cell attachment and mitosis. If the lens-forming ectoderm cells become firmly attached to each other so that no lateral cell movement can occur, and if mitosis increases the mass of these cells, the sheet of cells will buckle inward. Microfilament bundles and extracellular materials have been observed at the outer surface of the lens-forming ectoderm. These may limit the expansion of the cell layer so that folding will occur.

The lens begins to form as a thickening in the epidermis, called the *lens placode*. The lens placode then invaginates to form the *lens vesicle*. Fiber cells form from lens epithelial cells. Within the fiber cells, proteins called *crystallins* form. These proteins allow the lens to focus light upon the retina. Contact of the lens with the presumptive retina of the optic vesicle appears to induce lens fiber differentiation (Figure 10–5).

LENS REGENERATION. The lens can regenerate in vertebrates such as salamanders. If the normal lens is removed from such animals, a lens regener-

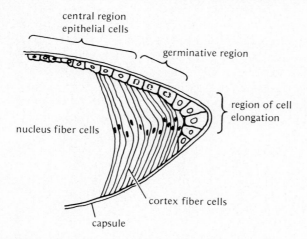

Figure 10–5. Lens-fiber formation. Epithelial cells elongate into fiber cells (cortex fiber cells). Nucleus fiber cells are those that formed during early lens development. From Papaconstantinou, *Science* 156 (1967): 338. Copyright 1967 by the American Association for the Advancement of Science.

ates from the dorsal rim of the iris (Figure 10–6). Thus, a lens, which is usually an epidermal derivative, can regenerate from the iris, which is a neural derivative. The events involved in lens regeneration include the loss of pigment granules in the iris and cell division. The cells then undergo typical lens formation, forming epithelial cells and lens fibers. If the lens is removed from the eye and then inserted back into the same eye, no new lens is formed. The iris does not begin to undergo change. These results suggest that the lens itself inhibits the iris from forming another lens. Only mature lenses and not undifferentiated lenses can inhibit the regeneration of a lens from the iris. When lens proteins from mature lenses are separated by gel electrophoresis, two of them will inhibit lens regeneration from the iris when placed in the eye. Thus, lens protein from the mature, differentiated lens may be released into the eye fluid and block formation of additional lenses from the iris. These experiments suggest that the salamander eye has evolved an elegant means of regenerating a lens from cells that normally do not form one. Along with this development has evolved a means of preventing the formation of additional lenses in a normal eye.

EYE DEVELOPMENT AND VIRAL DAMAGE. Before leaving the lens, let us mention how studies on lens development have led to an understanding of an important birth defect in human beings, namely blindness caused by German measles (rubella) virus. It has long been known that if a pregnant woman is infected with German measles virus during the first three months of pregnancy, her newborn child may be blind. If, however, she contracts the disease after the first trimester of pregnancy, blindness rarely occurs in the newborn child. It is known that the eye of the human embryo develops during the first six weeks. At six weeks, the lens vesicle has closed and the lens fibers have differentiated. It has been suggested by Robertson, Blattner, and Williamson that the lens and other organs that arise from ectodermal invaginations are susceptible to the harmful effects of certain viruses only as long as they are open to the exterior (e.g., in the cup stage or earlier). Once they close, viruses can no longer enter (Figure 10–7). Saxen's group in Finland showed that this was indeed true for the lens cataract induced by German

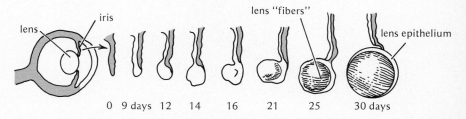

Figure 10-6. Lens regeneration from iris of salamander eye. From N. K. Wessells, *Tissue Interactions and Development*, Menlo Park, Calif.: W. A. Benjamin, 1977, p. 170, based on experiments in R. W. Reyer, *Quart. Rev. Biol.* 29 (1954).

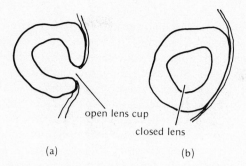

open lens cup

closed lens

(a)   (b)

Figure 10-7. Susceptibility of the lens to virus infection. Studies by Saxén's group indicate that the embryonic lens is only susceptible to infection by German measles virus before the closure of the lens cup.

measles virus in human embryos. Using chick embryos and human embryos obtained from therapeutic abortions, these workers showed that the virus-sensitive period was limited to the open stage of the lens vesicle. Closed lenses could be infected only if virus was surgically introduced into the lens vesicle. Thus, a careful study of the developing lens has led to important information regarding a major congenital abnormality.

OTHER EYE STRUCTURES

The cornea, the transparent covering of the front of the eye, is derived from epidermal ectoderm that lies over the anterior portion of the eye and from mesoderm (mesenchyme) cells that lie under the ectoderm. The optic cup—and especially the lens—induce and maintain the cornea. The mesenchyme of the cornea is continuous with the sclera (the protective coat that is seen as the white of the eye). The epidermal portion of the cornea is continuous with the skin or eyelid epithelium. The cornea becomes transparent by the dissolution of pigment granules in the prospective cornea. The lens and eye cup stimulate corneal development. If the eye cup is removed, the cornea does not develop. The lens alone can be transplanted under other epidermis. This epidermis then loses its pigment and differentiates into cornea. The eyeball also seems to be needed to maintain corneal transparency. If the eyeball is removed, chromatophores invade the cornea and the cornea becomes similar to normal skin.

The shape of the cornea is important in allowing light waves to bend properly. The cornea is curved by pressure exerted against it by the eye fluid. The coats around the eye include the outermost protective white of the eye (the sclera) and the layer between the sclera and the retina (the choroid coat). The choroid coat is pigmented and contains blood vessels that nourish the eye. The choroid coat and the sclera develop from mesenchyme that accumulates around the eyeball. They may form as a result of factors from the pigmented retina coat or from the neural retina.

SUMMARY

In summary, the eye develops by a complex series of reciprocal interactions among the optic vesicles, epidermis, and mesenchyme. The nature of the "inducers" is unknown. Some of the relevant mechanisms have been described. The mechanisms that cause the optic cup and lens vesicle to form include cell-shape changes caused by cytoplasmic microtubules and microfilaments, differential cell adhesiveness, and mitosis. Experiments by Roth's group suggest that selective adhesive recognition may be involved in the proper hookup of optic nerve fibers with the brain. Our understanding of the molecular basis of eye morphogenesis is still in its infancy, but these mechanisms will eventually be elucidated.

SECTION 2   DEVELOPMENT OF NERVOUS INTEGRATION AND BEHAVIOR

How do nerves that originate in the central nervous system get to specific end organs, the multitude of sense organs and muscles in our body? How do they innervate them? How does behavior develop? These questions are among the most absorbing problems in developmental biology today—and also among the most poorly understood. Hypotheses dealing with these questions have been proposed, and some concrete experimental evidence is improving our understanding. Here, we shall briefly survey these problems and the potential solutions to them.

HYPOTHESES OF NERVE-END ORGAN HOOKUP

A variety of hypotheses have been proposed to explain how nerves get to and hook up with specific end organs. A few will be described here. We can categorize these hypotheses as the chemical, electrical, mechanical, and adhesive recognition models of nerve growth and end-organ innervation. It should be stressed that a complete theory would probably be a combination of more than one of the models described. Thus, it is unlikely that any one of the hypotheses will turn out to be the whole answer. We shall see why as we examine the experimental evidence for each model.

CHEMICAL MODEL

The chemical (chemotaxis) model suggests that nerves grow along a gradient of specific chemicals, usually toward the source. We confronted the question of chemotaxis once before, in Chapter 5. There, we saw that known chemicals given off by female gametes of certain plant forms specifically attract male gametes of those species. Except in one major area, there is little concrete experimental evidence in support of a chemical model of directional nerve growth.

The exception is a substance (protein) called *nerve growth factor*. This protein has been purified from mouse submaxillary glands. It stimulates massive nerve outgrowth from ganglia (Figure 10-8). Nerve growth factor is present in many sources, such as mouse sarcoma tumor, snake venom, and mouse salivary glands. Nerve growth factor was discovered when Bueker implanted mouse sarcoma 180 tumor tissue into the body wall of two to three day chick embryos to see if nerves innervated rapidly growing tissue. The tumor tissue stimulated massive nerve outgrowth from adjacent ganglia. Rita Levi-Montalcini suggested that the tumor contains a nerve growth factor. In early attempts to determine if the factor was a nucleic acid or a protein, it was subjected to a snake venom that contains the enzyme phosphodiesterase, which degrades nucleic acid. In control experiments, the snake venom itself was used to see if nerve outgrowth occurred. To the surprise of the investigators, the venom was about one thousand times as potent as the tumor homogenate in promoting nerve outgrowth.

The salivary glands of mice, which are similar to the glands that produce snake venom, had a nerve growth factor activity up to ten thousand times that of the tumor homogenate and ten times that of snake venom. Tiny quantities of nerve growth factor (0.5 micrograms per gram of body weight) caused a four- to six-fold increase in the number of developing sympathetic ganglia (clusters of nerve cells in the spinal cord area).

Antibodies against nerve growth factor were prepared by Cohen and Levi-Montalcini by injecting purified factor into rabbits. When this antibody was injected into newborn rats, mice, rabbits, and kittens, it prevented the growth of sympathetic chain ganglia (nerve cell clusters in the spinal cord area). These experiments suggest that a substance similar to nerve growth factor exists in the embryo and that it is responsible for the growth and maintenance of sympathetic and sensory nerve cells.

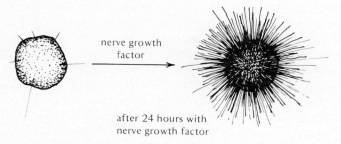

nerve growth
factor

after 24 hours with
nerve growth factor

Figure 10-8. The effect of nerve growth factor on a chick embryo sensory ganglion. A seven-day-old embryo sensory ganglion was cultured *in vitro* in a medium containing nerve growth factor. From experiments by Rita Levi-Montalcini and B. Booker.

How does nerve growth factor work? Does it influence the direction of nerve growth, or does it stimulate growth in general? We cannot as yet answer these questions completely. What does nerve growth factor do to cells that respond to it? First, large quantities of microtubules and microfilaments (see Chapter 15) appear in the cytoplasm. These elements may play a role in nerve growth and differentiation and in the cytoplasmic flow in nerve axons. In addition, RNA synthesis, protein synthesis, lipid synthesis, and glucose oxidation are enhanced in such cells.

Nerve growth factor, like insulin, appears to bind to specific surface receptors on nerve cell membranes. This can be observed by radioactively labeling nerve growth factor and measuring the rate at which it binds to the cell surface. It may be that nerve growth factor, like many hormones, primarily affects the cell membrane. By binding to the surface receptor, nerve growth factor may set off a series of reactions that are transported to the cytoplasm and nucleus by some sort of chemical messenger, such as cyclic AMP (adenosine monophosphate). Recent work suggests that nerve growth factor may be a protein-degrading enzyme (a protease). As yet, however, there is no conclusive evidence regarding the sequence of events by which nerve growth factor stimulates nerve growth. There is also no solid evidence in support of the contention that nerve growth factor is responsible for the highly directional nerve growth required for hooking up specific nerves with specific end organs. It may be that nerve growth factor is responsible for nerve cell growth and development, but not for guiding nerves to their specific end organs.

ELECTRICAL MODEL

Is it possible that nerves grow toward an end organ because of electrical currents or gradients between the outgrowing nerve and the end organ? There is no conclusive evidence to support this model. Some experiments, however, do suggest that nerves respond to electricity. If direct electric current is applied to the amputated forelimb stumps of laboratory rats, some regeneration occurs. Cartilage, bone marrow, bone, blood vessels, muscles, and nerves develop in this tissue. This suggests that electric currents can stimulate nerve growth.

When nerve explants were placed in a culture dish and an electric current was directed through the culture medium, the nerves always grew in the direction of the current. A more careful repeat of this experiment, however, was performed by Weiss. He found that the nerves grew in the direction of the current even after it was turned off! An examination of the culture plates revealed that the current oriented the molecules of the culture gel, forming fine grooves in the culture medium in the direction of the electric current. The nerves did not follow the electric current itself; they followed the grooves formed in the substratum. These experiments, therefore, provide no evidence for an electrical model of nerve growth. They do not, however, rule out the possibility that small electric currents may play some role in guiding nerve growth in the body, but evidence for this contention is still lacking.

MECHANICAL MODEL

The results of Weiss's experiments support a mechanical model of nerve guidance. Weiss extended these experiments by preparing little hollow frames of different shapes filled with blood plasma. As the plasma clotted, lines of tension were set up in directions that corresponded to the shape of the frames. Weiss found that nerves always grew out along these lines or grooves. He termed such nerve growth *contact guidance*. The nerves appear to grow according to mechanical guidelines. They use these guidelines as road maps to get to where they are going.

An extension of this model to animals suggests that nerves use blood vessels as road maps to get where they are going. There is little doubt that nerves, *in vivo*, do use mechanical guidelines such as blood vessels. Nerves can be observed to follow blood vessels and oriented tissues in the body. We can conclude that the mechanical model is an important part of the explanation for directed nerve growth.

It is unlikely, however, that this model is the whole story or that contact guidance explains all aspects of nerve-end organ hookup. Too many highly specific nerve hookups can be found in end organs to be explained by contact guidance alone. For example, nerves from the dorsal part of the retina hook up to very specific regions of the ventral part of the optic tectum of the brain, and nerves from the ventral part of the retina hook up to very specific regions of the dorsal portion of the optic tectum. The hookup points in the optic tectum are extremely close together. Several models have been proposed to explain such highly specific nerve hookups. One of these will be described here. Some of the others will more appropriately be developed in the behavior section.

ADHESIVE RECOGNITION MODEL

Roth and his colleagues proposed that nerves find specific regions in their end organs by adhesive recognition. That is, the nerve chemically "recognizes" its final resting place by the adhesion between the surfaces of the growing nerve cells and the surfaces of the cells in the target region of the end organ. This adhesion keeps the nerve in its proper place and prevents it from moving on to another area. The specific molecular groups responsible for such adhesive recognition may turn out to be sugar-binding enzymes on one cell surface that attach to sugar chains on the other surface. Roth and his coworkers have obtained some evidence in support of such an enzyme-substrate recognition. Adhesive recognition and its role in morphogenetic events such as nerve-end organ hookups is dealt with in detail in Chapter 12.

DEVELOPMENT OF BEHAVIOR

Of all the functions that develop in the embryo, fetus, and newborn, behavior is perhaps the most complex. Our understanding of the development of behavior is still in its infancy and far from the molecular level which has been reached in other areas. As yet, we are still at the experimental embryology level of investigation. There is little doubt, however, that in time we shall understand the development of behavior at the molecular level as we now understand such processes as early egg activation, cell differentiation, and so on.

Basic questions in the development of behavior are: Do the nerves that innervate end organs have some kind of prior molecular information about the target region? Or do they learn where they are or what they must do only after they get there and innervate the end organs? These questions are important to the story of behavior because they deal with the problem of what influences what in establishing the nerve pathways essential for behavior.

Sperry and Miner performed experiments that shed light on these problems. They rotated a large strip of skin on the flank of a frog tadpole 180°. Thus, once dorsal (back) skin was now in a ventral (belly) position and ventral skin was in a dorsal position (Figure 10–9). The skin was allowed to heal into place with this new orientation, and the tadpole was allowed to metamorphose into a frog. When the dorsal part of the original skin (now lying ventrally) was irritated, the frog scratched its back. When the ventral part of the original skin (now lying dorsally) was irritated, the frog scratched its belly. Thus, the frog responded according to the old orientation of the skin, not the new. This response could result from two possibilities:

1. The skin attracted the appropriate nerve. For example, belly (ventral) skin attracted a belly nerve, even though the skin was in a dorsal position.

2. The skin altered the way in which the nerve reacted—that is, it instructed the nerve about what kind of skin it really was.

When Sperry and Miner examined the innervation of the reversed skin, they found that the nerves normally found in the specific region innervated the

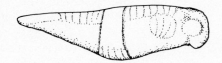

Figure 10–9. Reversal of skin on tadpole flank. From experiments by Sperry and Miner, 1951.

(reversed) skin in that region. In other words, a normal back (dorsal) nerve innervated the belly skin that was now on the back, and a normal belly nerve innervated the back skin that was now on the belly. Thus, originally ventral axons that now innervate the back skin moved to the stomach area seem to acquire dorsal properties. That is, they realign their interconnections with other neurons in the central nervous system (spinal cord) so that the evoked motor impulses go to nerves that innervate dorsal muscles. This suggests possibility number two, namely that the skin (end organ) can instruct nerves as to what they should do. Steinberg, however, suggests another alternative. An anatomically dorsal nerve may have some physiologically ventral fibers, and an anatomically ventral nerve may have some physiologically dorsal fibers. Even so, the skin (end organs) must play some role in activating the proper nerve fiber.

We should stress that in many cases of nerve regeneration, the regenerating nerves appear to know exactly where to go, almost as if they were part of a printed circuit. Behavioral evidence for this comes from experiments with amphibian eyes. If the eyes of frogs or salamanders are rotated 180° and the optic nerves are then cut, the optic nerves regenerate from the eyes to the visual centers in the brain, the optic tecta. Eventually, vision is restored. The animal, however, behaves as though every part of its visual field were upside down. Thus it appears that the regenerating nerve fibers reached their original end stations in the optic tecta. This has been confirmed by direct anatomical observations with regenerating optic nerves in fish. Nerves, therefore, do seem to know exactly where to go (Figure 10–10).

The concept that early behavior develops as a result of nerve–end organ hookup and not from learning is supported by a variety of additional studies. Many muscles in the embryonic chick become active only after nerves hook

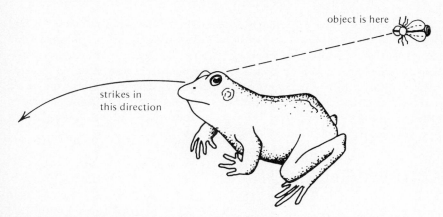

object is here

strikes in this direction

Figure 10–10. Behavior of frog after 180° rotation of eyes. After R. W. Sperry, in S. S. Stevens, ed., *Handbook of Experimental Physiology* (New York: John Wiley, 1951), pp. 236–80.

up to these muscles. Thus, the nerves appear to influence the muscle, enabling it to function. Also the intricate movements involved in the hatching of the chick appear precise and integrated right from the beginning without any "practice" at earlier stages.

Similarly, if salamander embryos are reared in an anesthetic drug that prevents muscle activity, they develop but are paralyzed. However, when the larvae are returned to water without the drug, they immediately behave like normal swimming larvae raised entirely without the drug. Thus, all of the early behavioral activities were not learned; they began immediately as soon as the muscles were allowed to act. This implies that certain behavior is controlled by the inherent qualities of the nerve-muscle hookup. Likewise, in the chick embryo, some muscular movements occur before any functional sensory nerve connections exist. These movements appear to be totally controlled by motor nerve-muscle hookups.

Interesting experiments have been performed with kittens suggesting that there is a time in development when the vision nerves are plastic and can be specified by environmental influences. After this period of plasticity, the nervous pathways are fixed and learning cannot undo the specification that occurred during the plastic period. The plastic period appears to exist in the one- to two-month-old kitten. A similar plastic period may appear in the two- to four-year-old human child. If kittens are reared during the plastic period in environments consisting only of vertical stripes or horizontal stripes, they do not cope well with visual detection tasks in the real world. The cats raised with horizontal stripes walk into vertical objects, such as chair legs, apparently without detecting them. The cats raised with vertical stripes do not respond to horizontal visual cues, such as the chair seat. The animals could not detect anything made of lines that were not present during the earlier plastic period.

Such phenomena are clearly important to human development. For example, if a two- to four-year-old child has a severe squint or astigmatism, permanent visual problems can occur because the child's brain pathways are fixed at that age. Such problems may be uncorrectable after the plastic period. Many types of behavior in human beings are probably initiated by pathways that develop during the plastic period. Exposing a child to a variety of learning experiences during the two- to four-year-old period may be the best way to assure that the child will learn well in later years.

SECTION 3   HEART DEVELOPMENT

The heart pumps blood through the blood-vessel system, thereby supplying the body tissues with food and oxygen and removing wastes that accumulate in these tissues. Obviously it is an organ we cannot do without. The development of the heart involves a series of cellular migrations, fusions, and specific differentiation—indeed, a multitude of the morphogenetic mechanisms described in Chapter 12. Let us examine some of the experiments that led to our present understanding of heart embryogenesis. First, however, we shall summarize the events that occur in heart development.

HEART EMBRYOGENESIS: SUMMARY OF EVENTS

The major events that occur during heart development in the chick embryo can be summarized as follows:

1. The heart begins to differentiate as two vesicles in the hypomere mesoderm on either side of the developing foregut. We shall show later how these vesicles originate.

2. As the ectoderm and endoderm fold to form the head fold of the embryo, lifting the anterior end of the embryo off the yolk, the two heart rudiments are brought together in the ventral midline, and they fuse. Each of the rudiments consists of an inner lining, the *endocardium*, and an outer muscle layer, the *myocardium* (Figure 10–11). These events occur between the 25th and 30th hour of incubation in the chick embryo, or between the 7- and 20-somite stage. In human embryos, similar events occur during the third week, or by about the 8-somite stage.

3. The paired heart vesicles begin to fuse at the anterior (head) end, and continue to fuse in the posterior direction. The first region formed after fusion is the most anterior portion of the heart, the *truncus* or *conus arteriosis*, which leads to the ventricle, the thick-walled muscular pumping chamber. Next to form is the atrium, the heart chamber that delivers blood to the ventricle. The last part to form is the *sinus venosus*, the heart chamber that receives venous blood.

4. The heart tube bends to form an S-shape. The heart begins to beat just after the paired heart rudiments begin to fuse, just before the conus arteriosus forms.

The amphibian heart forms in a way rather similar to that of the chick. However, the amphibian embryo does not fold to bring the heart areas together. Instead, the prospective heart mesoderm cells on each side of the embryo converge ventrally (Figure 10–12). Mesenchyme-like loose cells appear to come from the right and left regions of the ventrally converging hypomere splanchnic mesoderm. These cells form the endocardium, the

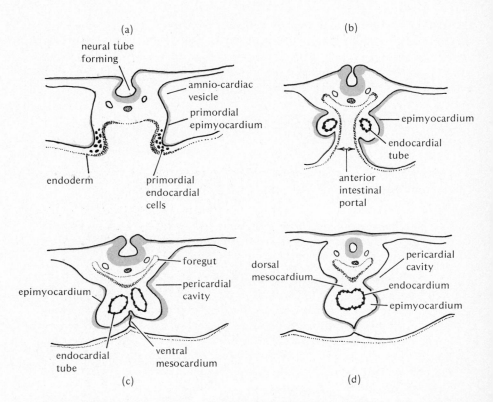

Figure 10–11 (a)–(d). Cross sections of heart development in 25- to 30-hour chick embryos. From T. W. Torrey, *Morphogenesis of the Vertebrates* (New York: John Wiley).

inner heart lining, as a thin-walled tube. The endocardial tube divides at both ends. At the anterior end are the two ventral aortae; at the posterior end are the two vitelline veins that bring the first venous blood to the heart. As the hypomere regions from each side converge and fuse ventrally, the splanchnic mesoderm envelops the endocardial tube, forming the muscular myocardium. As in the chick, the heart tube then twists, forming an S-shape. Two common cardinal veins form by the union of veins from the head and posterior regions of the body. Along with the vitelline veins these enter the most posterior heart chamber, the sinus venosus. This chamber leads into the thin-walled atrium, which in turn joins with the thick-walled ventricle and conus arteriosus as in the chick (Figure 10–13). This S-shape is found in all vertebrates. Later in embryogenesis, however, the hearts of different vertebrate groups develop differently, sometimes forming multiple chambers that adapt the hearts to the physiological needs of specific animal groups.

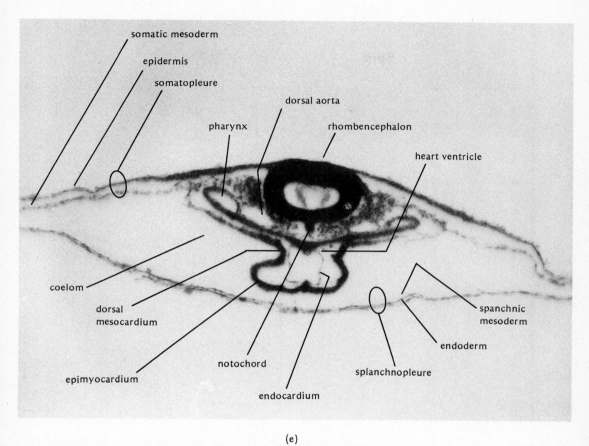

(e)

Figure 10–11 (e). Chick embryo, 33 hrs., cross section, showing heart.

EXPERIMENTAL ANALYSIS OF MECHANISMS OF HEART DEVELOPMENT

The analysis of chick heart development was accomplished by elegant experiments by Mary Rawles, Robert DeHaan, and others. Their experiments illustrate how radioactive labeling, microsurgery, and transplantation can be used to elucidate the development of a major organ. Portions from all parts of the prestreak blastoderm of chick embryos were isolated in culture. Beating tissue developed from these portions, showing that heart-forming potential is not restricted to any localized region of the early epiblast. In the

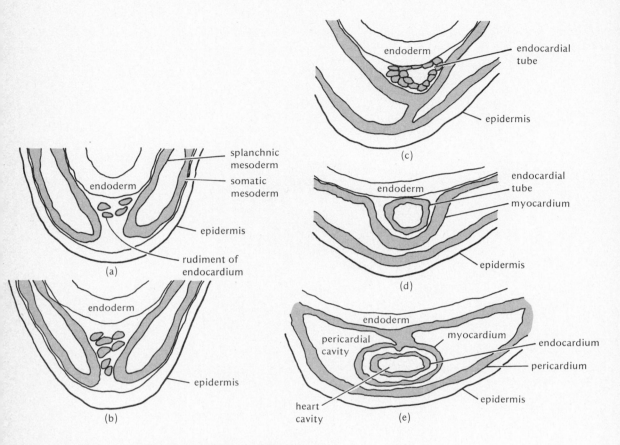

Figure 10-12. Development of the amphibian heart.

primitive streak stage, only cells from the posterior half of the blastoderm develop into beating tissue when isolated in culture, showing that the heart forming region becomes more localized as development proceeds. Mary Rawles showed that certain fragments from the blastoderm of a head-process stage embryo would give rise to heart tissue when cultured on the vascular chorioallantoic membrane of an older embryo. (The head-process stage is the stage when notochordal cells move forward from Hensen's node.) The heart-forming regions appeared to be localized in the epiblast at either side of the midline. These cells seemed to move through the primitive groove and take up residence on either side of the head process (Figure 10-14).

De Haan and his colleagues Rosenquist and Stalsberg performed a beautiful series of experiments designed to determine more accurately the fates of very specific regions of the blastoderm in forming the heart. These workers transplanted fragments of blastoderm from a radioactively labeled donor

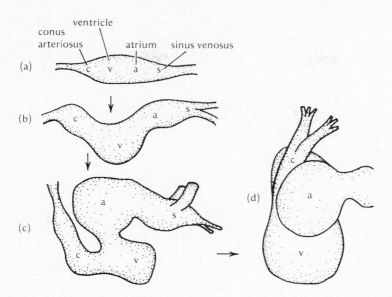

Figure 10–13. Twisting of frog heart rudiment. From Joy B. Phillips, *Development of the Vertebrate Anatomy* (St. Louis: C. V. Mosby, 1975).

embryo to an unlabeled recipient. The fate of many such specific fragments can be followed by autoradiography and by carbon marking (using specks of carbon to mark specific areas) and microcinematography. By these methods, DeHaan et al. determined that when the primitive streak is well developed, heart-forming cells lie in paired regions about midway down the length of the streak, extending from the midline about halfway to the edge of the

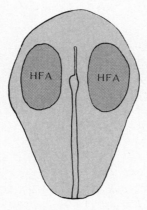

Figure 10–14. Heart forming area (HFA) in the head-process stage of the chick blastoderm.

embryo. The cells enter through the primitive groove. By the end of the head-process stage, the heart-forming mesoderm is organized in two separate regions. The most anterior parts of the heart-forming regions fuse to give rise to the most anterior part of the heart, the conus arteriosus. The middle parts fuse to give rise to the middle part of the heart, the ventricle. The posterior parts fuse to give rise to the most posterior parts of the heart, the atrium and sinus venosus (Figure 10–15). This brings us back to the summary given previously. The experimental analysis has helped us understand the events that led up to the formation of the two fusing mesodermal vesicles with which we began our summary.

How does the heart-forming mesoderm get to where it is going? DeHaan and colleagues transplanted a radioactively "hot" square of endoderm-mesoderm into a "cold" head-process embryo, as done previously. After a few hours the embryo was sectioned and examined by autoradiography. The results showed that the radioactive mesoderm moved forward relative to the endoderm (Figure 10–16). As will be seen shortly, the heart-forming mesoderm appears to use the associated endoderm as a road map to guide it to where it is going.

We can now ask some questions: Do inductive interactions like those found in other systems occur during heart development? What is the relationship between the heart-forming mesoderm and the closely associated endoderm? Some evidence for inductive interactions in heart formation exists in the amphibian system. If the endoderm is removed from newt embryos, the heart never develops. Thus, the intimate contact between the

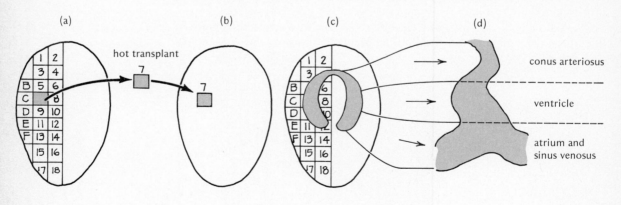

(a)   (b)   (c)   (d)

hot transplant

conus arteriosus

ventricle

atrium and
sinus venosus

donor embryo

1 μc ³H
thymidine
2 hours

recipient embryo

Figure 10–15. Experimental determination of chick heart fate map. "Hot" explant from (a) is transplanted to "cold" recipient (b). (c) and (d) show the fate map of the chick heart-forming regions projected upon a chick heart, as developed from the experiments shown in (a) and (b). After R. L. DeHaan, *Ann. N.Y. Acad. Sci.*, 1965, and R. L. DeHaan, in *The Emergence of Order in Developing Systems*, M. Locke, ed. (New York: Academic Press, 1968), p. 208.

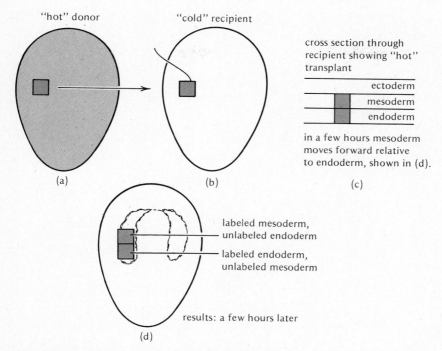

"hot" donor   "cold" recipient

cross section through
recipient showing "hot"
transplant

ectoderm
mesoderm
endoderm

in a few hours mesoderm
moves forward relative
to endoderm, shown in (d).

(a)   (b)   (c)

labeled mesoderm,
unlabeled endoderm

labeled endoderm,
unlabeled mesoderm

results: a few hours later

(d)

Figure 10-16. Forward migration of heart-forming mesoderm relative to endoderm in chick embryo. After experiments by DeHaan and colleagues.

endoderm and the heart-forming mesoderm seems to be required for heart development in the amphibian embryo.

Certain agents can be used to separate the preheart mesoderm and endoderm in the chick embryo. These substances include sodium citrate and ethylene diaminetetraacetic acid (EDTA), which bind divalent cations. If head-process stage chick embryos are treated with sodium citrate, the endoderm will be almost completely removed, leaving the ectoderm and mesoderm relatively intact. Many small twitching heart vesicles form in these embryos, but they do not migrate together to form a heart. These experiments suggest that the endoderm is the specific substratum (cell layer) needed to guide the migrating prospective heart cells to the heart region. If a piece of endoderm with the precardiac mesoderm attached is removed from the embryo and cultured *in vitro*, the mesoderm migrates along the endoderm surface and differentiates into a twitching heart vesicle. The endodermal cells in the precardiac area apparently elongate, furnishing an oriented substratum for migration. Adhesive gradients may exist in the endoderm that help the prospective heart cells move in a specific direction. For example, the preheart cells may move in the direction of increasing "stickiness" of the associated endoderm. Clearly, the interactions that occur between the heart-forming mesoderm and the endoderm are still not well understood.

A final question about heart development is: What causes and controls

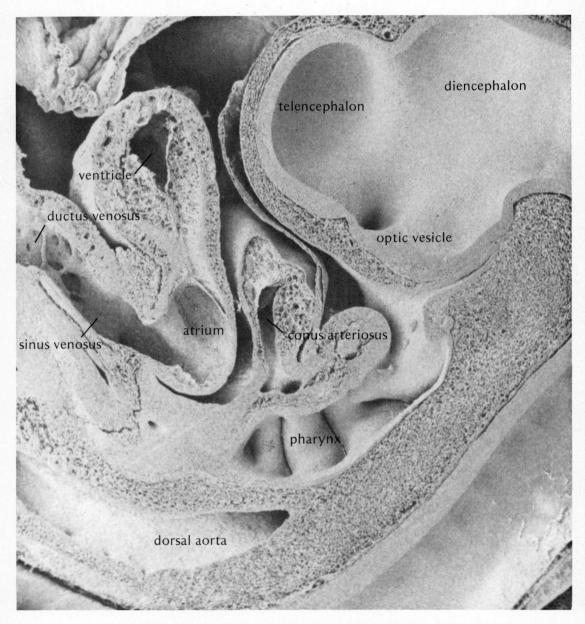

Figure 10–17. Seventy-two-hour chick embryo showing heart chambers. Courtesy of Peter Armstrong.

heart beating? It has been shown that the heart muscle cells differentiate after the cells have stopped dividing. The chick heart begins to beat at the 10-somite stage in the chick embryo. Beating begins irregularly in the ventricular region and spreads to the rest of the heart as the other regions differentiate. It stabilizes at a rhythmic rate of about 35 beats per minute, and gradually increases to about 85 beats per minute as the ventricle loops. After sixty or so hours of incubation, the chick heart rate reaches the typical embryonic rate of about 115 beats per minute.

A chick heart that is contracting at about 115 beats per minute can be cut into three sections, the sinoauricular part (sinus venosus plus atrium or auricle), the ventricular part, and the conus part. Only the sinoauricular portion will beat at 115 contractions per minute; the other two fragments will revert to the slower, early embryo rate. This experiment suggests that the sinoauricular region acts as a "pacemaker" that spreads the faster beating rate to the other parts of the heart. The factors controlling heart rate, however, seem to be very complex. Experiments with chick cells have shown that, in culture, the pulsation rate is inversely related to the size of the cell cluster (with seven-day heart cells). A larger cell cluster beats more slowly than a smaller cell cluster. It is therefore unlikely that pacemaker cells alone govern the absolute rate of whole heart beating. Other factors, such as the volume of heart tissue present, also play a role. In addition, DeHaan has shown that potassium ion concentration plays a key role in controlling heart cell beating. A lower potassium ion concentration apparently increases the percentage of cells that beat.

In summary, a variety of mechanisms are involved in heart development. Prospective heart-forming mesoderm cells in the chick embryo migrate on an endoderm substratum, eventually forming two mesodermal vesicles on either side of the developing foregut. These heart rudiments fuse in an anterior to posterior direction to form the early heart. The newly formed heart tube twists into a looped structure with specific chambers. This twisting may be caused by the migration of sheets of cells and by changes in cell shape (see Chapter 12). Finally, after cell division stops, the heart-muscle cells differentiate and spontaneous beating begins. Figures 10–17, 10–18 and 10–19 are scanning electron micrographs of the heart and associated structures in the developing chick embryo. Much remains to be learned about heart development, but we have come a long way in our understanding of this area of embryology, thanks to the work of Rawles, DeHaan, and others.

SECTION 4   LIMB DEVELOPMENT AND REGENERATION

LIMB DEVELOPMENT

The development of limbs has been extensively studied in this century, and a great deal of interesting information has surfaced. Much of what we know

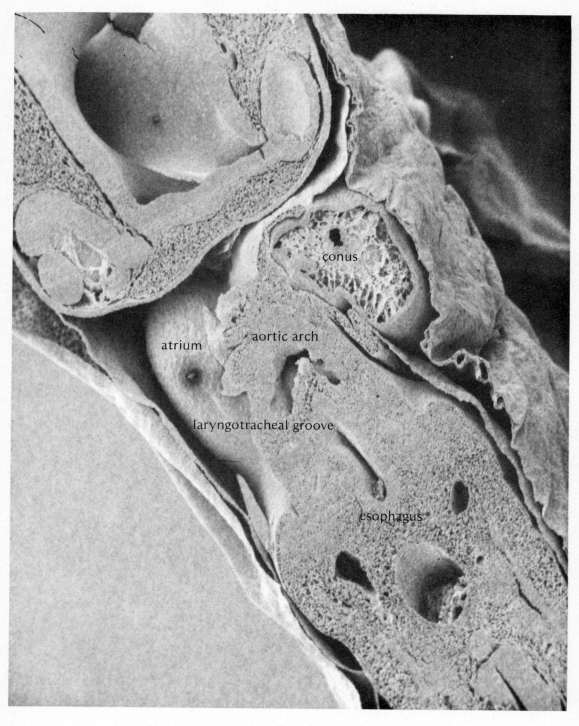

Figure 10-18. Seventy-two-hour chick embryo showing heart chambers. Courtesy of Peter Armstrong.

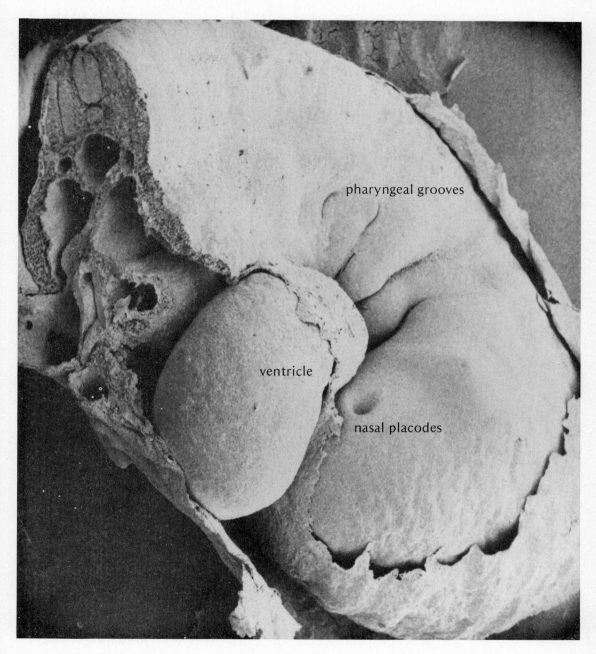

Figure 10–19. Seventy-two hour-chick embryo showing heart chambers. Courtesy of Peter Armstrong.

came from experiments involving operations on the developing limbs of vertebrates. We shall stress some of these experiments here. In this section we shall examine basic limb development in vertebrates and some of the hypothetical mechanisms controlling it. We shall then look briefly at the intriguing process of limb regeneration. Some vertebrates can regenerate entire limbs after the normal limbs are amputated. An understanding of this phenomenon may lead to methods of stimulating limb regeneration in man. As will be seen, we are on the threshold of medically useful developments in this area.

A LIMB FATE MAP

Before we examine the processes involved in limb development, we shall briefly consider the origin of the limb buds and the nature of a limb bud fate map. Recall from Chapter 9 that limbs are derived from the somatic mesoderm of the hypomere. Cells from the somatic mesoderm accumulate in the limb-forming regions and migrate to a position just below the epidermis. The epidermis above this accumulation thickens in many vertebrates. This thickening is called the *apical ectodermal ridge* (Figure 10–20). Thus, the limb bud consists of a core of somatic mesoderm covered by epidermis; the latter often differentiates into an apical ectodermal ridge. Figure 10–21 summarizes the formation of the limb bud.

Figure 10–22 shows a fate map developed by Saunders for a chick forelimb. Fate maps such as this are worked out by labeling different portions of the limb bud with carbon particles or with radioactive tracers, and following the fate of these marked regions in the adult limb. A similar procedure was used by Vogt to construct whole embryo fate maps (see Chapter 7). Note that all of the limb bones are derived from the mesoderm portion of the limb bud. We shall return to the respective roles of the mesoderm and the ectoderm in limb formation later in our story.

THE LIMB-FORMING FIELD

The limb bud and surrounding area is an example of an *embryonic field*, an area of the embryo that possesses a set of specific properties. Because this subject has been most thoroughly examined in relation to the limb system, we shall preface our discussion by defining an embryonic field and explaining why the limb bud area qualifies as one. In this way, we shall begin to understand the limb bud and its properties.

Embryonic fields are areas of the embryo that possess the following four properties:

1. The potency of tissue to form a given structure is more widespread in the embryo than the prospective structure itself.

2. The expression of this potency decreases with distance from the prospective structure.

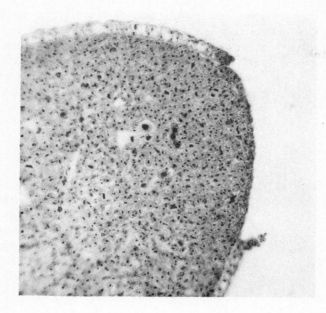

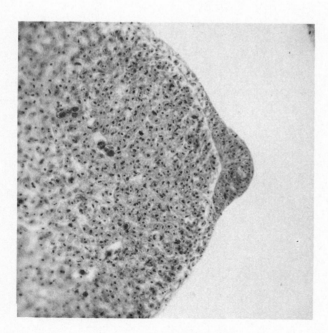

Figure 10-20. Tip of chick embryo wingbud. (a) Apical ectodermal ridge has been removed. (b) Apical ectodermal ridge is present. From Saunders, *J. Exp. Zool.* 108 (1948): 363–403. Courtesy of J. Saunders.

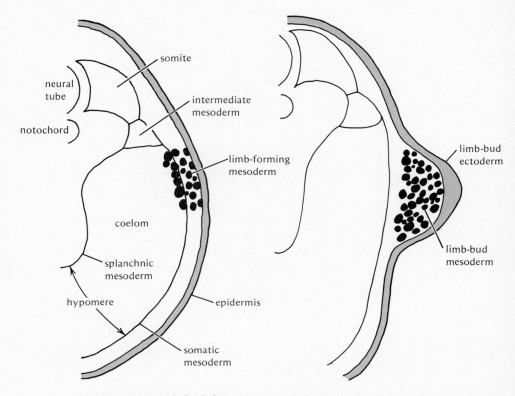

Figure 10–21. Limb bud formation. Limb-forming mesoderm is derived from the somatic mesoderm of the hypomere.

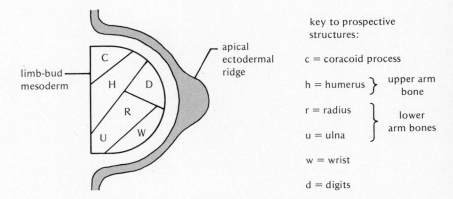

Figure 10–22. Fate map of chick forelimb. Saunders inserted carbon particles in the limb bud mesoderm and traced these marks to develop limb fate map. A similar fate map can be developed for the hindlimb, including the femur (upper leg), the tibia and fibula (lower leg), the ankle, and the toes.

3. Parts of the primordium of the structure (the embryonic tissue bud that gives rise to the structure) can yield the complete structure.

4. Augmentation of the primordium can give rise to the complete structure.

Amphibian limb buds and eye, ear, and heart primordia are examples of embryonic fields. That is, these primordia possess the four properties given above. This was demonstrated by numerous experiments indicating that a given primordium area has each specific field property. A brief examination of the four field properties indicates that an embryonic field is an area that regulates itself. In other words, if some pieces of the area are removed, a complete structure can still form. Thus, cells in an embryonic field must be able to recognize their own position and their own fate. They must also be able to change their fates in response to a disruption of the normal primordium. We shall begin by examining a number of interesting experiments showing that the limb-forming area is an embryonic field. After we get a "feel" for limb-forming area, we shall turn to a number of experiments that elucidate the roles of the different portions of the limb bud in controlling limb formation.

The first field property is that the potency to form a given structure is more widespread than the prospective structure. The amphibian embryo is choice material for experiments on the limb-forming area because operations on embryos are easily accomplished, the embryos themselves are readily available, and they develop in plain water. Byrnes, Braus, and Balinsky performed experiments on frog embryos that support the first field property. If a normal forelimb bud is removed from the embryo, a normal limb still develops from the surrounding tissue. Thus, the potency to form an amphibian limb is indeed more widespread than the limb bud itself. Field property one is therefore satisfied.

The second field property is that the expression of the potency to form the structure decreases with distance from the prospective structure. Balinsky used abnormal inducers (such as ear vesicle or nose rudiments) to induce limb formation on the flank between the normal forelimb and hindlimb sites in the amphibian embryo. The underlying mesoderm cells accumulated under the epidermis to form a limb bud, and an additional limb developed. Balinsky found that the further away from the normal limb-forming site, the more difficult it was to induce a limb with abnormal inducers. The area equidistant from the normal forelimb site and normal hindlimb site was least able to form a limb (Figures 10-23, 10-24). In addition, limbs produced closer to the normal forelimb site were forelimb-like (four digits and thin). Limbs formed closer to the normal hindlimb site resembled hindlimbs (five digits and thick). In between, the limbs that did form had characteristics of both forelimbs and hindlimbs. Thus, the expression of the

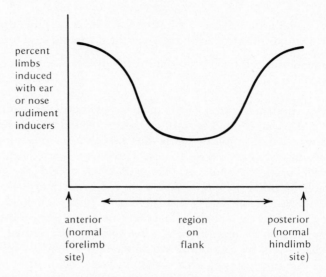

Figure 10-23. Limb induction in newts with abnormal inducers. Balinsky found that it was easier to induce limbs closer to the normal limb forming sites. Limbs formed nearest the normal forelimb site resembled forelimbs, limbs formed nearest the normal hindlimb site resembled hindlimbs. Limbs formed between the two sites possessed hindlimb and forelimb characteristics.

potency to form an amphibian limb decreases with distance from the prospective limb. Thus, field property two is also satisfied.

Field property three is that parts of the primordium can yield the complete structure. Experiments that support this property in amphibians come from the work of Harrison, Swett, and others. These workers removed portions of the limb bud as shown in Figure 10-25. When half—or even

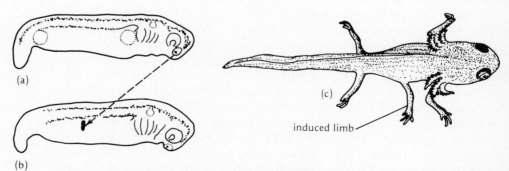

Figure 10-24. Induction of limbs with artificial inducers. (a)–(c) Limb induction in newt with nose rudiment, (a) and (b) show operation. Position of normal limb buds shown in (a). Induced limb shown in (c). From: I. B. Balinsky, *An Introduction to Embryology* (Philadelphia: W. B. Saunders, 1975), p. 387.

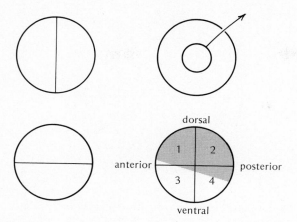

Figure 10–25. Parts of limb bud can give rise to complete limb. Harrison and Swett found that halves (a) and (b), the periphery (c) or single quadrants (d) could give rise to complete limbs. The shading in (d) indicates the best limb-forming ability.

three-quarters—of the limb bud was removed, the remaining portion could form a complete, normal limb. When the limb bud was divided into quadrants, certain quadrants formed limbs more often than did other quadrants. Parts of the amphibian limb bud can yield a complete limb. Thus, field property three is satisfied.

The fourth field property is that augmentation of the primordium (increasing the amount of tissue) can produce the complete structure. If an extra piece of amphibian limb mesoderm is stuffed under the ectoderm along with the normal mesoderm, a single complete limb is still formed. Field property four is thus satisfied.

The limb-forming area is therefore truly an embryonic field because experiments have shown that it has all four field properties. How do cells in the limb-forming area adjust to all of the operations needed to form a normal limb? This problem is not yet fully solved. French, Bryant, and Bryant, however, recently developed a theory explaining how cells in fields such as limb buds assess their positions and therefore regulate the development of the structure. They suggest that each cell has some sort of molecular information giving its position on the radius of a circle and around the circle—a pattern, in short, like polar coordinates (see Figure 10–26). Experiments that will eventually prove or disprove the "circle" theory of positional information are being performed. The molecular nature of the positional information is also under investigation. Many experiments have already provided support of the "circle" rule. For example, if an X-irradiated amphibian limb stump is provided with a complete piece of nonirradiated skin (epidermis and dermis) containing a full circle of cells from the area, regeneration occurs. If the skin contains few cells from the circumference of the circle, regeneration fails.

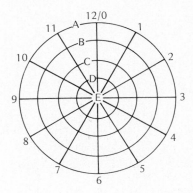

Figure 10-26. Polar coordinates in a positional information field. Each cell is assumed to have information with respect to its position on a radius (the line from A through E) and its position around the circle (0 through 12). Positions 12 and 0 are identical, so the sequence is continuous. From S. J. Bryant and L. E. Iten, *Devel. Biol.* 50 (1976): 212.

MECHANISMS OF LIMB DEVELOPMENT

The limb bud consists of two major parts: the limb-bud mesoderm and the ectodermal covering, the apical ectodermal ridge. Let us begin by investigating the following questions: What is the role of the limb-bud mesoderm in limb development? What is the role of the apical ectodermal ridge?

Some investigators believe that the apical ectodermal ridge induces the underlying limb-bud mesoderm to differentiate into limb bones (Saunders-Zwilling model of limb development). Others believe that the apical ectodermal ridge does not induce mesodermal differentiation, but that it serves as a specific protective boundary over the mesoderm (Amprino hypothesis of limb development). As we shall see, there is no solid proof that the apical ectodermal ridge has an inductive function. Indirect evidence suggests that induction may occur, but at this point in time the question remains unresolved. We shall come back to the apical ectodermal ridge shortly. First, let us turn to an interesting related question.

What determines whether a hindlimb or forelimb is formed from a limb bud? In other words, what determines the property of "limbness"? The answer is unambiguous: The limb mesoderm determines if the limb formed will be fore or hind. The following experiment demonstrates this fact. Saunders and Zwilling transplanted chick embryo mesoderm from a hindlimb (leg) bud to a position under the apical ectodermal ridge of the forelimb (wing) bud. They also did the reciprocal experiment of transplanting mesoderm from the wing bud to a position under the apical ectodermal ridge of a leg bud. In the first case (leg mesoderm under wing bud) a leg (with scales and claws) developed. In the second case (wing mesoderm under leg bud) a wing developed (with feathers, etc.). Thus limbs are mesodermally specific.

In chick embryos, the apical ectodermal ridge is essential for limb development. Nonlimb ectoderm cannot replace the apical ectodermal ridge. If the ridge is removed from a limb bud, or if it is replaced by nonlimb ectoderm, all further limb outgrowth stops. Additional evidence for the importance of the apical ectodermal ridge comes from Zwilling experiments with the chick "wingless" mutation. This mutant forms wing buds, but wings fail to develop. If the mutant wingless mesoderm is replaced with mesoderm from a normal nonmutant wing bud, the wing fails to develop. The mutant "wingless" epidermis does not possess an apical ectodermal ridge. Thus, even if normal mesoderm is present under "wingless" mutant epidermis, no apical ridge develops and no wing develops.

Other experiments also show the importance of the apical ectodermal ridge in limb development. If a second apical ectodermal ridge is grafted onto a limb bud alongside the original ridge, the distal portions of the limb that develop are doubled (that is, two bones are formed side by side). Similarly, prospective thigh mesoderm develops into toes when grafted right below the apical ectodermal ridge of a wing. Although the "legness" of the grafted mesoderm still remains, the wing apical ectodermal ridge dramatically influences the thigh mesoderm, causing it to form more distal structures, toes, at the end of the wing.

It should be noted that the mesoderm also exerts an influence on the apical ectodermal ridge. Recall the "wingless" chick mutant. If "wingless" mutant mesoderm is stuffed under the apical ectodermal ridge of a normal chick wing bud, the ridge degenerates. If a thin sheet of mica is placed between the ridge and the mesoderm of a normal chick limb bud, the ridge flattens. These experiments suggest that the limb mesoderm "maintains" the apical ectodermal ridge, perhaps by releasing some chemical factor.

Apical ectodermal ridges can be placed in a small porous basket. If such a basket is added to a culture dish containing limb bud mesoderm, the ridges prevent the death of mesoderm cells. In the absence of ridges, the mesoderm cells begin to die. The factor that prevents mesoderm cell death may be the same factor that influences limb mesoderm outgrowth in the intact limb bud. Similar culture experiments also indicate that the limb-bud mesoderm produces a factor that prevents the flattening of the apical ectodermal ridge. As we are now in the age of biochemistry, the molecular nature of these factors will probably be elucidated within the next decade or so.

What other factors are involved in limb development? In Chapter 12, "Mechanisms of Morphogenesis," we shall examine some of the interesting mechanisms that control the development of form in embryos. It is appropriate to mention here that cell death plays an important role in limb development. The digits on your hands and feet (your fingers and toes) develop as the result of cell death in the regions between the digits. Cell death also helps shape the upper arm, forearm, and elbow region. Let us just mention one experiment here to give us some insight into the role of cell death in limb development. If cells are taken from the portion of a stage 17 chick embryo wing bud that is destined to die at stage 24 and grafted to the somite area of the embryo or placed in culture, the grafted cells will die when the embryo reaches stage 24 (or when the cells would have reached stage 24 in the culture experiments). This suggests that although stage 17 embryo cells look normal, they are "programmed to die" and will die at the right time. The stage 17 cells, however, can be prevented from dying by grafting to the dorsal side of the limb bud. If the cells programmed to die are taken from a later (stage 22) embryo, however, they cannot be saved even by transplanting them to the dorsal portion of the limb bud. Thus a "death clock" is set by stage 17 but it can be turned off by certain environmental influences. Once the embryo reaches stage 22, however, the "death clock" counts down and cannot be turned off, no matter what environment the cells are placed in. In cases where the "death clock" can be turned off, limb mesoderm seems to provide a factor that can change the program. The nature of the "death clock" and the factors influencing it are not well understood. We shall, however, get additional insight into cell death as a morphogenetic mechanism in Chapter 12.

Limbs develop in a proximo-distal sequence. In other words, the limb regions closest to the body differentiate first and the digits differentiate last. Cell density is greatest in the proximal regions and lower near the growing tip because cells can move away from each other more easily at the tip. It is known that when cells are densely packed (contacting each other on all sides), growth and movement of such cells may cease. This is termed contact inhibition of growth and movement. When growth (cell division) ceases, differentiation often begins. Thus, the closeness of cell contact in the different regions of the forming limb may play an important role in controlling the program of limb differentiation. The mechanism by which cell contact inhibits continued growth and movement is not well understood. It is, however, the topic of a great deal of current investigation. As an aside, we might mention that one of the reasons why tumor cells continue to grow and spread is that tumor cells are less inhibited by contact with each other than nontumor cells are. Thus, the mechanism of contact inhibition seems to be defective in tumor cells. We shall return to this topic in Chapter 16.

LIMB AXES. How does limb orientation come about? An area of limb-bud mesoderm near the posterior junction between the limb bud and the body plays a key role in determining limb orientation. This area is called a *zone of polarizing activity* (ZPA) (Figure 10–27). If such an area is transplanted beneath the apical ectodermal ridge of a chick wing bud, an additional wing develops in this area. The posterior side of the new wing always faces the implanted ZPA. The ZPA appears to determine the anterior-posterior axis of the limb; it also appears to stimulate its outgrowth.

The dorsal-ventral axis of the limb is (at least in part) determined by limb-bud ectoderm. The limb-bud mesoderm can be removed, dissociated

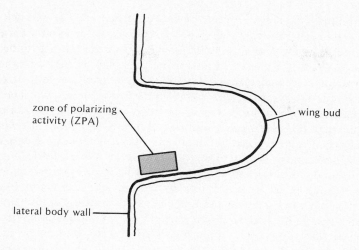

Figure 10–27. Location of the zone of polarizing activity in a chick wing bud. Based on findings by J. W. Saunders.

into single cells, and then repacked and placed under the ectoderm. The limb formed from such an implant has a dorsal and ventral side but no anterior-posterior differences. ZPA cells would have been totally dispersed in the implant, making an anterior-posterior axis impossible to develop. The dorsal and ventral (top-bottom) surfaces that develop in such an implant are located next to the original dorsal and ventral sides of the ectodermal cover over the mesodermal implant. Thus, the ectoderm plays a role in determining the dorsal-ventral axis of the developing limb. The means by which the ZPA and the ectoderm determine limb orientation are not yet understood.

LIMB REGENERATION

Why can adult newts regenerate legs while adult human beings cannot? The answer to this question would bring us a long way toward developing a technique for regenerating amputated human limbs. The answer is not yet in hand, but a great deal of exciting information has recently become available on parts of this puzzle. Before we look into this work, let us briefly summarize the regeneration abilities of various vertebrates.

Newts and salamanders (urodele amphibians) have remarkable powers of regeneration. These organisms can regenerate entire limbs, jaw parts, gill parts, eye parts, and tails. Frogs (anuran amphibians) can regenerate normal limbs only in the tadpole stage. Adult frogs cannot regenerate limbs. Lizards can regenerate tails; fish can regenerate fins but not tails; birds can regenerate parts of the beak. Thus, the only adult vertebrates that normally regenerate limbs are the newts and salamanders.

Mammals cannot regenerate limbs after simple amputation, but they can regenerate many internal tissues. For example, a large portion of the liver can be removed, and regeneration will restore it. In newly born opossums, a limb can regenerate if a piece of brain tissue is implanted in the limb prior to amputation. It should be noted that the infant opossum limb is not fully differentiated and is in the same state as the frog tadpole leg when it begins to lose regenerative ability. Three important points should be kept in mind. The first is that the presence of additional nervous tissue can stimulate limb regeneration, not only in infant opossums, but also in other organisms such as frogs. The second is that less differentiated tissue is more capable of regeneration than more differentiated tissue. The third point is that limb regeneration can be stimulated in organisms that do not normally regenerate limbs. This raises the hope that someday it might be possible to stimulate limb regeneration in humans. We shall return to the topic of stimulating limb regeneration after we examine what actually occurs during limb regeneration in vertebrates.

SEQUENCE OF EVENTS IN LIMB REGENERATION

The following sequence of postamputation events occurs in the newt limb:

1. The wound is covered by local epidermal cells that spread over the cut surface by active amoeboid movement.

2. The muscle and cartilage just below the covering cells dedifferentiate. The activity of degradative enzymes increases, and the bone and cartilage matrix disintegrate. Muscle cells and cartilage cells lose their differentiated appearance and become transformed into embryo-like cells.

3. These dedifferentiated, embryo-like cells accumulate under the epidermal covering of the wound and, together with the epidermal covering, are called the regeneration *blastema* or regeneration bud. The undifferentiated cells that make up this blastema come from surrounding tissue, not from distant parts of the body. This was shown by labeling experiments: Radioactively labeled limb cells (muscle, cartilage, bone, epidermis, and nerve Schwann cells) all dedifferentiated and took part in forming the regeneration blastema. Also, local limb X-ray irradiation blocked regeneration; whereas whole body irradiation (the limb shielded from the X-rays) did not. These results suggest that local limb cells play the major role in forming the regeneration blastema.

4. The blastema cells begin active division and growth.

5. Division and growth decrease and the cells begin to differentiate. This includes the synthesis of muscle proteins and cartilage matrix by the newly formed muscle and cartilage cells, respectively. As mentioned earlier, when cell division and growth cease, conditions favorable for differentiation develop. New limb bones and muscle form, and regeneration is complete. A summary of these events is given in Figure 10–28.

Let us now look briefly at experiments that will help us understand some of the basic mechanisms operating in regenerating limbs. One important principle is that if a limb is amputated, regeneration involves structures distal to the cut (Figure 10–29). Thus, if the cut is through the lower arm, the parts of the lower arm distal to the cut, the wrist and the digits, will regenerate. If

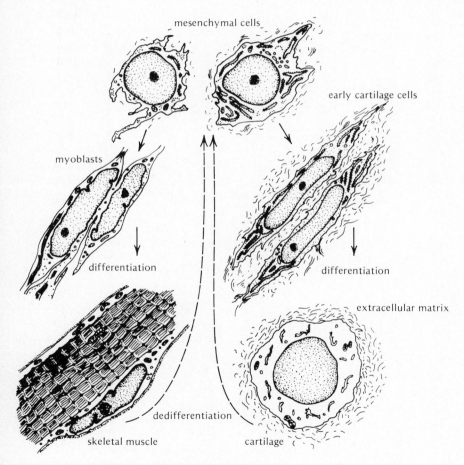

Figure 10–28. Cytology of limb regeneration. Cartilage and muscle apparently dedifferentiate into mesenchymal cells early in regeneration. The mesenchymal cells then form the muscle and cartilage of the newly regenerated limb. From E. D. Hay in D. Rudnick, ed., *Regeneration* (New York: Ronald Press, 1962), pp. 177–210.

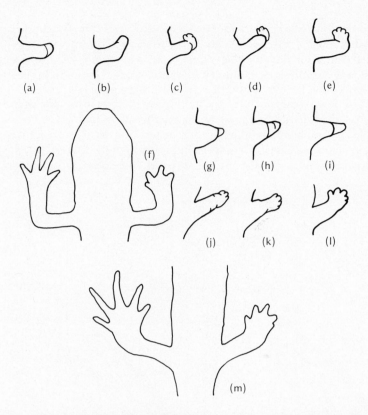

Figure 10-29. Regeneration of the newt *Triturus cristatus*. (a)–(f) Consecutive stages of the regeneration of a forelimb amputated above the elbow. (g)–(m) Consecutive stages of the regeneration of a hindlimb amputated above the knee. From G. Schwidefsky, *Roux. Arch.* 132 (1934): 57–114.

the cut is through the upper arm, the parts of the upper arm distal to the cut, the lower arm, the wrist, and the digits will regenerate. The cells in the area of the cut surface must have information that allows the regeneration of only the distal (missing) structures. In our earlier discussion, we noted that French, Bryant, and Bryant formulated a theory in which cells can recognize their positions on a circle. Perhaps the cells in the cut region recognize what needs to be made distal to the cut by their ability to recognize their place on a circle around the cut area. Future work will determine whether this phenomenon does truly play an important role in limb development and regeneration.

Other important sets of experiments were those of Zwilling and Searls. Zwilling dissociated wing-bud mesoderm into single cells, mixed the cells up, and grafted this group of cells under the ectoderm on the body of a chick embryo. A normal limb formed with the bone and muscle in the right places. Did the mesoderm cells "sort out," the precartilage cells moving to the center of the limb bud and the premuscle cells moving to the periphery? Or did their positions determine what they would form?

Searls and Janners labeled chick embryos with tritiated thymidine. They then removed either the central core of the limb bud (prospective cartilage) or the peripheral portion of the limb bud (prospective muscle). Each labeled region was implanted into the limb bud of a cold (unlabeled) chick embryo at right angles to the long axis of the developing bones. Thus, each "hot" implant spanned both prospective muscle and prospective cartilage in the cold limb bud (Figure 10–30). What happened? The tissue that formed from the hot implant became cartilage in the central, cartilage-forming region of the host bud, and became muscle in the peripheral, muscle-forming region. Thus, in the developing limb bud, the fate of prospective muscle and cartilage cells is not irreversibly fixed but can be changed by the environment around the cells. Cell position is a key factor in controlling differentiation.

What about regenerating adult limbs? When the cartilage and muscle dedifferentiate, do cells originally derived from cartilage redifferentiate only into cartilage, or can these cells redifferentiate into muscle? Do cells origi-

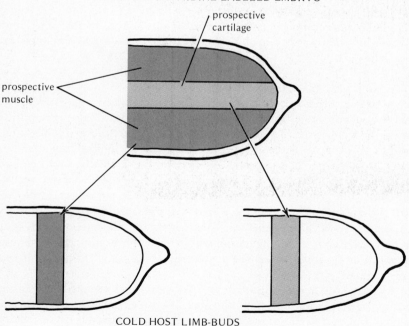

Figure 10–30. Experiment showing effects of cell position on tissue differentiation in limb buds. In both (a) and (b), the implant forms muscle in the peripheral regions of the limb buds and cartilage in the central core regions. Modified and redrawn from R. L. Searls and M. Y. Janners, *J. Exp. Zool.* 170 (1969): 365.

nally derived from muscle redifferentiate only into muscle, or can these cells form cartilage? The answer to this question is still a bit unclear. An experiment, however, suggests that although salamander cartilage can dedifferentiate, the resulting cells redifferentiate mainly into cartilage. Muscle cells that dedifferentiate from muscle tissue, however, can apparently form both muscle and cartilage. Before amputation, radioactively labeled salamander muscle, epidermis, cartilage, or cartilage plus perichondrium (the connective tissue around cartilage) was grafted into irradiated host salamander limbs. (The host limbs were irradiated to discourage the participation of host cells in the regeneration process.) The experimental results were quite interesting. Only muscle was able to generate all of the component tissues in the regenerated limb (cartilage, perichondrium, the connective tissues of joints, fibroblasts, and muscle). Cartilage generated cartilage, perichondrium, and connective tissue, but not muscle. Epidermis apparently took no part in the regeneration at all. What do these results mean? They suggest that muscle can dedifferentiate and redifferentiate into many tissues, while cartilage can redifferentiate only into cartilage and connective tissue. Note, however, that muscle contains several cell types, including myoblasts, connective tissue, and blood elements. Thus, it is still unclear if any cells in the adult limb can really redifferentiate into totally different cell types. At any rate, the fates of cells in the early limb bud are not irreversibly fixed. In the adult limb, however, it remains to be seen how fixed are the fates of the component cell types.

STIMULATION OF REGENERATION

Let us now return to the problem of stimulating limb regeneration. We already alluded to one important factor that stimulates limb regeneration, namely, nervous tissue. Recall that limb regeneration occurred in the infant opossum when nervous tissue was implanted into the limb. In salamanders, nerves also appear to be important in regeneration; normally, limbs will not regenerate after the nerves have been removed. In the embryo, however, the salamander limb can develop even if nerves are kept out of the developing limb. If these nerveless limbs are amputated and nerves are kept out of the area, the limbs will still regenerate. Thus, a limb that had had no nerves could regenerate without them, but a limb that has had nerves must have them for regeneration. So limbs appear to develop an "addiction" to nerves if they were originally present.

Exactly how nerves stimulate regeneration is not yet well understood. An important factor seems to be the quantity of nervous tissue present, not the type. Regeneration is stimulated by most types of nervous tissue and does not appear to depend on whether the nerves are sensory or motor. In fact, if a limb nerve is moved from its normal position in the newt and placed under the skin in a region not too far from the normal limb, a new limb can form in this area. Nerves, therefore, exert a profound influence on regeneration.

What do nerves do to stimulate regeneration? If nerves are removed from a regenerating limb, the RNA concentration, DNA concentration, protein synthesis, and synthesis of extracellular glycosaminoglycan (a kind of glycoprotein molecule) decrease. Singer isolated and partially purified a basic protein from whole brains or brain nerve endings that stimulates protein synthesis in regenerating limbs. This material may turn out to be the nerve "factor" that stimulates regeneration. Nerves may also stimulate regeneration by stimulating blood vessel growth to the area. Increased blood supply to the regenerating limb may directly promote development by supplying nutrients, oxygen, hormones, and other substances to the regeneration blastema.

Schwann sheath cells surrounding nerves can give rise to the entire regeneration blastema and to the entire regenerated limb. Salamander limbs can be irradiated with X rays so that regeneration does not occur. Regeneration will occur, however, in such an irradiated stump if an unirradiated nerve is implanted in the amputation site. By taking nerves from a pigmented salamander and implanting them in the irradiated stump of a nonpigmented salamander, Wallace showed that the entire regenerated limb arises from the donor nerve Schwann sheath cells and from fibroblasts, connective tissue cells. Thus, not only can nerves stimulate regeneration, but nerve sheath cells and fibroblasts can apparently give rise to all of the bone and muscle cells of the regenerated limb!

Cell damage at the wound surface plays an important role in stimulating regeneration. A newly regenerated limb can form in newts even if no amputation has occurred. A tight ligature around a limb or the introduction of cartilage breakdown products under the skin can trigger partial or complete limb regeneration. Products from damaged cells, therefore, may stimulate regeneration. This conclusion is supported by experiments in which beryllium nitrate is applied to the amputation surface. This salt suppresses regeneration, possibly by binding substances that stimulate regeneration as they are released from damaged cells. The beryllium treatment must be carried out immediately after amputation, or it will not inhibit regeneration. This suggests that the salt must bind the products from damaged cells immediately, or the products will start the regenerative process. Another amputation of the beryllium-treated stump allows regeneration to occur unless beryllium is applied again.

ELECTRIC CURRENTS, NERVES AND REGENERATION. Several studies suggest that natural electricity in organisms plays a role in growth, morphogenesis, differentiation, and regeneration.

Electric currents are generated in living organisms by the flow of ions such as sodium and calcium. Investigations into such electricity began about 1843, when the German physiologist E. Du Bois-Reymond measured steady current flow in wounded human skin. In the 1960s, Lionel F. Jaffe at Purdue

University measured electrical gradients in eggs and early embryos of the marine brown alga, *Fucus*. In the 1970s, Jaffe and his student Richard Nuccitelli developed a sensitive vibrating probe apparatus to record the current entering or leaving cells or tissues suspended in media. The probe device is a rapidly vibrating platinum electrode that oscillates between two points in the medium without penetrating the cells. It detects the tiny differences in voltage between these positions produced by the current under study. This probe has measured small steady electric currents in various plants and animals, including root hairs, pollen tubes, amoeba pseudopodia, *Fucus* embryos, and developing amphibian and insect systems.

Jaffe, Vanable, Jr., and Borgens at Purdue used the vibrating probe to study current flows during amphibian limb regeneration. They found that a large current density, 50–100 microamperes per square centimeter, leaves the stump end of salamander limbs after amputation and returns to the undamaged portions of the limb and body.

Jaffe, Vanable, and Borgens eliminated sodium uptake by removing it from the artificial pond water used to maintain the animals or by treating their skin with a drug that blocked the uptake of sodium from the pond water. When sodium uptake was blocked, the current leaving their limb stumps was greatly reduced or eliminated. These results suggested that the skin surrounding the limb stump was a source of currents driven out of the stump end, and that sodium ion movement was a source of the current.

Jaffe and colleagues now blocked the current efflux with current-blocking drugs or by sewing a flap of whole skin over the stump end in place of the normal wound covering. Regeneration was inhibited. Another observation begins to make sense in light of these findings. Children can regenerate amputated finger tips if the finger is severed between the last joint and the nail, and if the fingertip is not sewn closed. Regrowth may include new fingernail and fingerprints. Surgical closure of the fingertip places skin over the stump, and would block the efflux of current just as in the amphibian work. Indeed, natural current flows peaking at about 22 microamperes per square centimeter have been measured leaving the amputated fingertips of children.

Adult frogs, unlike salamanders or newts, do not regenerate amputated limbs. Most of the current flowing from a cut stump is shunted beneath the skin, leaving very little current in the stump tissue. This effect occurs in frogs but not in salamanders because frogs possess subdermal lymph spaces (large sinuses filled with conductive body fluids) that effectively short-circuit the current. Salamanders do not possess these sinuses; neither do frog tadpoles. Frog tadpoles can regenerate limbs, but then lose this capability as they metamorphose into adult frogs. It is interesting that the precise area on their limbs where loss of regenerative capacity occurs is the area where the subdermal lymph sinuses begin to appear.

This work suggests that one might stimulate regeneration by altering the current that flows into stump tissue. Borgens and colleagues implanted small batteries under the skin on the backs of frogs with amputated limbs. They then placed the electrodes so that the current flow through the stump was similar in direction, magnitude, and duration to that in the salamander stump (see Figure 10–31). Control frogs had dead battery implants. The electrically treated limb stumps regenerated to varying degrees, although the limbs were very abnormal. New bone, muscle, nerve, and bone precursors were all present past the point of amputation. The control stumps showed only scar tissue and fracture callus. Borgens and colleagues also placed electrodes so that the current flow was in the opposite direction. The amputated limbs did not regenerate. Thus, a weak current (0.0000002 amperes) initiated limb regeneration. The next question is, how did the current do this?

Marcus Singer and colleagues found many more nerves in the stumps of salamanders and newts than in the stumps of frogs. When nerve tissue was removed from salamander and newt stumps, the regenerative ability was lost. Singer added nerve tissue to a frog's forearm stump by rerouting the sciatic nerve bundle from the frog's leg to its stump (Figure 10–32). The limb regenerated. Other experiments showed that when an electric field was applied to frog stump tissue as discussed previously, nerve growth within the stump was enhanced. Perhaps electric current stimulates regeneration by increasing the nerve supply to the stump. Mizell and others have shown that nerve tissue stimulates regeneration in newborn opossums and lizards; thus

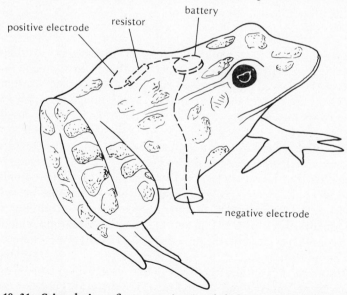

Figure 10–31. Stimulation of regeneration in adult frogs using a battery implanted under the skin. Two electrodes are positioned to deliver current through the stump. The current is designed with a polarity and magnitude resembling the natural current flow through a salamander stump. Based on work of Borgens, Vanable, and Jaffe.

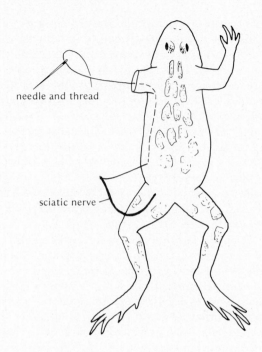

needle and thread

sciatic nerve

Figure 10–32. Procedure for rerouting sciatic nerve from frog's leg to forearm stump. Based on experiments by Marcus Singer.

this effect is not limited to amphibians. Robinson, McCaig, and Hinkle found that when an electric field was applied to a culture of embryonic nerve cells isolated from the clawed frog *Xenopus*, they grew toward the negative pole (cathode). They speculated that a natural electric field helped guide embryonic nerves from the developing spinal cord to their destination in developing muscle.

Borgens, Cohen, and Roederer applied about 10 microamperes of steady current to two small electrodes positioned about 4–5 centimeters apart next to the spinal cord of larval lampreys. The spinal cord was then completely severed between the electrodes. The regeneration of spinal cord nerves was greatly enhanced, as compared with controls receiving no current but treated identically in all other ways. It appears that electric curent plays an important role in both nerve growth, and limb regeneration, phenomena that appear to be interrelated.

Currents can move cellular components (especially ions) in the cell membrane and within the cell cytoplasm. They may also concentrate the molecules involved in cellular recognition and the reception of hormone signals in specific regions of cells. Exactly which molecules are responsible for the growth-inducing effects of current, however, is still poorly understood.

Some work, however, from Singer's group and others provides insights into the mechanism by which nerves stimulate regeneration. Singer's group isolated a crude protein fraction from the supernatant of a 140,000 × G centrifugation of newt brain homogenates. When the stumps of denervated limbs were treated with this fraction, they incorporated radioactive amino acids into protein and tritiated thymidine into DNA at rates similar to nondenervated control stumps. Similar results were obtained with both newt and chicken brain extracts by Choo and colleagues. Previous work by Mescher's group, Singer's group, and others showed that protein synthesis, RNA synthesis, and DNA synthesis in denervated stumps were substantially lower than those in nondenervated control stumps. The active factor in the nerve extracts might be a 13,000 molecular weight basic protein called *fibroblast growth factor*. This protein is released by proteolytic enzymes from myelin, the major component of nerve sheaths. Mescher and Gospodarowicz showed that fibroblast growth factor stimulates mitosis in cells in culture and also stimulates DNA synthesis when infused into the stumps of denervated limbs of adult newts. Fifty micrograms of this factor increased the mitotic rate in the denervated stumps to about 70% that in control stumps from nondenervated limbs, a 5-fold increase over the rate in untreated denervated limb stumps.

Clinically useful techniques have already been developed from this basic research. Brighton and Friedenberg at the University of Pennsylvania School of Medicine and Bassett of Columbia University have used electric current sources to treat fractures that failed to heal properly. Induced currents increased successful healing by as much as 85% in these difficult cases. Eventually, electric current will probably be useful in treating spinal cord injury as well as bone damage.

To sum up, the limb bud has remarkable self-regulating properties that enable it to form a limb even after a massive mutilation. Adult limbs can regenerate in newts and salamanders, but not in most other vertebrates. The sequence of events in limb regeneration includes cell migration, dedifferentiation of limb tissues, and redifferentiation of newly regenerated limb tissue. Implanted nerve tissue, products of cell damage, and electric current can stimulate limb regeneration in systems that do not usually regenerate limbs. Work in this field offers hope that eventually it will be possible to stimulate limb regeneration in human beings.

SECTION 5   UROGENITAL ORGANOGENESIS

We now turn to the development of the urogenital system. In addition to describing the embryology of this system, we shall examine the mechanisms that play important roles in its development.

What is the urogenital system? Aren't we talking about two systems—

the urinary, or kidney, system and the reproductive system? In fact, we are. However, we shall examine the development of these systems as a single system because their embryogenesis is interrelated. In fact, as you will see, the structures that belong to one system at one stage in development sometimes become a part of the other system later on.

Before we begin, let us define more carefully what we mean by the urogenital system. The urogenital system consists of the kidneys and accessory structures involved in the formation of urine, together with the gonads and accessory structures utilized for reproduction. We should mention that the kidneys are by no means the only organs involved in excretion. Some fish excrete salt through their gills. Salt is excreted by the nasal glands of some marine birds and reptiles, and by the rectal glands of sharks and rays. Skin glands (sweat and mucous glands) excrete salt, nitrogenous wastes and water. The liver, lungs, and salivary glands are other structures involved in the regulation of body fluids and the excretion of wastes. Thus, although we will focus on the kidney here, we must keep in mind that the kidney is not the only organ involved in eliminating wastes and maintaining the body fluids.

KIDNEY

We shall begin our study of urogenital development by examining the embryology of the kidney in the more primitive vertebrates. We shall then work our way up to the advanced vertebrate systems. We shall discuss the development of the genital structures after our discussion of kidney development. The interrelationships between the structures used for excretion and reproduction will be shown.

In Chapter 9 we noted that the kidneys and gonads are derived from the intermediate mesoderm, or mesomere. This is the region of mesoderm between the somite and the hypomere. Let us now examine the development of the functional kidney of aquatic embryos and larvae: namely, the kidney of embryonic fish and tadpoles. This kidney is called the *pronephros*. Some of the structures of this kidney become part of the urogenital system of higher vertebrates, as we shall see later.

PRONEPHROS

The pronephros, the functional kidney of fish and amphibian embryos, develops from the intermediate mesoderm (mesomere). The specific area that gives rise to the pronephros is a region consisting of segmented portions of intermediate mesoderm, called *nephrotomes*. The cells in the nephrotomes separate and form an internal cavity, the *nephrocoel*. *Pronephric tubules* thus form from the nephrotomes and contain internal nephrocoels. The mechanisms by which tubules form are unclear, but they appear to involve the types of morphogenetic cellular rearrangements discussed in Chapter 12. As shown in Figure 10–33, the pronephric tubules are continuous with the coelom by

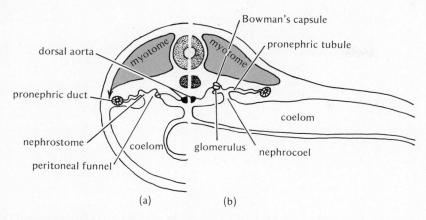

Figure 10–33. Diagram of cross section through embryos. (a) Anamniote (fish or frog embryo). The glomerulus is considered external, since the branch of the aorta pushes into the side of coelom. (b) Amniote (bird, reptile, or mammal embryo). The glomerulus is surrounded by Bowman's capsule and is considered internal. Most amniotes and anamniotes possess internal glomeruli. From J. B. Phillips, *Development of Vertebrate Anatomy* (St. Louis: C. V. Mosby, 1975), p. 434.

ciliated funnels (*nephrostomes*) and fuse at the other end to form the *pronephric duct*. This duct elongates in a posterior direction and fuses with the cloaca. The pronephric duct thus directly transfers wastes into the cloaca for elimination.

A network of fine blood vessels (*glomus*) is associated with the ciliated funnels of the pronephric tubules. Wastes pass through the lining of these blood vessels into the coelomic area right beside the openings of the pronephric funnels (nephrostomes). The wastes are picked up by these funnels, enter the pronephric tubules, and are transported to the cloaca via the pronephric duct.

The pronephros is functional in fish embryos and in amphibian larvae. A pronephros also forms in reptiles, birds, and mammals, but it never becomes functional.

MESONEPHROS

As the larval fish and amphibians become adults, the pronephric tubules disintegrate. Thus, an important mechanism involved in the formation of the adult kidney is differential death of the embryonic pronephric tubules. The pronephric duct, however, remains intact. The nephrotomes in the intermediate mesoderm now aggregate and form a second set of tubules. This set of tubules forms posterior to the pronephric tubules and connect to the pronephric duct after the first set disintegrates. The new tubules are the *mesonephric tubules*. The old pronephric duct is now called the *mesonephric*, or *Wolffian, duct* (Figure 10–34), since the mesonephric tubules have attached to

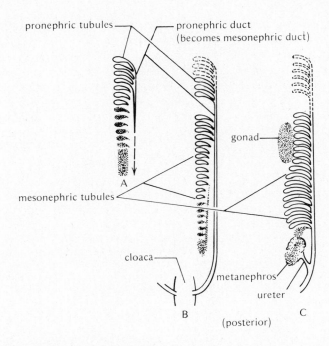

Figure 10-34. Development of kidney in vertebrates. (a) Pronephric duct becomes mesonephric duct. (b) and (c) Relationship of pronephric duct, mesonephric duct, and ureter to kidney development. Modified from Burns, "Urogenital System," in Willier, Weiss, and Hamburger, eds., *Analysis of Development* (Philadelphia: W. B. Saunders, 1955).

it. Waddington and O'Connor found that the pronephric duct appears to induce the formation of mesonephric tubules. If the pronephric duct is blocked from growing posteriorly and does not reach the area where the mesonephric tubules form (the mesonephrogenic mesoderm), well-developed tubules never form. Thus one important mechanism of mesonephric tubule morphogenesis is induction by the pronephric duct. The nature of induction is explored in Chapter 11.

The dorsal aorta branches into fine vessel clusters called *glomeruli* that contact the mesonephric tubules. The portion of each tubule that contacts the glomerulus expands and invaginates to form *Bowman's capsule*. Wastes are filtered from the blood in the glomeruli through Bowman's capsules on the mesonephric tubules and into the tubules proper. Each Bowman's capsule plus its glomerulus is a functional renal unit called a *Malpighian body*. This is more complex than the situation in the pronephros, in which wastes could be picked up from the coelom area by the nephrostomes of the pronephric tubules. Kidney function in the mesonephros (and also in the more advanced *metanephros*) is not confined to simply filtering wastes from the glomeruli into Bowman's capsules. Many additional processes occur in the more advanced kidneys, including the reabsorption of nutrients, hormones, and other substances from the kidney tubules back into the blood.

The mesonephros is the functional kidney of adult fish and amphibians. Some individuals term the mesonephric kidney an *opistonephros* because it resembles both the mesonephros and the metanephros. It resembles the metanephros in that it lacks nephrostomes and in that the mesonephric tubules are connected to large collecting ducts.

It should be stressed that the mesonephric kidney is not only present and functional in adult fish and amphibians, it is also the functional kidney of embryonic reptiles and birds and of mammalian embryos whose wastes are not adequately removed by the mother (such as the pig embryo). The mesonephros does not function as an excretory organ in mammalian embryos from which the mother (through the placenta) effectively removes wastes. Let us now examine the development of the metanephros, the functional kidney of adult reptiles, birds, and mammals.

METANEPHROS

A new duct, the *ureter*, forms as an outgrowth at the point where the mesonephric duct joins the cloaca. The outgrowth begins as an evagination, the *ureteric bud*. The developing ureteric bud grows and branches into the metanephrogenic mesoderm, the posterior portion of the mesomere nephrotome that gives rise to the metanephric kidney. Much as in the development of the mesonephros, the metanephric tubules differentiate from the metanephrogenic mesoderm as a result of induction by the ureter (Figures 10–34, 10–35). This is shown by the fact that metanephric tubules do not develop if the ureteric bud does not enter the metanephrogenic mesoderm. The metanephros is complex, and includes Bowman's capsules and associated glomeruli just as the mesonephros does (Figure 10–33).

In summary, the pronephros, mesonephros, and metanephros develop in the intermediate mesoderm or mesomere. Each consists of tubules connected to a main excretory duct. We shall now turn to the genital system. It will become apparent that some structures in the embryonic kidney system switch roles and become incorporated into the genital system. Some of these mechanisms will be reexamined in the chapter on mechanisms of morphogenesis, Chapter 12.

GENITAL SYSTEM

The gonads function in reproduction, the perpetuation of the species. The germ cells formed in the gonads directly combine to form the new organism. In Chapter 9 we learned that the gonads are of mesodermal origin. What

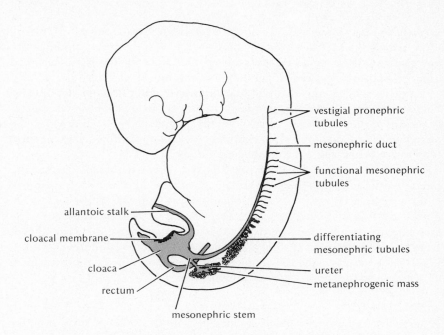

vestigial pronephric
tubules

mesonephric duct

functional mesonephric
tubules

allantoic stalk

cloacal membrane

differentiating
mesonephric tubules

cloaca

ureter

rectum

metanephrogenic mass

mesonephric stem

Figure 10–35. (a) Sagittal section through mammalian embryo, showing the location of three types of kidneys. From J. B. Phillips, *Development of Vertebrate Anatomy* (St. Louis: C. V. Mosby, 1975), p. 439.

about the germ cells? What is their origin? We shall examine this question first, then move on to discuss gonad development.

ORIGIN AND MIGRATION OF THE PRIMORDIAL GERM CELLS

The *primordial germ cells* are the cells that, after many stages of development, give rise to functional gametes. At first one would assume that these cells arise in the developing gonad. This is not the case. These cells apparently arise in different areas in different organisms, then migrate to the gonad area or are transported to this area via the bloodstream. Furthermore, the primordial germ cells are not necessarily of mesodermal origin.

The following is a simple summary of the origin of the primordial germ cells in some groups of vertebrates (Figure 10–36):

Amphibians: Primordial germ cells originate in the vegetal endoderm of the early frog embryo. In salamanders, they originate in the hypomere of the mesoderm.

Birds and Reptiles: Primordial germ cells originate in the extraembryonic endoderm.

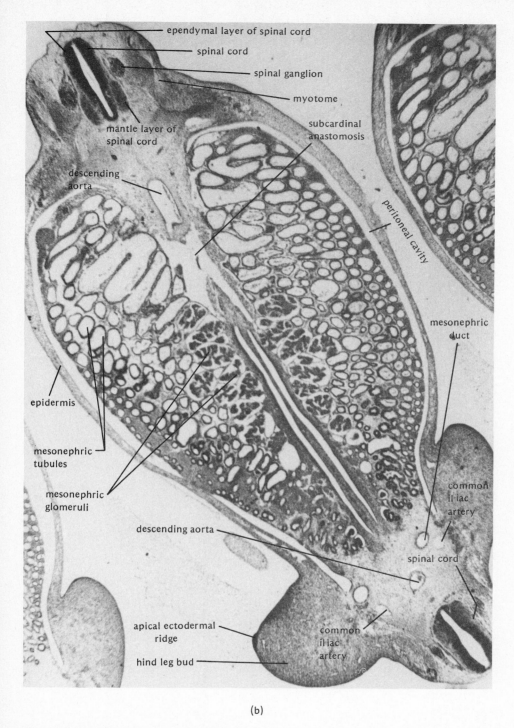

ependymal layer of spinal cord

spinal cord

spinal ganglion

myotome

mantle layer of
spinal cord

subcardinal
anastomosis

descending
aorta

peritoneal cavity

mesonephric
duct

epidermis

mesonephric
tubules

mesonephric
glomeruli

descending aorta

common
iliac
artery

spinal cord

apical ectodermal
ridge

hind leg bud

common
iliac
artery

(b)

**Figure 10–35.(b) Cross section through 10 mm pig embryo, showing
mesonephros.**

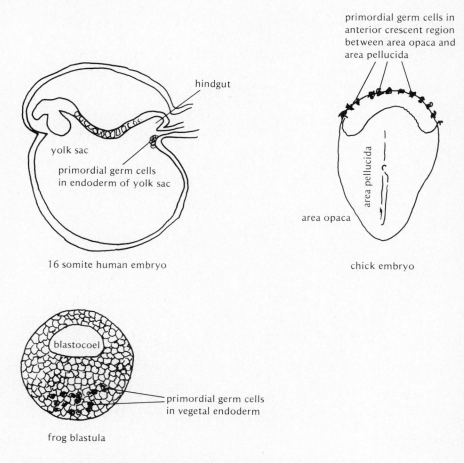

Figure 10-36. Origin of primordial germ cells in some vertebrates.

Human Beings: Primordial germ cells originate in the yolk sac endoderm.

In some invertebrates, however, primordial germ cells are distinguishable before germ layer formation. These cells should not be grouped as being derived from a specific germ layer.

How does one identify the origin of the primordial germ cells? Several types of experiments led to the conclusions above. For instance, the area thought to give rise to the germ cells can be irradiated with ultraviolet light or with X-rays. The area can be excised from the embryo, or killed with a hot needle. In some systems, primary germ cells can be distinguished in the early embryo by using special stains. In the former experiments, when the germ cells are killed, the gonad develops without germ cells. In any case, one can identify the region of the early embryo that gives rise to the germ cells, and it is not the gonad.

A second question involved is: How do the primordial germ cells get to the gonad? The mechanisms involved are not fully understood. In birds, primordial germ cells enter the blood vessels for transport to the gonad area. In frogs, primordial germ cells migrate in the dorsal endoderm to the gonad area. In mice and humans, the movement of the primordial germ cells from the endoderm to the gonad area can be traced by staining the embryo for alkaline phosphatase, an enzyme localized in the primordial germ cells. Primordial germ cells get to their destination by migration or blood transport. This is reminiscent of the ways in which tumor cells spread in the body—by invasion (migration) or metastasis (blood transport). Interesting comparisons of this sort between tumor and embryonic cells will be discussed in Chapter 16.

To say that primordial germ cells get to their destination by mechanisms such as migration or blood transport is only part of the explanation. How does a migrating primordial germ cell know where to stop? Is it possible that specific adhesive recognition occurs between the primordial germ cell and the region of destination? That is, do primordial germ cells stop because they specifically stick to the gonad area? The mechanism of specific adhesive recognition will be described in more detail in Chapter 12.

In summary, primordial germ cells in vertebrates appear to originate in the endoderm or mesoderm. However, these cells should perhaps not be classified as germ layer derivatives per se; instead they should be considered as special cell types apart from the germ layers. The mechanisms by which germ cells arrive at their destination in the genital region include migration and blood transport. Specific adhesive recognition may help determine where the germ cells stop. Let us now turn to gonad development.

GONAD DEVELOPMENT

In the previous section we noted that the primordial germ cells originate outside the gonad area and migrate there by such mechanisms as migration and selective cell adhesion. Where is this gonad region? How does the gonad develop? In the following discussion, keep in mind that the gonads and associated structures usually develop in pairs, just as the kidneys do.

As indicated in Chapter 9, in vertebrates the gonads arise in the posterior intermediate mesoderm—the region right above the posterior splanchnic mesoderm of the hypomere. The primordial germ cells migrate into this region of mesoderm, which then thickens into a structure called the *germinal ridge*. Some researchers feel that there is no conclusive evidence that the promordial germ cells that migrated into the germinal ridge area give rise to the gametes. These researchers hold that the primordial germ cells simply stimulate gonad development, and that the gametes arise from the gonad mesoderm. The experiments discussed in the last section, however, suggest that the primordial germ cells are indeed gamete precursors. No proof to the contrary has yet materialized.

INDIFFERENT GONAD

There is some controversy about the events that occur in the early development of the gonads. The surface of the germinal ridge, the *germinal epithelium*, contains some of the primordial germ cells. There is some evidence that the germinal epithelium buds off to form the *primary sex cords*. Others believe that the primary sex cords develop from strands of mesoderm from neighboring areas. In any case, as seen in Figure 10–37, the primitive gonad now consists of an outer germinal epithelium (or *cortex*) and an inner region, the medulla, made up of the primary sex cords. Note that at this stage in development, the gonad is neither an ovary nor a testis; instead, it is called an *indifferent gonad*, and this stage the *indifferent stage*.

The indifferent stage is characterized not only by the indifferent gonad, but by the presence of all the ducts and tubules present in adult males and adult females. In genetic males, certain parts of the gonad and certain ducts degenerate before adulthood. In genetic females, other parts of the gonad and certain other ducts and tubules do so. Thus, the genital system of males and females develops as the result of death in some structures and growth in others. Let us examine this absorbing story in a little more detail.

GENETIC FEMALES

In genetic females, the primary sex cords of the gonad degenerate and the inner *medulla* becomes reduced. The outer region, or cortex, develops instead. The primordial germ cells contained in the cortical region cluster in groups

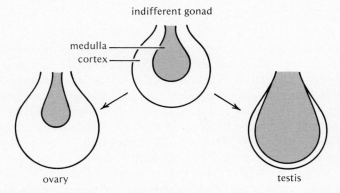

Figure 10–37. Differentiation of the indifferent amphibian gonad into an ovary or testis. Relative predominance of cortex or medulla is shown. From R. K. Burns (after Witschi) in *Survey of Biological Progress I* (New York: Academic Press, 1949).

and become surrounded by follicle cells that protect and nourish the developing oocytes. These clusters of cells are called *secondary sex cords* or *nests of oogonia*. The female gametes or eggs develop from oogonia that undergo maturation as described in Chapter 4.

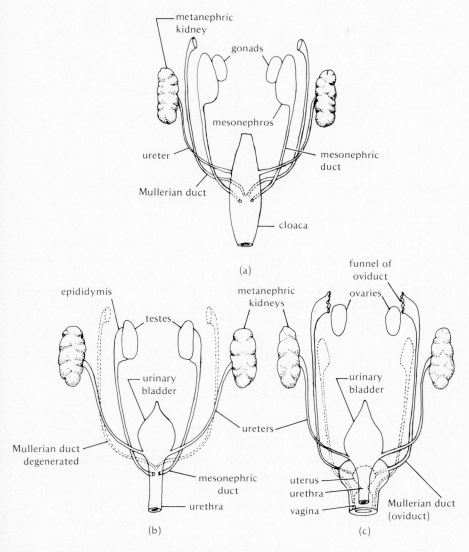

Figure 10–38. Diagram showing the transformations of the genital ducts in mammalian embryos from an indifferent stage (a) to the male (b) and female (c) conditions. From B. I. Balinsky, *An Introduction to Embryology*, 4th ed. (Philadelphia: W. B. Saunders, 1975), p. 437.

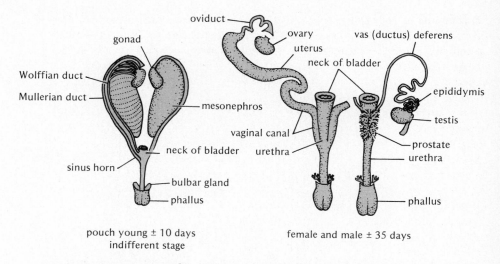

Figure 10–39. Sexual differentiation in young opossums. (a) Indifferent stage (about 10 days). (b) Female and male embryos (about 35 days). From R. K. Burns, *Survey of Biological Processes I* (New York: Academic Press, 1949), 233–266.

The urogenital ducts present at the indifferent stage of sexual development include the mesonephric duct with its tubules and Bowman's capsules (described earlier) and the *Müllerian ducts*. The Müllerian ducts develop in the intermediate mesoderm, grow posteriorly, and fuse with the cloaca. In genetic female reptiles, birds, and mammals, the mesonephric ducts and associated tubules degenerate. The Müllerian ducts possess a funnel-shaped anterior opening close to the ovary, called the *ostium tubae*. As eggs are ovulated, they fall into the ostium tubae and are carried down the Müllerian ducts. Since the Müllerian ducts transport eggs, they are the *oviducts* of the female (Figures 10–38, 10–39).

Thus, in genetic female vertebrates, the primary sex cords of the gonad degenerate and the cortex of the gonad becomes the important gamete-producing region. Secondary sex cords, or nests of oogonia, develop in the cortex of the ovary. The mesonephric ducts and tubules degenerate in reptiles, birds, and mammals; the Müllerian ducts persist, forming the functional oviducts. Clearly, the adult gonad with its associated structures develops as a result of the degeneration of some embryonic regions and the persistence and development of others. After considering the development of the male gonad and accessory structures, we shall briefly examine some of the mechanisms that appear to control genital development.

GENETIC MALES

In genetic males, the inner region of the gonad, the medulla, develops. This region contains the primary sex cords. The primordial germ cells in the primary sex cords develop into spermatogonia (see Chapter 4). Spermatogonia form the male gametes, the spermatozoa. As development proceeds, the primary sex cords become hollow structures called the *seminiferous tubules*, which contain the primordial germ cells. The tubules connect to adjoining cell groups (*rete cord cells*), which also hollow out to form a network of tubules called the *rete testis*. The tubules of the rete testis connect to the mesonephric tubules, and then join the mesonephric duct. Sperm are therefore released from the seminiferous tubules into the ducts of the testis and finally into the mesonephric duct that is connected with the cloaca (Figures 10–38, 10–39). Here is a perfect example of how some ducts and structures originally associated only with the urinary system are incorporated into the reproductive system. We should also note that in genetic males, the Müllerian ducts degenerate after the indifferent stage of sexual development. In addition, remember that the mesonephric duct maintains both a urinary function and a reproductive function in fish and amphibians. In reptiles, birds, and mammals, however, the mesonephric duct carries only sperm and therefore is called the sperm duct or *ductus deferens*. The urinary function is taken over by the metanephros with its associated ureter.

To summarize, the medulla of the gonad develops in genetic male vertebrates, and seminiferous tubules form from the primary sex cords. The mesonephric ducts persist and act as sperm ducts, and the Müllerian ducts degenerate.

MECHANISMS OF GONAD DEVELOPMENT

In Chapter 12 we shall examine examples of mechanisms that are of general importance in controlling morphogenesis. Here, we shall look into the mechanisms that control some of the changes that occur during genital development in each sex. We noted that in both genetic males and genetic females, structures of both male and female reproductive systems are present at the indifferent stage of sexual development. After the indifferent stage, some of these structures degenerate and others develop, depending upon the specific sex of the individual. What causes these changes? The answer to this question is not fully understood.

Sex hormones probably control some of these changes. As will be seen in Chapter 12, such hormones can control the level of lysozomal enzymes in cells. Lysozomal enzymes are degradative enzymes (cathepsins, acid phosphatase, ribonuclease) carried in cell digestive organelles called lysozomes. The levels of these enzymes increase in the sexual ducts that degenerate. For example, before the Müllerian ducts degenerate in genetic males, the level of these enzymes increases in these structures. These enzymes may directly degrade the structures.

It was suggested above that sex hormones influence the levels of these enzymes in the structures. Injections of male hormones (androgens) into genetic females causes the Müllerian ducts to break down. If, however, female hormones (estrogens) are inoculated into genetic males the Müllerian ducts persist (Figure 10–40). In any case, it is unclear whether these hormones

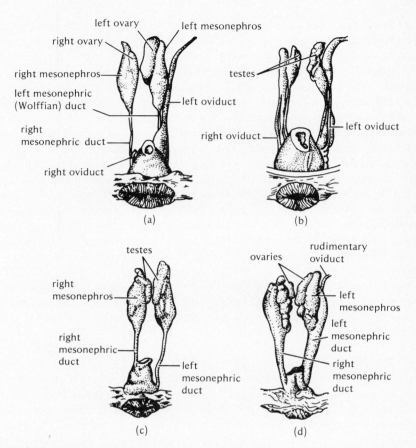

Figure 10-40. Effect of sex hormones on sex-duct development in chick embryos.
(a) Normal female embryo, no added hormones, eighteen-day incubation. (b) Male embryo treated with female hormone. Both oviducts present and enlarged. (c) Normal male embryo, no added hormones, seventeen-day incubation, no oviducts. (d) Female embryo treated with male hormone. Oviducts mostly absent, mesonephric ducts enlarged. From Willier, Weiss, and Hamburger, *Analysis of Development* (Philadelphia: W. B. Saunders, 1955). © 1955 by the W. B. Saunders Co., Philadelphia, Pa.

play a key role in the early sexual differentiation of the embryo. Some evidence suggests that they do. For example, Frank Lillie observed twin cattle with intermingled fetal blood supplies. If a genetic male and a genetic female embryo developed together, the genetic female becomes masculinized. The male twin apparently releases male sex hormones that modify the female. In another experiment, a piece of a gonad of one sex was transplanted near the developing gonad of an embryo of the other sex. The genital structures of the recipient were modified in the direction of the sex of the donor gonad. These experiments, however, do not prove that sex hormones control genital differentiation in the early embryo. Future work will definitively indicate the factors that control the differentiation of genital structures in the embryo.

H-Y ANTIGEN. The testis is organized into seminiferous tubules, which produce sperm. These tubules contain germ cells in various stages of meiosis and Sertoli cells that support and nourish the developing gametes. Ohno and colleagues have elucidated some of the mechanisms that control the formation of seminiferous tubules in male mammals. Apparently, a specific antigen, H-Y antigen, plays a key role in testis morphogenesis.

H-Y antigen is a protein embedded in the plasma membrane of male cells containing a Y chromosome. Males of all the mammal species examined possess the H-Y antigen, whereas female cells do not. In Ohno's experiments, a young mouse testis was disaggregated into single cells by the proteolytic enzyme trypsin. The cells were then treated either with a specific antibody that removes H-Y antigen from the cells, or with an antibody that does not

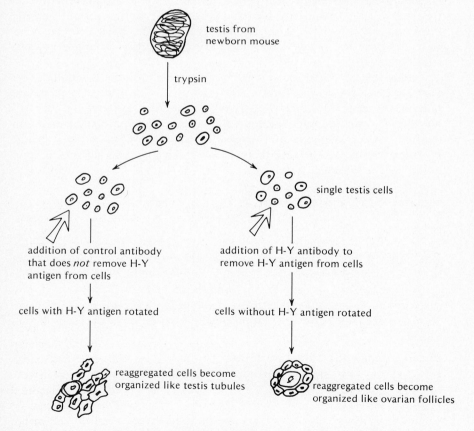

Figure 10–41. The importance of H-Y antigen in controlling testis morphogenesis.
Based on experiments by Ohno and colleagues (*J. Am. Med. Assoc.* 239 (1978: 217–220).

remove the antigen. Each batch of testis cells was then reaggregated in rotary culture. The reaggregated cells possessing H-Y antigen formed structures resembling normal seminiferous tubules. The reaggregated cells without H-Y antigen formed structures resembling the organization of ovaries (Figure 10–41). These ovary-like structures consisted of single primordial germ cells surrounded by follicle-like structures formed from Sertoli cells. This experiment strongly suggests that H-Y antigen is directly involved in testis morphogenesis and organization. One would speculate that the Y chromosome in males codes for H-Y antigen, that this antigen becomes embedded in the cell surface, and that it causes an indifferent gonad to form the tubule organization of a testis. Exactly how H-Y antigen works, however, is not well understood.

The H-Y antigen (or one very similar) has been identified in many organisms, not just in mammals. In *Xenopus laevis*, the male possesses the XX genotype and the female the XY genotype. Here, the H-Y antigen is associated with females. It appears to cause the indifferent gonad to form ovarian tissue. It is not difficult to understand how H-Y antigen could become evolutionarily associated with the heterogametic sex. How it could cause testis organization in one case and ovary organization in another is less straightforward.

Some recent studies, however, have cast doubt on the contention that H-Y antigen is responsible for sex determination. In one study, a male mouse mutant was reported to be H-Y negative. In another, at least one-third of all XY mice of a specific strain developed as phenotypically normal females with ovaries, even though these XY females possessed H-Y antigen. Thus, H-Y antigen may not be absolutely necessary for inducting testis formation in the undifferentiated gonad.

SECTION 6 DEVELOPMENT OF THE IMMUNE
SYSTEM

We fight off infectious diseases and probably even such diseases as cancer with our immune system. This system contains cells that travel throughout our tissues, seeking out and destroying foreign cells and substances. It also produces antibodies that specifically combine with foreign invaders of all sorts. Let us first see how this system develops and then briefly discuss how it works.

DEVELOPMENT OF TISSUES AND
CELLS OF THE IMMUNE SYSTEM

CHICK EMBRYO
In the chick embryo, two glands are associated with the early development of cells that respond to foreign substances (antigens). These organs are the

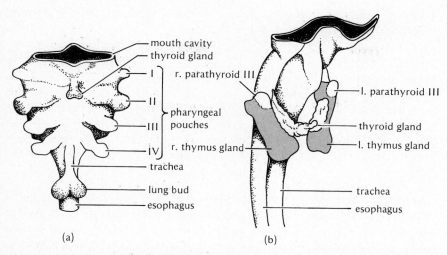

Figure 10–42. Derivation of the thymus. (a) Parts of the third and fourth pharyngeal pouches (endoderm) that migrate down to the chest cavity (b) form the thymic lobe. From G. L. Weller, *Contrib. Embryol. Carneg. Inst.* 24 (1933): 93.

thymus gland and the *bursa of Fabricius*. The thymus gland, as mentioned in Chapter 9, develops from the pharyngeal pouches (Figure 10–42). The bursa of Fabricius develops as an outpocketing of the hindgut (Figure 10–43).

The stem cells that give rise to the cells of the immune system are derived from the yolk sac wall in the chick embryo. They have darkly staining basophilic cytoplasm. The stem cells enter the rudiment of the thymus after six to seven days of incubation in the chick embryo. The stem cells give rise to white blood cells called *lymphocytes*. Large numbers of medium and small lymphocytes can be observed in the thymus gland after 12 days of incubation.

Stem cells originating in the yolk sac enter the bursa of Fabricius after

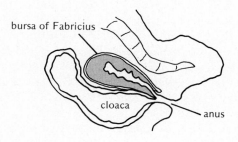

Figure 10–43. Bursa of Fabricius in chicken. From Burnet, *Self and Not Self* (New York: Cambridge University Press, 1969).

daltons), is not found until hatching. We shall return to these and other antibodies shortly.

MOUSE EMBRYO

In the mouse embryo, stem cells that "seed" the organs of the immune system also originate in the yolk sac wall. These cells travel from the yolk sac to the fetal liver and finally to the bone marrow. As in the chick embryo, the lymphoid stem cells of the mouse embryo can be identified by their large size and darkly staining cytoplasm. The stem cells enter the thymus after eleven days of gestation. After sixteen days, lymphocytes can be observed in the fetal liver.

ORIGIN OF T AND B LYMPHOCYTES

We saw that two major glands in the chick embryo become seeded with lymphoid stem cells, the thymus and the bursa of Fabricius. Similarly, the mouse thymus is seeded by lymphoid stem cells. A bursa of Fabricius has not been identified in the mouse, but tissue with a similar function is present. The lymphocytes produced by the thymus and bursa in the chick embryo have different functions. The thymus lymphocytes (*T cells*) play a major role in *cellular immunity*. Cellular immunity involves a direct attack by a T cell on a foreign cell. The bursa lymphocytes (*B cells*) are active in *humoral immunity*. Humoral immunity involves the synthesis of free soluble antibodies that combine with foreign antigens.

T cells mature in the thymus from the stem cells and become able to respond to foreign cells (become competent). These T cells then enter the circulatory system. B cells (in the chick embryo) mature in the bursa from the stem cells that originally seeded it. Around the time of hatching, B cells from the bursa enter the circulatory system and seed the bone marrow, spleen, and lymph nodes. The bone marrow is the major source of proliferating B cells in the adult. The origin of the B cells that confer humoral immunity in the mouse embryo is not well understood. It is assumed that the B cell population developed from stem cells somewhere in the body (possibly in the fetal liver) and then seeded the bone marrow. The likely source of B cells in adult mammals is thus the bone marrow.

The role of the bursa and thymus in conferring immunity has been shown by surgical removal experiments. If the bursa is removed from a chick embryo before the B cells were released for seeding other organs, the immune response fails to develop. After the B cells are released from the bursa, however, this gland can be removed without affecting the immunity response. Similarly, if the thymus of, for example, a mouse embryo is removed before the T cells have been released (at birth), the immunity is decreased. After birth when T cells have already been released into the circulation, the thymus can be removed with little effect on immunity. What do we mean by immunity? How do T and B cells function?

FUNCTION OF T AND B CELLS IN THE IMMUNE RESPONSE. We have seen that T and B lymphocytes mature in the thymus and (in birds) the bursa respectively. These cells are released into the circulatory system and seed other lymphoid tissues. We also mentioned that the T cells play a major role in cellular immunity, while the B cells form circulating antibodies (humoral immunity). How do the T and B cells respond to foreign antigens during the immune response? A very large variety of substances might elicit an immune response. How can the cells of the immune system respond to these substances in a specific manner? Let us begin with the B cell system responsible for humoral immunity.

HUMORAL IMMUNITY

B CELL CLONES

B cells synthesize antibodies in response to antigenic stimulation. Mature antibody-secreting B cells are called *plasma cells*. Antigenic substances (substances able to elicit the production of antibodies) include proteins, glycoproteins, and polysaccharides that are folded in complex molecular patterns (see Chapters 14 and 15). The B cells recognize some of these patterns as foreign, and synthesize an antibody that specifically combines with the antigen.

Several hypotheses have been developed to explain the specific response of B cells to antigenic stimulation. One of the most widely accepted of these was the *clonal selection hypothesis* developed by Sir Macfarlane Burnet and others (Figure 10-44). As we shall discuss later, this hypothesis is not restricted to B cells. The hypothesis assumes that each B cell can produce only one kind of antibody, but that each B cell has all the genes needed for producing all types of antibody. A B cell is restricted or determined for a single response. In other words, only one set of genes that regulate the production of one kind of antibody can be activated. Other genes for other antibodies remain inactive throughout the life of the cell.

The fate of B lymphocytes is thus fixed so that they can produce only one type of antibody. When an antigen comes along that can elicit a response in a given determined lymphocyte, this antigen interacts with the B cell surface. Such interaction occurs because the committed lymphocyte has already synthesized a small amount of specific antibody, and this antibody was incorporated into the surface membrane of the B cell. The interaction between the specific antigen and the surface antibody stimulates the lymphocyte to divide, resulting in a clone of cells that possess the specific surface antibody and are committed to its synthesis (Figure 10-45). In this way, a large population of specific antibody-forming cells can be produced to respond to a given antigen. Large quantities of antibody are thus produced, permitting an effective immune response upon antigenic stimulation.

There is good evidence that the clonal selection hypothesis is correct. For example, if lymphocytes are exposed to two different antigens, and then

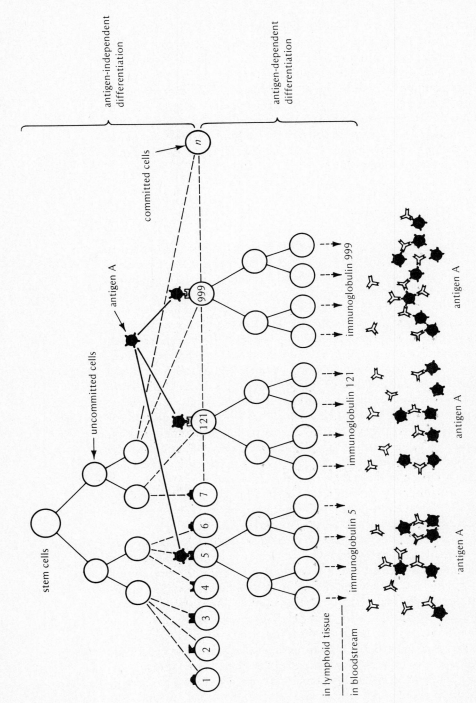

Figure 10-44. Clonal selection theory of antibody formation. From G. Edelman, The structure and function of antibodies, *Scientific American 223* (1970): 34. Copyright © 1970 by Scientific American, Inc. Used with permission.

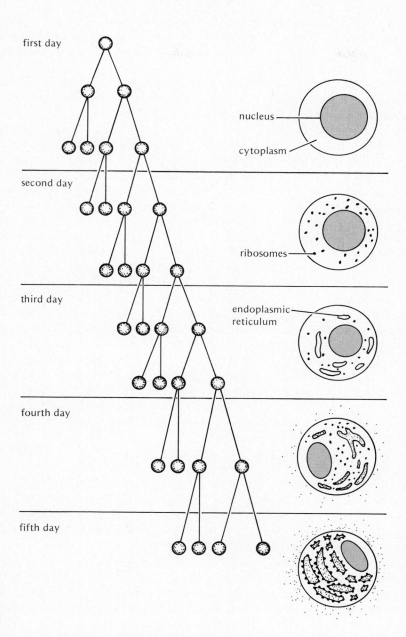

first day

nucleus

cytoplasm

second day

ribosomes

third day

endoplasmic
reticulum

fourth day

fifth day

Figure 10-45. Development of B plasma cells. The mature plasma cell has extensive endoplasmic reticulum and a great deal of antibody. From G. J. V. Nossal, How cells make antibodies, *Scientific American*. Copyright © 1964 by Scientific American, Inc. Used with permission.

twelve to thirteen days of incubation. These stem cells are similar to those that entered the thymus at six to seven days. After seventeen days of incubation, many lymphocytes are present in the bursa, as observed earlier in the thymus.

After fourteen days of incubation, the production of antibody by lymphocytes can be detected. The type of antibody produced is termed *IgM*, or *immunoglobulin M*. It has a molecular weight of about 900,000 daltons. Another major type of antibody, *IgG* or *immunoglobulin G* (molecular weight, 150,000 single B cells are placed in microdrops, each microdrop contains antibody against either antigen 1 or 2, but not both. Thus, one B cell can only respond to one antigen.

Humoral immunity is, however, not totally dependent upon B cells and their descendants. If the thymus is removed at an early stage, the production of circulating antibodies in response to some antigens was less effective, but not absent. In addition, if B cells alone are inoculated into animals that were previously irradiated to destroy their immune systems, the animals showed no effective reaction to antigens. Inoculation of B plus T cells, however, permitted effective antibody production by the B cells. Thus it seems that T cells and perhaps other white blood cells such as *macrophages* assist the B cells in their role of producing specific antibodies. At present it is believed that a strong humoral immune response involves several cell types. Macrophages receive antigenic stimulation and in some way process it. A signal from the macrophages is then sent to T cells. The T cells then interact with B cells, stimulating them to produce large quantities of specific antibody. The basis of these interactions is not well understood.

Now that we have surveyed the cell types involved in antibody production, a brief examination of antibodies is appropriate. We shall then move on to look at T cell mediated immunity.

ANTIBODIES

We briefly mentioned two types of circulating antibodies, IgM and IgG. Another type of blood antibody is *IgA*. Other classes of antibodies (*IgD* and *IgE*) are minor components of the serum and will not be described here. IgM has a molecular weight of about 900,000 daltons, while the molecular weight of IgA, like IgG is about 150,000 daltons.

IgG is the major immunoglobulin in the serum of immunized adult mammals such as humans. Each IgG molecule is composed of four polypeptide chains held together by disulfide bonds and noncovalent associations (see Figure 10–46). Two of these chains, called *light chains*, are identical and have a molecular weight of 23,000 daltons. The other two, also identical, are called *heavy chains*. The heavy chains have a molecular weight of 53,000 daltons and are thus more than twice the size of the light chains.

The important functional sites on a specific IgG molecule, the antigen-combining sites, possess unique amino acid sequences not found on an IgG

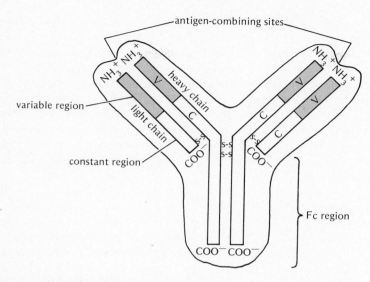

Figure 10-46. Immunoglobulin molecule. The antigen combining sites are located between the V regions of the light and heavy chains. From Hood, Weissman, and Wood, *Immunology* (Menlo Park, CA: Benjamin/Cummings, 1978), p. 5. Used with permission.

molecule made against a different antigen. These sites, the *variable regions*, consist of part of the light and heavy chains. The remaining portion (the *constant region*) has generally identical amino acid sequences.

The variable regions of the antibody molecule account for the specificity of antibodies in combining with antigens. How do B cells make antibodies whose polypeptide chains consist of a constant plus a variable region? This is an important puzzle that is not yet fully understood. One would expect that a single gene codes for a single polypeptide chain. But how can a single gene code for a constant amino acid sequence plus a variable amino acid sequence in one polypeptide chain?

Two genes code for each polypeptide chain in immunoglobulin molecules. The experimental evidence that supports this contention includes molecular hybridization work (Chapter 13). For example, light chain messenger RNA has complementary base sequences with two different DNA fragments of embryonic cells. Also, by sequencing DNA it has been shown that variable region genes are not adjacent to constant region genes.

How are polypeptide chains with products from two separate genes constructed in antibody-forming cells? Genes or gene products could be combined at the DNA, RNA, or polypeptide level. Recent work suggests that some of the combining occurs at the RNA level. Apparently, high molecular weight nuclear RNA undergoes processing that removes the RNA between the variable and constant immunoglobulin message sequences. This processing brings the variable region and constant region RNA message sequences together. This combined message is then transported to the cytoplasm for translation into immunoglobulin polypeptide.

Variable and constant region genes may also be combined at the DNA level. This would explain the mechanism by which a given B cell synthesizes only one type of specific immunoglobulin molecule in response to only one type of antigen. The rearrangement of the light and heavy variable and constant genes in the DNA could activate the transcription of the combined variable-constant gene sequence. This could be a critical step in the differentiation of antibody-forming cells. An excellent discussion of the problem of antibody synthesis and antibody-forming cell differentiation is found in Hood, Weissman, and Wood (1978).

CELLULAR IMMUNITY

As indicated earlier, two systems of immunity protect vertebrates against foreign invaders: the humoral response and the cellular response. The humoral immune response is based upon antibodies formed by B plasma cells. The antibodies circulate around the body in the blood serum. The cellular immune response is mediated by the T cells.

T cells, like B cells, have antigen receptors on their surfaces. The B cell receptors are known to be specific antibodies. The molecular nature of the T cell receptor, however, is not known. Like B cells, each T cell carries only one kind of surface antigen receptor that responds to only a few closely related antigen determinants. Antigen stimulated T cells give rise to *killer T cells* that directly kill foreign cells. Other types of T cells have other roles, such as helping B cells differentiate and proliferate.

The clonal selection hypothesis accounts for the large-scale response of T cells to specific antigens just as it does for that of B cells. The initial encounter with the antigen causes T cell clones to proliferate and differentiate. The specific antigen probably interacts with the surface of the T cells that have receptors for that antigen. The proliferating T cells form a clone of T cells that respond to the specific antigen. The B cell response is to synthesize specific antibodies, the T cell response (killer T cells) is to directly attack the invading cell. In both T and B cells, some "memory" cells retain the capacity for continued proliferation upon encountering the given antigen a second time. The other cells are short-lived and die a few days after carrying out their task, whether it be forming an antibody (B plasma cells), or killing the invading cells (killer T cells).

Maturation of T cell lines involves establishing a microenvironment in the embryonic thymus that stimulates maturation, seeding the thymus by T cell precursors that arise in the yolk sac, proliferating the T cell precursors and differentiating them into mature T cells in the thymus, and releasing the T cells into the blood stream for migration to the peripheral lymphoid tissues (Figure 10-47).

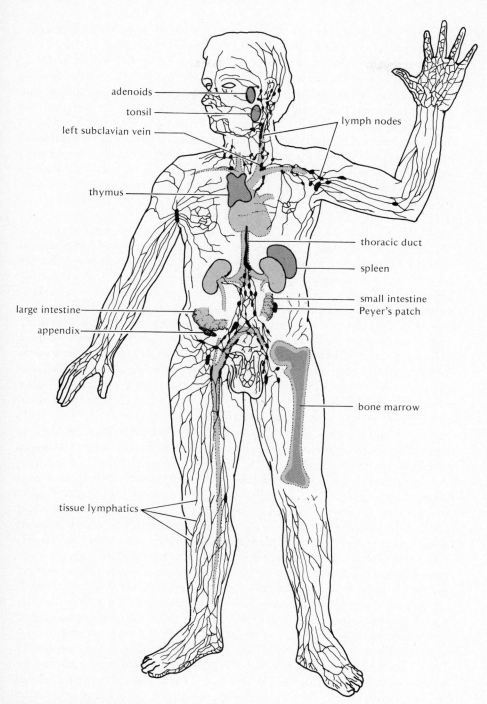

Figure 10–47. Human lymphoid system. From N. K. Jerne, The immune system, *Scientific American* 229 (1973): 52. Copyright © 1973 by Scientific American, Inc. Used with permission.

Before we conclude our discussion we should mention that a T cell can be distinguished from a B cell by detecting specific cell surface markers. For example, in mice, antigens Ly-4 and Pc-1 are specific for the B cell line. Ly-4 is found on young B cells; whereas Pc-1 is found only on differentiated B cells (plasma cells). Thy-1 is a membrane glycoprotein antigen found on T line cells, but not on B cells. Many T cells (but not B cells) carry the antigen Ly-1 or Ly-2 and Ly-3.

Antigens that enter the body through the upper respiratory and gastrointestinal tract are filtered through the local lymph nodes and through specialized lymphoid organs such as the tonsils, adenoids, appendix, and Peyers patches (Figure 10–47). Antigens that enter the bloodstream are filtered out by macrophages, the large white blood cells that line the blood vessels in the spleen, lungs, and liver. Lymphocytes (T and B cells) pass through the lymphoid organs and back into the bloodstream. In the lymphoid organs the T cells and B cells contact antigens and macrophage-processed antigens. T cells reside in the diffuse surface cortex of the lymph nodes; whereas B cells migrate to discrete regions in the lymph nodes called follicles. B cells and T cells thus recognize specific regions of the lymph nodes. Contact with antigen causes the T and B cell clones to proliferate and differentiate.

When stimulation by specific antigens causes the T and B cells to mature, large numbers of killer T cells and antibodies produced by B-plasma cells are distributed by the lymph vessels to the bloodstream and body tissues. The killer T cells accumulate in large numbers on the blood vessel walls near the sites where the specific antigens invaded the tissues. The killer T cells crawl through the blood vessel walls and enter the tissues, where they attack the cells or other entities that contain the antigen. Most B-plasma cells remain in the lymph node, secreting antibody that enters the circulation; most killer T cells enter the circulation and travel to distant sites to attack the foreign antigens (Figure 10–48).

In summary, the development of the immune system involves a complex series of steps that form active cells (such as B-plasma cells and killer T cells) that respond to specific antigens. These final cell types are derived from B and T cells, which in turn arise from stem cells in specific microenvironments such as those in the thymus, bursa of Fabricius, bone marrow, and fetal liver. Clones of B-plasma cells, killer T cells, and other associated cell types form in response to antigenic stimuli. The antigen interacts with specific surface receptors in the B and T cell lines. Each B cell responds to only one type of antigen and can only synthesize antibody against this antigen. The mechanism by which a given B cell is programmed to synthesize only one kind of antibody may involve the translocation of variable and constant region genes at the DNA level to bring them close to each other. This could commit the cell to exclusive production of a particular immunoglobulin molecule by activating the transcription of the complete antibody genes that are adjacent to each other. Such DNA modification during cell differentiation in the

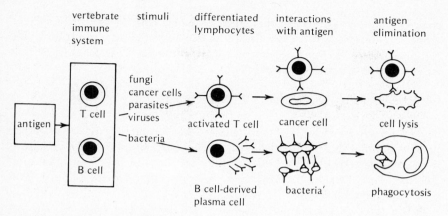

Figure 10–48. Reactions that occur upon stimulating T and B cells. From Hood, Weissman, and Wood, *Immunology* (Menlo Park, CA: Benjamin/Cummings, 1978), p. 2.

immune system is extremely important. DNA modification may be as important as simple activation and repression of genes in the differentiation of many different cell types.

In Chapter 12, we shall see that cell surface receptors can often move laterally in the plane of the cell membrane. Antigens that bind to T and B cell surface receptors could move them together, in some way triggering events in the nucleus that lead to clone proliferation and specific antibody synthesis (in B cells). This idea is supported by the finding that multivalent molecules called lectins (see Chapter 12) can stimulate the differentiation of T and B cells. These lectins bind to cell surface molecules, drawing them together (Figure 10–49).

The sequence of events in the differentiation of antibody-producing cells can be summarized as follows. B cell precursors first manufacture heavy chains, then light chains, according to the mechanisms described so far in this

(a) immunoglobulin or lectin receptors with random lateral mobility

(b) patching with divalent or multivalent antigen or lectin

(c) movement of patches toward the cell pole

(d) capping at the cell pole

Figure 10–49. Movement of antigen receptors (such as immunoglobulins) in the cell surface of a lymphocyte caused by the binding of multivalent antigens. From Hood, Weissman, and Wood, *Immunology* (Menlo Park, CA: Benjamin/Cummings, 1978), p. 34.

chapter. The heavy and light chains combine, and IgM and IgD are displayed at the cell surface. When a specific antigen binds to the surface antibody, the cell proliferates, forming a clone of mature B lymphocytes that are specialized to synthesize the specific antibody (see Figure 10–50).

GENETICS OF ANTIBODY DIVERSITY

The immune system displays a mechanism of differentiation that may also be present in other developmental systems. Not long ago, this mechanism was not taken seriously by the scientific community because of its unusual nature. The data, however, are now in hand, and this rather remarkable process is now firmly established.

We noted earlier that both the heavy and the light polypeptide chains of

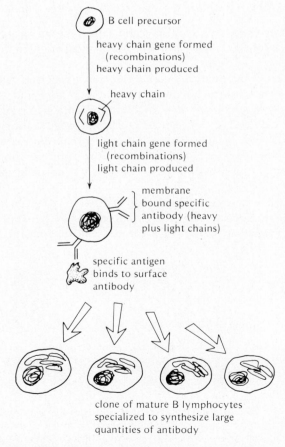

B cell precursor

heavy chain gene formed
(recombinations)
heavy chain produced

heavy chain

light chain gene formed
(recombinations)
light chain produced

membrane
bound specific
antibody (heavy
plus light chains)

specific antigen
binds to surface
antibody

clone of mature B lymphocytes
specialized to synthesize large
quantities of antibody

Figure 10–50. The differentiation of antibody-producing cells.

immunoglobulins contain both a variable region and a constant region. Messenger RNA coding for the light and heavy chains contains sequences for both the constant and variable regions. Thus, the presence of both regions in a light or heavy chain is not the result of joining separately synthesized constant and variable amino acid segments. Instead, both regions are coded in each messenger RNA molecule. How are both a constant and a variable gene sequence incorporated into a single mRNA? The answer is that two separate DNA sequences (constant and variable genes) are joined together. A single mRNA containing both of these sequences is then transcribed from that joined sequence. Let us consider this so-called gene joining or *jumping genes* mechanism in more detail.

A human light chain is formed as follows: The variable region of the light chain is coded in so-called V and J sequences of DNA, as shown in Figure 10–51. The constant region, in turn, is coded by a C gene. There are about 150 alternative V sequences, each separated by a short intervening

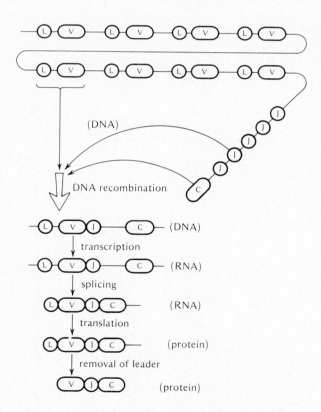

Figure 10–51. The formation of an antibody light chain, showing how genes are recombined to allow vast diversity in the structure of antibodies.

sequence from a leader (L) sequence. The L and V segments are separated by a long noncoding sequence of DNA from 5 J (joining) sequences. The J sequences are separated from a single C gene by an intervening sequence (Figure 10–52).

During the development of lymphocytes, one of the V genes with its L sequence is recombined with one of the J sequences and, together with the single C gene, forms an active light chain gene. This light chain gene is transcribed into a primary RNA molecule. The intervening sequences are then cut out, yielding a light chain mRNA that can be translated into the light chain precursor polypeptide. As the light chain passes through the cell membrane, the leader is cleaved away. This leader, which is hydrophobic (water-repellent), may help transport antibody through the cell membrane.

If one of about 150 or so V genes is joined with one of 5 J genes, 150×5 or 750 different genes for a light chain variable region can be formed. Additional variation comes from the different places where the V and J genes can be joined. If we assume that such alternative joining sites increase diversity about tenfold, then the total number of possible V–J combinations is about 750×10, or 7500.

Heavy chain diversity is brought about by the same mechanisms. The

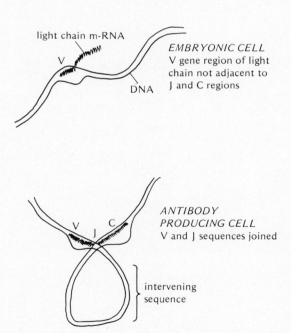

light chain m-RNA

EMBRYONIC CELL
V gene region of light chain not adjacent to J and C regions

DNA

ANTIBODY PRODUCING CELL
V and J sequences joined

intervening sequence

Figure 10–52. Hybridization technique using specific light chain messenger RNA. This technique permits the visualization of antibody genes with the electron microscope.

heavy chain variable region, however, consists of a D gene along with V and J segments. Thus, the heavy chain diversity is even greater than that of light chains. According to estimates, as many as 80 heavy chain V genes, 6 J genes, and 50 D genes exist in humans. Thus, $80 \times 6 \times 50$, or 24,000, combinations can form. Assuming about 100 gene combination points, there are about $24,000 \times 100$, or 2.4 million possible different heavy chains.

Thus, there are about 7500 possible light chains and 2.4 million possible heavy chains, yielding a total of 7500×2.4 million, or 18 billion possible human antibodies. Thus, some 300 separate genetic segments in embryonic DNA can give rise to 18 billion different antibodies. Mutations in the variable region genes would increase the diversity even more.

The gene shuffling that we have described might play a role in developing diversity in systems other than the immune system. Investigation of the role of such "jumping genes" in other systems will surely yield new insights into the mechanisms of embryonic differentiation.

HYBRIDOMA TECHNOLOGY AND MONOCLONAL ANTIBODIES

Monoclonal antibodies, because of their exquisite specificity and homogeneity, have been lauded as having great potential in many areas, including the analysis of embryonic development and the diagnosis and treatment of cancer. The specificity and homogeneity of these antibodies arises because they are derived from descendants of a single cell (hence the term monoclonal). Conventional antibodies are less specific and homogeneous because they are made by many different cells in an organism.

The discovery of monoclonal antibody technology is an important story that should be described here. Georges Köhler and Cesar Milstein fused a normal lymphocyte with a myeloma cancer cell, creating a hybrid cell (*hybridoma*) that combined the two important properties of both parents. The hybridoma produced the single specific antibody characteristic of its lymphocyte parent; it also grew continuously like its cancerous myeloma parent. In this way, antibody against a specific antigen was continuously produced by descendants of the original hybrid cell.

More specifically, Köhler and Milstein inoculated a mouse with sheep red blood cells, an extremely effective antigen. Later, they removed the spleen of the mouse and mixed the sensitized lymphocytes with P3 myeloma cells. Some of these hybridomas and their descendants constantly synthesized antibody against sheep red blood cells. This discovery became an immunologist's dream: It is now possible to produce constant, pure sources of single antibodies that can be used to detect the presence of specific antigens and, possibly, to treat diseases such as cancer. Pure, homogeneous antibodies can now be made in large quantity, something that was impossible with conventional immunological techniques.

READINGS AND REFERENCES

EYE DEVELOPMENT

Coulombre, A. J. 1965. The eye. In R. L. DeHaan and H. Ursprung, eds. *Organogenesis*, pp. 219–51. New York: Holt, Rinehart and Winston.

Jacobson, M. 1968. Development of neuronal specificity in retinal ganglion cells. *Develop Biol.* 17: 202–18.

Karkinem-Jaaskelainen, M., and Saxén, L. 1976. Advantages of organ culture techniques in teratology. In J. D. Ebert and M. Marois, eds., *Tests of Teratogenicity in Vitro*, pp. 275–284. New York: Elsevier-North Holland.

Lopashov, G. V., and Stroeva, O. G. 1961. Morphogenesis of the vertebrate eye. *Adv. Morphogen.* 1: 331–78.

Reyer, R. W. 1962. Regeneration in the amphibian eye. In D. Rudnick, ed., *Regeneration*, pp. 211–65. New York: Ronald Press.

Robertson, G. G., Blattner, A. P., and Williamson, R. J. 1964. Origin and development of lens cataracts in mumps infected chick embryos. *Am. J. Anat.* 115: 473–86.

Stone, L. S. 1959. Regeneration of the retina, iris and lens. In C. S. Thornton, ed., *Regeneration in Vertebrates*, pp. 3–14. Chicago: University of Chicago Press.

Wessells, N. K. 1977. *Tissue Interactions in Development*. CA: W. A. Benjamin. Menlo Park, CA.

DEVELOPMENT OF NERVOUS INTEGRATION AND BEHAVIOR

Barbera, A. J., Marchase, R. B., and Roth, S. 1973. Adhesive recognition and retinotectal specificity. *Proc. Nat. Acad. Sci. U.S.* 70: 2482–86.

Cohen, S. 1960. Purification of a nerve growth promoting protein from the mouse salivary gland and its neurocytotoxic antiserum. *Proc. Natl. Acad. Sci. U.S.* 46: 302–11.

Gaze, R. M. 1970. *The Formation of Nerve Connections*. New York: Academic Press.

Hamburger, V. 1968. Emergence of nervous coordination. Origins of integrated behavior. In M. Locke, ed., *The Emergence of Order in Developing Systems*, p. 251. New York: Academic Press.

Hamburger, V. 1962. Specificity in neurogenesis. *J. Cell. Comp. Physiol.* 60: 31 (suppl. 1).

Jacobson, M. 1970. *Development Neurobiology*. New York: Holt, Rinehart & Winston.

Jacobson, M. 1969. Development of specific neuronal connections. *Science*, 163: 543.

Jacobson, M. 1967. Retinal ganglion cells: Specification of larval connections in *Xenopus laevis*. *Science* 155: 1106.

Levi-Montalcini, R. 1964. Growth control of nerve cells by a protein factor and its antiserum. *Science* 143: 105.

Sperry, R. W. 1944. Optic nerve regeneration and return of vision in anuran. *J. Neurophysiol.* 7: 57.

Sperry, R. W. 1965. Embryogenesis of behavioral nerve nets. In R. L. DeHaan and H. Ursprung, eds., *Organogenesis*, New York: Holt, Rinehart & Winston.

Weiss, P. 1934. *In vitro* experiments on factors determining the course of the outgrowing nerve fiber. J. Exp. Zool. 68: 393–448.

HEART DEVELOPMENT

DeHaan, R. L. 1976. Cell coupling and electrophysiological differentiation of embryonic heart cells. In J. D. Ebert and M. Marois, eds., *Tests of Teratogenicity in Vitro*, pp. 225–32. New York: Elsevier-North Holland.

DeHaan, R. L. 1968. Emergence of form and function in the embryonic heart. In M. Locke, ed., *The Emergence of Order in Developing Systems*, New York: Academic Press.

DeHaan, R. L. 1965. Morphogenesis of the vertebrate heart. In R. L. DeHaan and H. Ursprung, eds., *Organogenesis*, pp. 377–420. New York: Holt, Rinehart & Winston.

DeHaan, R. L., and Sachs, H. G. 1972. Cell coupling in developing systems: The heart cell paradigm. In A. A. Moscona and A. Monroy, eds., *Current Topics in Developmental Biology*, vol. 7, pp. 193–228. New York: Academic Press.

Patten, B. M. 1958. *Foundations of Embryology*. New York: McGraw-Hill.

Rawles, M. E. 1943. The heart-forming areas of the early chick blastoderm. *Physiol. Zool.* 16: 22–42.

Wilens, S. 1955. The migration of heart mesoderm and associated areas in *Amblystoma punctatum*. *J. Exp. Zool.* 129: 579–606.

LIMB DEVELOPMENT

Amprino, R. 1965. Aspects of limb morphogenesis in the chicken. In R. L. DeHaan and H. Ursprung, eds., *Organogenesis*, pp. 255–84. New York: Holt, Rinehart & Winston.

Balinsky, B. I. 1974. Supernumerary limb induction in the anura. *J. Exp. Zool.* 188: 195–202.

Becker, R. D. 1974. The significance of bioelectric potentials. *Bioelectrochemistry and Bioenergetics* 1: 187–99.

Borgens, R. B., Vanable, J. W. Jr., and Jaffe, L. F. 1969. Small artificial currents enhance *Xenopus* limb regeneration. *J. Exp. Zool.* 207: 217–25.

Choo, A. F., Logan, D. M., and Rathbone, M. P. 1978. Nerve trophic effects: An in vitro assay for factors involved in regulation of protein synthesis in regenerating amphibian limbs. *J. Exp. Zool.* 206: 347–54.

French, V., Bryant, P. J., and Bryant, S. V. 1976. Pattern regulation in epimorphic fields. *Science* 193: 969–81.

Gospodarowicz, D., and Mescher, A. L. 1980. Fibroblast growth factor and the control of vertebrate regeneration and repair. *Ann. N.Y. Acad. Sci.*, 339: 151–74.

Harrison, R. 1918. Experiments on the development of the forelimb of amblystoma, a self-differentiating equipotential system. *J. Exp. Zool.* 25: 413–61.

Hay, E. D. 1965. Metabolic patterns in limb development and regeneration. In R. L. DeHaan and H. Ursprung, eds., *Organogenesis*, pp. 315–36. New York: Holt, Rinehart, & Winston.

Illingsworth, C. M., and Barker, A. T. 1980. Measurements of electrical currents during the regeneration of amputated fingertips in children. *Clin. Phys. Physiol. Meas.* 1: 87–89.

Iten, L. E., and Bryant, S. V. 1975. The interaction between the blastema and stump in the establishment of the anterior-posterior and proximo-distal organization of the limb regenerate. *Develop. Biol.* 14: 119–47.

Jabaily, J. A., and Singer, M. 1977. Neurotrophic stimulation of DNA synthesis in the regenerating forelimb of the newt, *Triturus*. *J. Exp. Zool.* 199: 251–56.

Mescher, A. L., and Gospodarowicz, D. 1979. Mitogenic effect of a growth factor derived from myelin on denervated regenerates of newt forelimbs. *J. Exp. Zool.* 207: 497–503.

Milaire, J. 1965. Aspects of limb morphogenesis in mammals. In R. L. DeHaan and H. Ursprung, eds., *Organogenesis*, pp. 283–300. New York: Holt, Rinehart & Winston.

Mizell, M. 1968. Limb regeneration: Induction in the newborn opossum. *Science* 161: 283–86.

Mizell, M., and Isaaco, J. J. 1970. Induced regeneration of hindlimbs in the newborn opossum. *Amer. Zool.* 10: 141–55.

Rubin, L., and Saunders, J. W. 1972. Ectodermal-mesodermal interactions in the growth of limbs in chick embryo. *Develop. Biol.* 28: 94–112.

Saunders, J. W., Jr., and Fallon, J. J. 1966. Cell death in morphogenesis. In M. Locke, ed., *Major Problems in Developmental Biology*, pp. 289–316. New York: Academic Press.

Searls, R. L., and Janners, M. Y. 1969. *J. Exp. Zool.* 170: 365.

Singer, M. 1974. Neurotrophic control of limb regeneration in the newt. *Ann. N.Y. Acad. Sci.* 228: 308–21.

Swett, F. H. 1937. Determination of limb axes. *Quart. Rev. Biol.* 12: 322–39.

Thorton, C. S. 1970. Amphibian limb regeneration and its relation to nerves. *Amer. Zool.* 10: 113–18.

Wallace, H. 1972. The components of regrowing nerves which support the regeneration of irradiated salamander limb. *J. Embryol. Exp. Morph.* 28: 419–35.

Wessells, N. K. 1977. *Tissue Interactions in Development*. Menlo Park, CA: W. A. Benjamin.

Zwilling, E. 1961. Limb morphogenesis. *Adv. Morphog.* 1: 329.

UROGENITAL MORPHOGENESIS

Blackler, A. W. 1962. Transfer of primordial germ cells between two subspecies of *Xenopus laervis. Jour. Embryol. Exp. Morph.* 10: 641.

Burns, R. K. 1961. Role of hormones in the differentiation of sex. In W. C. Young, ed., *Sex and Internal Secretions*, pp. 76–160. Baltimore: William and Wilkins.

Fraser, E. A. 1950. The development of the vertebrate excretory system. *Biol. Rev.* 25 : 159–87.

Gruenwald, P. 1952. Development of the excretory system. *Ann. N.Y. Acad. Sci.* 55 : 142–46.

Jost, A. 1965. Gonadal hormones in the sex differentiation of the mammalian fetus. In R. L. DeHaan and H. Ursprung, eds., *Organogenesis*, pp. 611–28. New York: Holt, Rinehart & Winston.

Lillie, F. R. 1916. The theory of the free-martin. *Science* 43: 611.

Ohno, S. 1978. The role of H-Y antigen in primary sex determination. *J. Am. Med. Assoc.* 239: 217–20.

Phillips, J. B. 1975. *Development of Vertebrate Anatomy*, pp. 434–39. St. Louis: C.V. Mosby.

Saxén, L. 1971. Inductive interactions in kidney development. In *Control Mechanisms of Growth and Differentiation, 25th Symposium*, Society of Experimental Biology. New York: Academic Press.

Silvers, W. K., Gasser, D. L., and Eicher, E. M. 1982. Serologically detectable male antigen and sex determination, *Cell* 28: 439–40.

Swift, C. H. 1914. Origin and early history of the primordial germ-cells in the chick. *Amer. J. Anat.* 15 : 483–516.

Willier, B. H., Weiss, P. A., and Hamburger, V., eds. 1955. *Analysis of Development*. Philadelphia: W. B. Saunders.

Witschi, E. 1948. Migration of the germ cells of human embryos from the yolk sac to the primitive gonadal folds. Carnegie Institution of Washington, publication 575, *Contributions to Embryology* 32: 67.

DEVELOPMENT OF THE IMMUNE SYSTEM

Auerbach, C. 1974. Development of immunity. In J. Lash and U. R. Whittaker, eds. *Concepts in Development*. Sunderland, MA: Sinauer.

Burnert, S. M. 1969. *Self and Not-Self*. London: Cambridge University Press.

Burnet, M. 1969. *The Clonal Selection Theory of Acquired Immunity*. London: Cambridge University Press.

Edelman, G. M. 1975. Antibody structure and molecular immunology. *Science* 180: 830.

Edelman, G. M. 1970. The structure and function of antibodies. *Sci. Amer.* 223: 34.

Greaves, M. F., Owen, J. T., and Raff, M. C. 1975. *T and B Lymphocytes: Origins, Properties and Roles in Immune Responses.* Amsterdam: Excerpta Medica.

Jerne, N. K. 1973. The immune system. *Sci. Amer.* 229: 52.

Hood, L. 1972. Two genes: One polypeptide chain—fact or fiction? *Fed. Proc.* 31: 179.

Hood, L., and Prahl, J. 1971. The immune system: A model for differentiation in higher organisms. *Adv. Immunol.* 14: 291–351.

Hood, L. E., Weissman, I. L., and Wood, W. B. 1978. *Immunology.* Menlo Park, CA: Benjamin/Cummings.

Katz, D. H. 1977. *Lymphocyte Differentiation, Recognition and Regulation.* New York: Academic Press.

Leder, P. 1982. The genetics of antibody diversity. *Scientific American,* 246, no. 5: 102–15.

Miller, J. F. A. P. 1966. Immunological function of the thymus. *Lancet* 2: 748.

Nossal, G. J. V., and Ada, G. L. 1971. *Antigens, Lymphoid Cells and the Immune Response.* New York: Academic Press.

Porter, R., and Knight, J., eds. 1972. *Ontogeny of Acquired Immunity,* CIBA Foundation Symposium. New York: Elsevier-North Holland.

Raff, M. C. 1974. Development and differentiation of lymphocytes. In T. J. King, ed., *Developmental Aspects of Carcinogenesis and Immunity,* pp. 161–72. New York: Academic Press.

Secarz, E., Herzenberg, L. A., and Fox, C. F., eds. 1977. *The Immune System II: Regulatory Genetics.* New York: Academic Press.

Szenberg, A., and Warner, N. L. 1962. Dissociation of immunological responsiveness in fowls with a hormonally arrested development of lymphoid tissues. *Nature* 194: 146.

Wade, N. 1982. Hybridomas: The making of a revolution. *Science* 215: 1073–75.

Weissman, I. L. 1967. Thymus cell migration. *J. Exp. Med.* 126: 291.

Weissman, I. L., Gutman, G. A., and Friedberg, S. H. 1974. Tissue localization of lymphoid cells. *Ser. Haematol.* 8: 482.

11

EMBRYONIC INDUCTION

The early embryo is a mass of similar cells that eventually become different. All are derived from the single fertilized egg, all contain identical genes. How do they become different? This is a major area of interest in developmental biology today. We already touched on this fascinating problem in previous chapters. In the chapter on cleavage, for instance, we saw that certain special types of cytoplasm or surface regions can be segregated to specific cells during the cleavage process. The interaction of cells with neighboring cells plays an important role in causing cells to differentiate. This was easily seen in neural tube differentiation, in which the prospective neural ectoderm must interact with the archenteron roof for neural differentiation to occur. This interacting system is a major topic of this chapter. The concept of embryonic induction originated in this system. Embryonic induction simply means that one tissue (inducing tissue) interacts with another tissue (responding tissue), causing the responding tissue to differentiate. This sort of interaction was also seen in organogenesis. The tip of the optic vesicle induces prospective lens ectoderm to differentiate into lens; the tip of the growing ureter induces metanephrogenic mesoderm to differentiate into metanephric kidney tubules. In short, inductive interactions between cells play a major role in causing them to differentiate. These inductive interactions often involve a chemical inducer that passes from the inducing tissue to the responding tissue. This inducer may activate specific genes that code for the specific proteins required for cell differentiation. In this chapter, we will focus on the nature of the interactions involved in embryonic induction, the factors responsible for "turning cells on" to differentiate. This will serve as an introduction to the later study of differentiation. In Chapters 13, 14, and 15, we shall examine what actually happens during differentiation at the nucleic acid, protein, and organelle levels.

ORIGIN OF THE INDUCTION CONCEPT

Let us describe a few of the experiments that led to the concept of embryonic induction. The first of these involved separating embryos at the 2-cell stage. Hans Driesch separated the sea urchin embryo at the 2-cell stage and found that each blastomere gave rise to a complete, normal (smaller) larva. E. B. Wilson did the same experiment with the *Amphioxus* embryo, with the same results. Each cell could give rise to an entire embryo.

Hans Spemann separated the amphibian embryo at the 2-cell stage. In about half the cases, at least one of the two blastomeres formed a complete embryo. In many cases, both blastomeres did so. It turned out that if the first cleavage plane passed through the *gray crescent* region of the zygote, then each blastomere gave rise to a complete embryo. The gray crescent is a surface region of the zygote that is exposed soon after fertilization when the outer pigmented surface of the zygote moves toward the point of sperm entry. We shall discuss the gray crescent later in this chapter.

What does this mean? The results suggest that the gray crescent contains something essential for normal embryonic development. The gray crescent gives rise to the dorsal lip of the blastopore. The dorsal lip invaginates into the embryo during gastrulation to form the roof of the archenteron (prospective notochord) that eventually lies under the prospective neural tube. Neural differentiation does not occur unless the archenteron roof underlies the prospective neural ectoderm. This is seen if one removes the outer membranes from the amphibian egg and grows the embryo in a solution of lithium chloride or in hypertonic solutions. Under these conditions, the prospective archenteron roof grows outward rather than inward. The archenteron roof does not underlie the prospective neural ectoderm, and the prospective neural ectoderm fails to differentiate into the neural tube.

In 1924, a set of beautiful experiments by Spemann and Mangold showed that the interaction of the archenteron roof with the prospective neural tube is essential for neural differentiation. (These experiments led to the award of a Nobel Prize.) Spemann and Mangold transplanted the dorsal lip of the blastopore of one species of newt early gastrula into the blastocoel of another species of newt early gastrula. Because the species differed in pigmentation, the origin of specific structures forming in the host embryo could be determined. The host embryo developed two neural systems, and in some experiments two entire embryos (Figure 11-1). In another experiment, the dorsal lip of one embryo was transplanted near the lateral lip of the blastopore of another embryo. The transplant invaginated and induced a second neural tube as in the previous experiments. The differences in pigmentation between the host and dorsal lip donor embryos showed that both neural tubes in the host embryo were derived almost entirely from host tissue. Thus, the transplanted dorsal lip did not give rise to a neural tube; instead, it induced the host embryo to form a neural tube. In cases where whole second embryos formed, the neural tubes were of host origin, but the donor dorsal lip differentiated into a variety of tissues that normally form from this region. Spemann and Mangold called the dorsal lip the *organizer* because it could cause a complete second embryo to form.

GRAY CRESCENT

About thirty minutes after fertilization in the frog embryo, the pigmented surface of the egg shifts toward the sperm entry point, exposing a region of underlying grayish surface cytoplasm. This event indicates a dramatic reorganization within the zygote. The exposed region of cytoplasm is called the gray crescent. As previously mentioned, the gray crescent is the precursor of the dorsal lip of the blastopore; which, in turn, is the precursor of the archenteron roof. We have seen that the archenteron roof (prospective

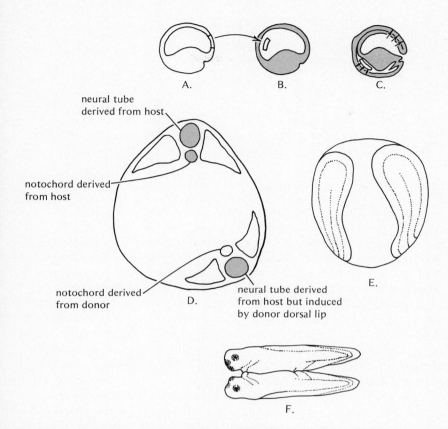

Figure 11-1. Transplantation of dorsal lip of blastopore. Dorsal lip from species (a) transplanted to blastocoel of species (b). (c) Dorsal lip from host (archenteron roof) and donor interacting with overlying ectoderm. (d) and (e) Two neural systems formed in host. (f) Two embryos formed in some hosts. After Spemann and Mangold, 1924 *Arch. Mikrosk. Anat. Entwmech.* 100: 599-638.

notochord) induces neural differentiation in the overlying ectoderm. An interesting question now arises. Can the gray crescent be transplanted into another embryo to induce neural differentiation? In other words, can the area that gives rise to the dorsal lip serve as the inducing tissue?

The answer is yes. Curtis transplanted the gray crescent from an 8-cell frog embryo to the prospective belly ectoderm region of a 1-cell embryo. The embryo cleaved. After gastrulation, a second embryo with neural tube and notochord developed at the site of the transplanted gray crescent (Figure 11-2). This gray crescent formed a dorsal lip and archenteron roof and induced the formation of a second neural tube (in addition to the neural tube induced by the embryo's own dorsal lip). The gray crescent represents only a surface region of the embryo. Clearly, important information is contained in

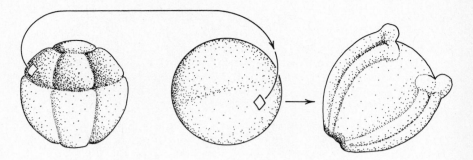

Figure 11-2. Gray crescent transplant. Gray crescent transplanted from 8-cell amphibian embryo (*Xenopus*) to the ventral margin of a 1-cell embryo induces a second embryo axis. After A. S. G. Curtis, *Endeavor* 22 (1963): 134.

this surface area, information that is essential in controlling differentiation. This is not unique to this system. In Chapter 6, we showed that the surface of sea urchin micromeres is different from that of the mesomeres and macromeres, and that these differences may be important in differentiation.

The ability of gray crescent cortex to induce neural differentiation is, however, the subject of controversy. Experiments by Malacinski and colleagues suggest that the gray crescent of the fertilized egg cannot promote or induce nervous system morphogenesis and differentiation. Instead, such inducing activity is progressively acquired, during development, by cells in the future dorsal region of the embryo.

Gray crescent cortex did not induce neural differentiation either in sandwich cultures of gray crescent cortex wrapped in early gastrula ectoderm or when the cortex was directly implanted into the blastocoel of a recipient blastula or gastrula. Also, dorsal cells from cleavage stages and early blastula stages showed little or no inducing capacity when implanted into the blastocoels of late blastulas or early gastrulas. Dorsal cells from late blastulas and early gastrulas, however, did show substantial inducing activity when implanted into blastocoels or when sandwiched in ectoderm. These experiments suggest that primary embryonic organizer activity gradually accumulates during early development in cells on the future dorsal side of the embryo. When the question of exactly when normal embryo tissue acquires neural inducing competence is resolved, researchers will be better able to identify the actual mechanism of neural induction.

Other experiments suggest that the gray crescent cortex is not the direct precursor of Spemann's organizer. Keller and colleagues have recently developed fate maps for the *Xenopus* embryo. These suggest that prospective notochord is formed not from surface gray crescent, but from internal cells. Experiments by Nieuwkoop and colleagues on the axolotl embryo also suggest that there is no direct topological succession between gray crescent cortex and dorsal mesoderm. Other work by Kirschner and Gerhart shows that double *Xenopus* embryos can be produced without transplanting gray crescent cortex to host embryos. Trauma alone can induce the formation of a second blastopore and, eventually, a double embryo.

Experiments performed by Gerhart and colleagues have shed new light on the role of the gray crescent as an inducer of dorsal structures in amphibian development. These workers rotated fertilized *Xenopus laevis* eggs and fixed them in the new upside-down orientation. The dense contents of the original vegetal hemisphere of the egg then flowed slowly in the direction of gravity (downwards). They found that the uppermost region of the egg becomes the site of dorsal development in the embryo. The egg thus selects its dorsal-ventral axis on the basis of its relationship to the gravitational field. These results were obtained with pre–gray crescent embryos. Gerhart and colleagues also found that post–gray crescent embryos are influenced by gravity, if the eggs are oriented and then subjected to centrifugal force. In this way they dissociated dorsal development from its normal topographical relationship with the grey crescent. A dorsal lip formed from areas other than the gray crescent, depending upon the embryo's orientation in the centrifugal field.

How do these results relate to Curtis's gray crescent transplant experiments? These workers showed that to implant the graft, Curtis had to remove the fertilization envelope of the host egg, which means that the egg was no longer free to control its orientation. Curtis's recipient eggs were inclined in such a way that gravity displaced the egg contents, thus forming a new embryonic axis. Since Gerhart and others produced such so-called twinning by merely regulating the gravitational force, they suggested that Curtis's results could easily be explained by similar gravitational effects with no dorsalizing activity present in the gray crescent.

The following working hypothesis was developed by Gerhart and colleagues to make sense of these conflicting experiments and to explain the basis for the dorsal inducing activity that resides in the dorsal lip of the blastopore. Initially, the sperm must determine the position of the prospective dorsal-ventral axis of the egg. This is shown by the gray crescent, blastopore, and the dorsal structures, which normally develop on the side of the embryo opposite the random entry point of the sperm. Since the egg cortex contracts toward the side of sperm entry, the sperm must orient this contraction in some way. How? Upon entering the egg, the sperm produces a large aster, presumably organized by its centriole. The microtubules of the aster redistribute the cytoplasmic contents slightly along the future dorsal-ventral axis. Drugs such as colchicine and vinblastine, which disrupt microtubules, block aster assembly and also cytoplasmic displacement. In addition, artificially activated *Xenopus laevis* eggs that lack a functional centriole do not exhibit these early displacements. Gerhart and coworkers suggest that this initial aster-driven asymmetry cues the direction of the cortical contraction, in turn leading to the major cytoplasmic redistribution that follows. We noted earlier that pre-crescent eggs were more sensitive to the axis-orienting effects of gravity than post-crescent eggs (these require centrifugation to displace the cytoplasmic contents). At the early pre-crescent stages, the cortical contraction can probably be cued by only a small, gravity-induced lateral displacement of cytoplasmic materials. In the post–gray crescent period, larger forces (generated by centrifugation) are required to reverse the cytoplasmic distribution produced by the aster-induced contraction and to create a new and opposite distribution.

The important dorsal-ventral regional differences arising in the fertilized egg before cleavage appear not to reside in the gray crescent cortex. Instead, they reside in the cytoplasmic materials, which are normally reorganized by the sperm aster and cortical contraction, and experimentally by gravitational force. While the cortex, including the grey crescent, contains some of the cellular components used by the egg to generate a dorsal cytoplasmic localization, it is not itself a lasting store of developmental information. The next questions, therefore, are: What are the important materials that undergo redistribution on the dorsal side of the fertilized egg? How do these materials form the inductively active vegetal cells that become part of the dorsal lip of the blastopore?

Before we move on to the induction process itself, let us note that what we have described for the amphibian also occurs in all the other vertebrates. That is, the neural tube is induced by the dorsal lip of the blastopore (archenteron roof) in the primitive chordate, *Amphioxus*, cyclostomes, bony fish, and amphibians. In reptiles, birds, and mammals, the structure that corresponds to the dorsal lip is a portion of the primitive streak, which serves the same inducing function. Thus, embryonic induction is a phenomenon of general importance in vertebrates and in other chordates.

TISSUE, STAGE, AND REGIONAL SPECIFICITY IN NEURAL INDUCTION

A question that now comes to mind is: Will *any* tissue respond to the archenteron roof and differentiate into neural structures? No. If a piece of dorsal lip is transplanted onto endoderm, no neural differentiation occurs. Only the ectoderm will form neural structures as a result of interaction with the dorsal lip. We say that only the ectoderm is *competent* to respond to induction by the dorsal lip. So, a tissue-specific competence for neural induction exists.

What about the developmental stage? Can ectoderm from any embryonic stage respond to the inducer by differentiating into neural structures? Again, no. Only gastrula-stage ectoderm is competent to respond to induction. A dorsal lip will not induce neural differentiation if grafted into a neurula. If a dorsal lip is transplanted to a blastula-stage embryo, the ectoderm differentiates into neural structures only after gastrulation is complete, when the normal host neural system develops. Thus, *stage-specific competence* for neural induction exists.

Does a dorsal lip taken from an early gastrula have the same inducing capacity as a dorsal lip taken from a later gastrula? Normally, the early dorsal lip invaginates and comes to lie beneath the anterior part of the prospective neural tube. A later dorsal lip comes to lie beneath the posterior part of the prospective neural tube. An early dorsal lip, or the anterior portion of the archenteron roof, can indeed be successfully transplanted into another embryo. There, it induces mostly head neural structures in the ectoderm (forebrain, eyes, nose rudiments, hindbrain, and ear vesicles). A late dorsal lip, or the posterior portion of the archenteron roof, induces mostly posterior structures, such as the spinal cord and associated trunk and tail organs. Thus, the dorsal lip or archenteron roof has *regional inducing specificity* (Figure 11–3).

CHEMICAL NATURE OF THE NEURAL INDUCER

Before we examine the work on the chemical nature of the elusive neural inducer, let us say at the outset that the molecular nature of the inducer (if it exists) is unknown. It is worthwhile, however, to look at some of the interesting approaches that have been taken in attacking this problem. This is not an exercise in futility, because, although the inducer has not been definitively identified, a far better understanding of embryonic induction has grown out of the investigations in this area.

Niu and Twitty attempted to isolate the neural inducer by growing dorsal lips in small drops of saline solution (Figure 11–4). The dorsal lips differentiated into mesodermal structures in these drops, showing that the conditions in the drops are conducive to the normal development of the dorsal lip tissue. The dorsal lips were cultured in these drops for about ten days, then removed. The researchers hypothesized that during the incubation time, the neural inducer would be released by the lips into the drops. After the dorsal lips were removed, ectoderm was placed in the "conditioned" drops. Induction occurred; the ectoderm differentiated into neural structures.

In control experiments, other tissues, such as ectoderm, were incubated in the drops before the responding ectoderm was placed to the drops. Neural induction did not occur in these cases. Only the dorsal lips released a factor into the drops that produced neural differentiation in the responding ectoderm. What was this factor? Was it the natural inducer? Many different enzymes were used to identify the chemical nature of the isolated "inducer." Only proteolytic (protein degrading) enzymes, such as trypsin, completely destroyed the inducing capacity of the drop. Other enzymes, such as ribonuclease, partially inactivated inducing activity in some experiments. It was concluded that the "neural inducer" in the drops was a protein, possibly a

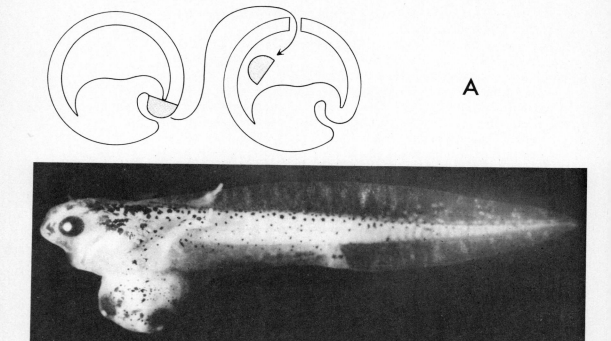

A

ribonucleoprotein. Unfortunately, although these experiments had great promise of solving the puzzle of the elusive inducer, little else has come from this work. It is quite difficult to obtain enough inducer, using the drop culture method, to carry out the biochemical experiments needed to purify and characterize a protein molecule. Modern microchemical methods, however, may offer means of purifying and characterizing the neural inducer even if large quantities are unavailable.

ABNORMAL INDUCERS

Further findings threw what looked like a monkey wrench into the induction problem. Apparently, many unrelated substances could induce neural differentiation in responding ectoderm. Killed dorsal lip, ground quartz, dragonfly lymph, killed guinea pig liver or bone marrow, live or dead human HeLa cells (cervical cancer cells)—all induced neural differentiation in responding ectoderm. Many of these experiments used the sandwich technique in which ectoderm is wrapped around the potential inducer and cultivated in fluid. What was going on?

Holtfreter found a possible solution to this enigma. A piece of isolated

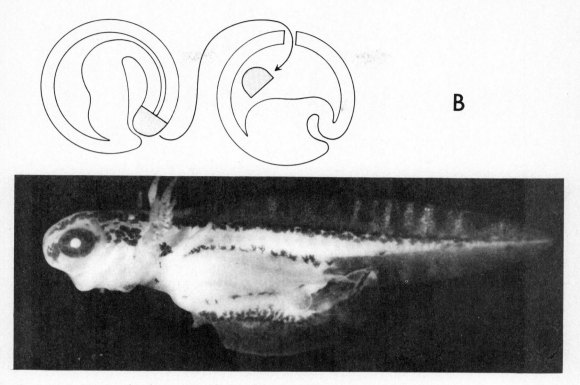

B

Figure 11–3. Regional inducing specificity of the dorsal lip. Early dorsal lip (early gastrula from amphibian) induces a secondary head in recipient embryo. Late dorsal lip (late gastrula from amphibian) induces a secondary trunk and tail in recipient embryo. Courtesy of Dr. L. Saxén.

prospective ectoderm would differentiate into neural structures when exposed for a short time to saline solutions that would slightly damage the tissue. If the pH of the solutions was lower than 5 or higher than 9.2, or if the ions in the solutions were altered, ectoderm often differentiated into neural structures. This work suggested that sublethal cytolysis was a cause of induction. That is, damage to cells in some way induces these cells to differentiate. Many of the so-called abnormal inducers may act in this way. On the other hand, the normal process of induction by the archenteron roof in the embryo may occur by some means other than sublethal cytolysis, since there is apparently no cell damage in the natural system. As you can see, our understanding of the chemical nature of neural induction is still in its infancy.

Some investigators feel that a simple mechanism may be responsible for neural induction in all normal and abnormal cases. One such mechanism is the sodium-uptake model. This model suggests that anything that causes neural induction does so by promoting sodium ion uptake by the ectoderm. Sodium is certainly required for induction to occur. If sodium ion is omitted from the

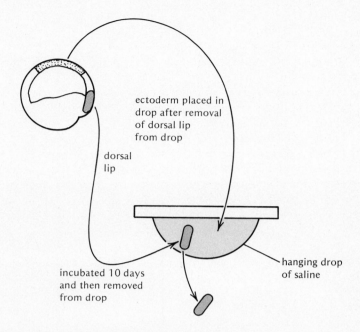

ectoderm placed in
drop after removal
of dorsal lip
from drop

dorsal
lip

hanging drop
of saline

incubated 10 days
and then removed
from drop

Figure 11-4. Attempt at isolating the neural inducer. Based upon experiments by Niu and Twitty, *Proc. Natl. Acad. Sci. U.S.*, 39 (1953): 985-89.

medium or reduced (from 88 mM to 44 mM), nerve cells fail to differentiate in ectoderm in the presence of dorsal lip. If ion uptake is a key first step to the induction puzzle, what do the ions do?

To sum up, many things can cause neural induction in responding ectoderm. The natural inducer may be protein or ribonucleoprotein. Many abnormal inducers may work by damaging the cells. The uptake of specific ions may be an important first step in the induction process. The future should provide us with more definitive answers to these problems.

MECHANISM OF INDUCER ACTION

We have not, as yet, definitively identified the chemical nature of the neural inducer. However, this is not the only important aspect of induction. Other key questions include: What is the nature of the response by the ectoderm to the inducer? What physical conditions are needed for induction to occur? Is contact between the archenteron roof and prospective neural tube needed for induction to occur? Let us examine these problems.

What is the nature of the response to the inducer? The response is controlled by genes in the responding ectoderm. Active genetic information from the dorsal lip is probably irrelevant to the response. In other words, the inducer provides permissive rather than instructive information to the responding ectoderm. Permissive refers to turning on the responding tissue's genes. Instructive refers to providing actual genetic information to the responding ectoderm. These conclusions are supported by the following experiment involving several species of salamanders.

Each species used has markers, such as certain head structures, not found in the other species. One of these structures is called a balancer. Only some salamander larvae have balancers.

The balancers are induced in the epidermis by the anterior portion of the archenteron roof. A balancer can be induced in a species that does not possess balancers by transplanting a piece of ectoderm from a species possessing a balancer to replace the ectoderm of the host. Thus, the archenteron roof of the species without a balancer can induce a balancer in ectoderm transplanted from a species competent to form the balancer. The reverse experiment was also performed, and showed that a balancer cannot be induced in ectoderm that is not competent to form a balancer. The ectoderm must possess the genetic code to be activated by the inducer. Thus, the inducer does not alter the genetic response of the responding tissue.

Another experiment along the same lines uses the RNA synthesis inhibitor actinomycin D. If newt gastrula ectoderm is exposed to an inducer, then separated from the inducer and immediately treated with actinomycin D, neural differentiation fails to take place. The ectoderm treated with actinomycin D remains healthy and forms epidermal structures, but not neural structures. This experiment suggests that the genes of the responding ectoderm must synthesize RNA (presumably messenger RNA) if neural differentiation is to take place in response to the inducer.

CELLULAR CONTACT IN EMBRYONIC INDUCTION

What physical conditions are needed for induction to occur? Is cell-to-cell contact between the archenteron roof and prospective neural tube required for normal neural induction? The question still remains an enigma.

One approach to investigating the need for cell-to-cell contact is to separate these two tissues with different artificial barriers. If induction is caused by a molecule passed from the archenteron roof to the overlying ectoderm, then a filter between them that allows molecular passage but not cell passage should not inhibit induction. Material that does not allow even small molecules to pass between the inducing and responding tissues should prevent induction.

In early experiments, membrane barriers were placed between the archenteron roof and ectoderm. Cellophane, which does not allow the passage to molecules, did indeed prevent induction. If permeable membrane filters were used instead, induction still occurred (Figure 11-5). These results suggested that cell contact was not needed for induction, and that the key was

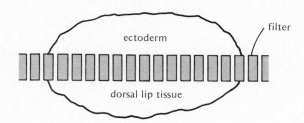

Figure 11-5. Separation of dorsal lip and responding ectoderm by porous filter permits induction to occur.

the passage of molecules between the inducing and responding tissue. Later, however, the permeable filters were examined with the electron microscope. Thin cell processes were sometimes observed in the filter pores. Thus, the cell-contact issue was not really resolved.

Toivonen and Saxén recently reinvestigated this problem using filters with tiny straight pores (0.1 μm) produced by shooting neutron beams through the filters. These filters were inserted between the inducing and responding tissues for a very brief time. The tissues were then separated and the ectoderm was cultured separately. Neural differentiation did indeed occur in the ectoderm. When the filter pores were analyzed under the electron microscope, no cell processes seemed to have entered the filter. This is perhaps the best experiment to date that shows that neural induction does not require contact between the cells of the archenteron roof and prospective neural tube. Thus, induction may occur in response to molecules diffusing from the archenteron roof to the responding ectoderm.

INDUCTION GRADIENTS

Although the chemical nature of the neural inducer is not known, indirect evidence suggests that two types of inducers in the archenteron roof may be active in inducing neural structures. Let us briefly look at some evidence suggesting that a "double gradient" is involved in the induction process.

Various abnormal materials were tested for inducing ability. Some materials, such as killed guinea pig liver, mainly induced forebrain structures in responding ectoderm. Other materials, such as killed guinea pig bone marrow, mainly induced trunk and tail structures in the responding ectoderm (Figure 11-6). These results suggested that more than one type of inducer exists, and that each inducer has a regional inducing specificity.

A mixture of forebrain inducer and tail inducer was then placed in contact with responding ectoderm. As one would expect, forebrain and trunk-tail structures were induced. However, hindbrain and spinal cord structures were formed as well (Figure 11-7). Thus, a mixture of a forebrain and trunk-tail inducer could induce intermediate structures. These experi-

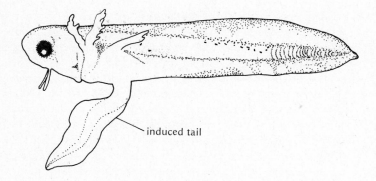

induced tail

Figure 11–6. Induction by abnormal inducer. Alcohol-treated (killed) guinea pig kidney tissue induces tail structures when transplanted into the blastocoel of a gastrula. After experiments by L. Saxén and S. Toivonen.

ments (by Toivonen and Saxén) suggested that in the natural system, two types of inducers exist in the archenteron roof. The forebrain inducer is concentrated in the anterior portion of the archenteron roof; the trunk-tail inducer, in the posterior portion. Where the two substances join (in the middle of the roof), middle structures are induced in the middle area of the responding ectoderm (Figure 11–7).

Tiedemann's group isolated an inducer that mainly evoked middle (hindbrain) structures. They separated this material by chromatography on DEAE cellulose into several fractions. One of the active fractions induced only forebrain structures in responding ectoderm; another active fraction induced only trunk-tail structures. No fraction induced much in the way of hindbrain structures. This is added evidence for the double gradient idea. There is no conclusive proof, however, that two or more inducers are present in the archenteron roof.

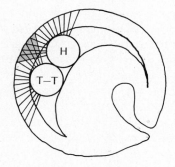

Figure 11–7. When a head structure inducer (H) plus a trunk-tail inducer (T-T) is implanted into a newt gastrula, it yields head, trunk-tail *and* middle structures. After Toivonen and Saxén, *Ann. Acad. Sci. Fenn. A.* 30 (1955): 1–29.

The experiments just mentioned suggest that two or more types of inducers exist in the archenteron roof, and that mixtures of these inducers are responsible for inducing intermediate structures in the responding neural ectoderm. A new question now arises: Can some differentiation of the neural tube result from the interactions of induced cells in the neural ectoderm? In other words, is the interaction between the roof cells and the archenteron cells the only interaction that takes place in neural induction? Or are there interactions between, for example, forebrain-induced ectoderm cells and trunk-tail or spinal-cord–induced cells, giving rise to intermediate structures such as the hindbrain? The former interaction can be termed *vertical interaction*, because it occurs between the archenteron roof and the neural ectoderm. The latter interaction can be termed *horizontal interaction*, because it occurs between the neural ectoderm cells themselves.

There is evidence suggesting that horizontal interaction contributes to neural differentiation. Again, this evidence is only indirect and inconclusive, but it is suggestive. Saxén, Toivonen, and Vainio placed one batch of amphibian gastrula ectoderm in contact with guinea pig liver (primarily an inducer of forebrain structures) and another batch in contact with guinea pig bone marrow (primarily an inducer of trunk-tail structures). These cultures were maintained for about 24 hours, which is time enough for induction but not for actual differentiation to occur. The inducers were removed from each ectoderm culture, and the ectoderm was dissociated into single cells. Some of the single ectoderm cells from each culture were allowed to reaggregate. In addition, ectoderm cells originally incubated with guinea pig liver were combined with cells originally incubated with guinea pig bone marrow. Reaggregated ectoderm cells from the guinea pig liver culture formed mainly forebrain and forebrain structures such as optic cups. Reaggregated ectoderm cells from the guinea pig bone marrow culture formed mainly spinal cord and trunk-tail structures. Aggregates of ectoderm cells from both cultures, however, formed large amounts of hindbrain and inner ear vesicles (Figure 11–8). These results suggest that the induced portions of the neural ectoderm do interact horizontally, and that these interactions lead to the differentiation of so-called middle neural structures.

The experiments just described were carried out using abnormal inducers: guinea pig liver and bone marrow. The following experiment, using natural inducing tissue, supports the conclusion that horizontal interaction plays a role in neural differentiation. A small number (less than 10) of amphibian gastrula ectoderm cells were cultured with dorsal lip. The ectoderm cells were then placed in a large uninduced mass of ectoderm. Neural differentiations occurred in the large ectoderm mass. The small number of induced ectoderm cells apparently interacted with the uninduced ectoderm mass, stimulating neural differentiation in the mass. Thus, interaction between ectoderm cells themselves seems to play a role in stimulating neural differentiation.

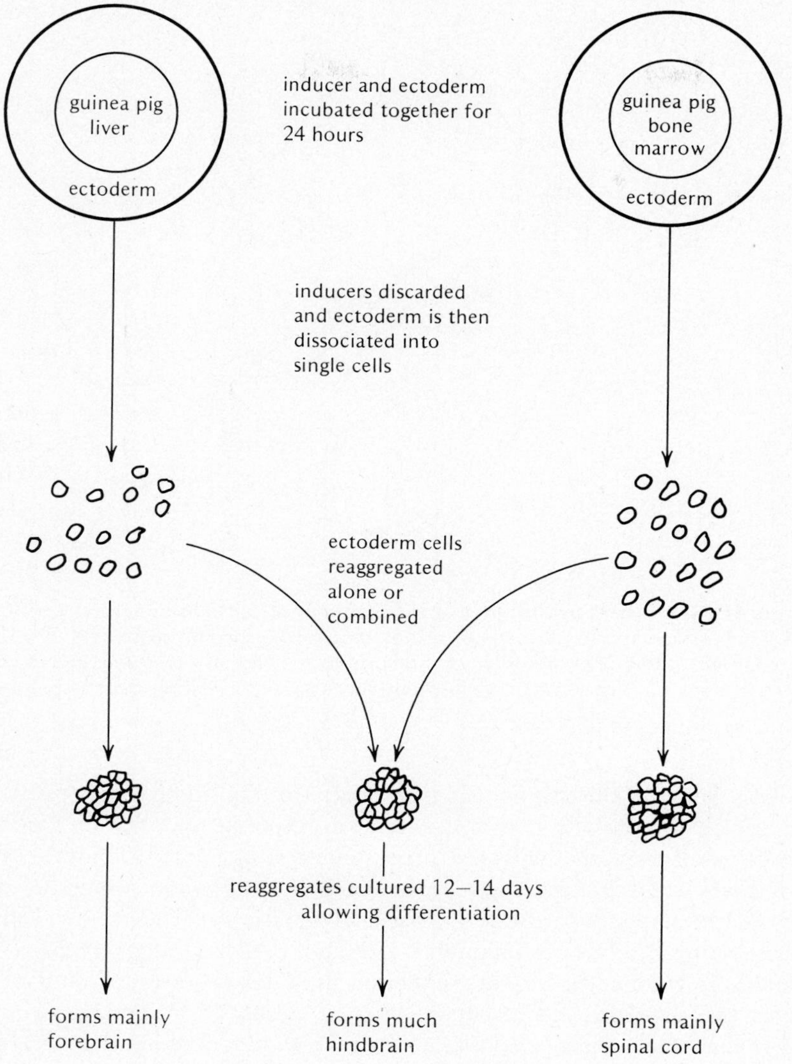
guinea pig liver
ectoderm
inducer and ectoderm incubated together for 24 hours
guinea pig bone marrow
ectoderm
inducers discarded and ectoderm is then dissociated into single cells
ectoderm cells reaggregated alone or combined
reaggregates cultured 12–14 days allowing differentiation
forms mainly forebrain
forms much hindbrain
forms mainly spinal cord

Figure 11–8. Experiment suggesting horizontal interaction in neural induction. Based on work by L. Saxén, S. Toivonen, and T. Vainio, *J. Embryol. Exp. Morphol.* 12 (1964): 333.

SUMMARY

In summary, neural induction is an example of how one cell interacts with another to cause differentiation. The inducer probably acts by turning on the

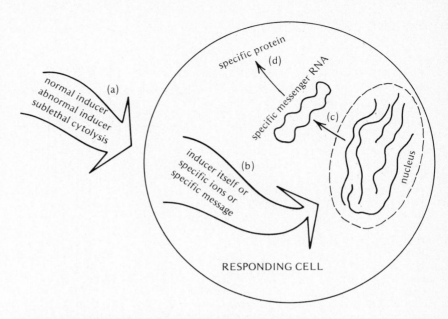

Figure 11-9. Hypothetical model of inducer action. (a) Inducer acts on responding cell. (b) Message reaches nucleus to begin transcribing specific messenger RNA (required, for example, for synthesizing neural protein). (c) Specific messenger RNA enters the cytoplasm. (d) The synthesis of a specific protein is required for differentiation.

specific genes in the responding tissue that are required for differentiation (Figure 11–9). We saw that only ectoderm responds to dorsal lip induction (tissue specificity), that only gastrula ectoderm responds to the inducer (stage specificity), and that specific regions of the dorsal lip induce specific structures in the responding ectoderm (regional specificity). The chemical nature of the natural inducer is unknown, although evidence suggests that it is a protein or ribonucleoprotein. Induction may not require cell contact between the inducing and responding tissues. Agents that damage cells and other abnormal agents can act as inducers. One model of induction suggests that inducing agents act by causing the responding tissue to take up specific ions such as sodium. Our understanding of embryonic induction is still in its infancy. The future should shed much light on this intriguing area.

Now that we have seen how interactions with neighboring cells can cause differentiation, we can turn to the process of differentiation itself (Chapters 13, 14, and 15). First we shall look at the forces that shape the embryo. It is these forces that get the cells to the right place at the right time so that processes such as induction and differentiation can proceed.

READINGS AND REFERENCES

Barth, L. G., and Barth, L. J. 1974. Ionic regulation of embryonic induction of cell differential in *Rana pipiens*. *Develop. Biol.* 39: 1–23.

Callers, J. 1971. Primary induction in birds. *Adv. Morphogen.* 8: 149–80.

Curtis, A. S. G. 1963. The cell cortex. *Endeavour* 22(87): 134–37.

Driesch, H. 1891. Entwicklungsmechanische Studien I–II. *Z. wiss Zool.* 53: 160–82.

Deuchar, E. 1970. Neural induction and differentiation with minimal numbers of cells. *Devel. Biol.* 22: 185–99.

Gerhart, J., Ubbels, G., Black, S., Hara, K., and Kirschner, M. 1981. A reinvestigation of the role of the grey crescent in axis formation in *Xenopus laevis*. *Nature* 292: 511–16.

Hara, K., Tydeman, P., and Kirschner, M. W. 1980. A cytoplasmic clock with the same period as the division cycle in *Xenopus* egg. *Proc. Natl. Acad. Sci. USA* 77(1): 462–66.

Holtfreter, J. 1968. Mesenchyme and epithelia in inductive and morphogenetic processes. In R. Fleischmajer, ed., *Epithelial-Mesenchymal Interactions*. Baltimore: William and Wilkins.

Keller, R. E. 1976. Vital dye mapping of the gastrula and neurula of *Xenopus laevis*. II. Prospective areas and morphogenetic movements of the deep layer. *Develop. Biol.* 51: 118–37.

Kirschner, M. W., and Gerhart, J. C. 1981. Spatial and temporal changes in the amphibian egg. *BioScience* 31: 381–88.

Malacinski, G. M., Chung, H. M., and Asashima, M. 1980. The association of primary embryonic organizer activity with the future dorsal side of amphibian eggs and early embryos. *Developmental Biology* 77: 449–62.

Nieuwkoop, P. D. 1973. The organization center of the amphibian embryo: Its origin, spatial organization and morphogenetic action. *Adv. Morphogenesis* 10: 1–39.

Nieuwkoop, P. D. 1977. Origin and establishment of embryonic polar axes in amphibian development. *Curr. Top. Devel. Biol.* 11: 115–32.

Niu, M. C., and Twitty, V. C. 1953. The differentiation of gastrula ectoderm in medium conditioned by axial mesoderm. *Proc. Natl. Acad. Sci. U.S.* 39: 985–89.

Saxén, L., and Toivonin, S. 1962. *Primary Embryonic Induction*. London: Logos Press.

Saxén, L., Toivonen, S., and Vainio, T. 1964. *J. Embryol. Exptl. Morphol.* 12: 333.

Spemann, H. 1938. *Embryonic Development and Induction*. New Haven: Yale University Press.

Tiedmann, H. 1968. Factors determining embryonic differentiation. *J. Cell Physiol.* 72 (suppl. 1): 129–44.

Toivonen, S., Tarin, D. and Saxén, L. 1976. The transmission of morphogenetic signals from amphibian mesoderm to ectoderm in primary induction. *Differentiation* 5: 19–25.

Weiss, P. 1950. Perspectives in the field of morphogenesis. *Quart. Rev. Biol.* 25: 177–98.

Wilson, E. B. 1904. Experimental studies on germinal localization. *J. Exp. Zool.* 1:1–72.

Yamada, T. 1962. The inductive phenomenon as a tool for understanding the basic mechanism of differentiation. *J. Cell. Comp. Physiol.* 60 (supp.): 49–64.

12

MECHANISMS OF MORPHOGENESIS

In previous chapters, we have attempted to explain the mechanisms that control developmental events. This chapter is devoted to a more in-depth examination of these mechanisms. Let us begin our study of the mechanisms of morphogenesis—the development of form in embryos—with a discussion of cell adhesion in embryonic systems. In the chapters on fertilization, gastrulation, neurulation, and organogenesis, we cited many examples of developmental events that are controlled, at least in part, by the ability of certain cells to adhere to certain other cells. After all, if the cells making up our stomachs didn't stick together in a very specific way, we wouldn't have stomachs. The same is true of all our other organs. We are the sums of many parts, each part consisting of tightly adhering cells. Cells in organisms, therefore, prefer to adhere to certain cells and not to others. Let us see if this preferential adhesiveness is also important in controlling embryonic cellular rearrangements. In addition, we shall examine cell adhesion at the molecular level. We shall treat the adhesion mechanism at length because it has been intensively investigated in recent years. Testable models have been developed to explain the molecular nature of adhesive recognition in embryos. We shall then discuss some of the other mechanisms involved in controlling the development of form in embryonic systems.

THE CELL SURFACE

The cell surface is where cell contact and adhesion take place. Cells in embryos move over one another and stick to one another at their surfaces.

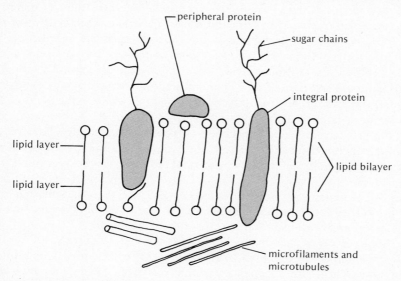

Figure 12–1. Fluid mosaic model of the cell surface. After Singer and Nicolson, *Science*, 175 (1972): 720–31.

Thus, before we examine embryonic cell rearrangements and associations, we shall look at the cell surface—the place where these associations begin.

Earlier (Chapter 5) we briefly described the cell surface as a lipid bilayer in which proteins and glycoproteins are imbedded. The cell surface has fluid properties that allow the proteins and glycoproteins to move laterally in the lipid. This movement is sometimes restricted by rod-shaped proteins, such as microtubules and microfilaments, that anchor some of the surface proteins from the cytoplasmic side of the membrane (Figure 12-1).

The cell surface, therefore, has fluid properties and is mosaic in nature—that is, it consists of various proteins interspersed in the lipid. This is Singer and Nicolson's fluid-mosaic model of the cell surface, currently the most popular and widely accepted model. We shall mention just a few of the important experiments that led to this current concept of the cell surface before we examine how the cell surface is involved in embryonic cellular rearrangements.

EXPERIMENTAL EVIDENCE FOR THE FLUID-MOSAIC MODEL OF THE CELL SURFACE

An elegant experiment by Frye and Edidin provided the basis for determining that the cell surface has fluid properties. These workers, using immunological techniques, developed complex stains that specifically bind to certain cell surface proteins. Thus, stains specific to the surface proteins of given cell types could be produced and coupled to fluorescent dyes. Two different cell types (a human cell and a mouse cell) were fused together, using Sendai virus to promote cell fusion. The surface proteins specific for each cell type were then observed using these complex stains and a fluorescence microscope. By watching the movement of each fluorescent stain bound on the surface of the hybrid cell, Frye and Edidin showed that the surface proteins of the human cell and mouse cell intermixed. These proteins intermixed at a rate suggesting that the proteins diffused laterally (moved sideways) in the plane of the cell surface (Figure 12-2).

Numerous subsequent experiments supported their results. Thus, investigators concluded that the cell surface is fluid in nature. We must stress, however, that under certain conditions (for example, low temperatures that solidify the lipid phase of the membrane) the movement of membrane proteins becomes restricted. In addition, evidence is accumulating that certain cell-surface proteins are anchored by cytoskeletal elements (microtubules and microfilaments) that restrict their movement.

Some of the proteins associated with the cell membrane are loosely associated and easily removed; whereas others are imbedded deeply in the lipid bilayer, sometimes spanning its entire thickness. The former are called peripheral membrane proteins and the latter are called integral membrane proteins. Integral membrane proteins can often be visualized using the

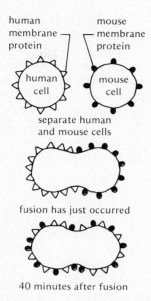

human
membrane
protein

mouse
membrane
protein

human
cell

mouse
cell

separate human
and mouse cells

fusion has just occurred

40 minutes after fusion

Figure 12–2. Lateral mobility of proteins in the membrane lipid. Frye and Edidin showed that 40 minutes after human and mouse cells were fused, their membrane proteins had completely intermixed.

freeze-fracture technique. In this technique, the membrane is split down the middle and one of the lipid bilayers is peeled away. The surface of the cleaved membrane is coated with a heavy metal that forms a replica of the fractured surface. The replica is examined with the electron microscope. Integral membrane proteins are observed as particles on the replica surface.

To sum up, the cell surface has fluid properties and is a mosaic consisting of lipid interspersed with proteins. The proteins can move laterally in the lipid.

SURFACE SUGARS

Figure 12–1 indicates that some of the proteins and lipids of the cell surface carry attached sugar chains. These sugar chains—and especially the terminal sugar residues of the sugar chains—reach farthest away from the cell surface and thus offer a first line of contact with approaching cells and molecules. Some of the molecular models proposed to explain the adhesive recognition of cells in embryonic systems emphasize the role of the surface sugars.

SELECTIVE CELL ADHESION AND TISSUE ASSEMBLY

We have seen specific examples of how specific cell adhesion plays a role in developmental events. For example, sperm-egg recognition appeared to depend upon specific adhesion between the gametes. Similarly, the filopodia of the secondary mesenchyme cells in the sea urchin gastrula preferentially adhere to the animal end of the embryo and play an important role in forming the gut tube. How does one investigate the role of cell adhesion in embryonic development? What happens if embryos are dissociated into single cells and the cells are permitted to reassociate? In other words, can we study the mechanisms controlling embryonic cellular rearrangements by taking embryos apart and seeing if they can put themselves back together? The answer is yes. Let us see what such experiments can tell us about the mechanisms of morphogenesis. We shall begin with sponges and then move on to embryonic systems.

SPONGE CELL SELECTIVE ADHESION

H. V. Wilson found that sponges could be disaggregated into single cells by pressing them through silk cloth. When the single sponge cells were allowed to settle in a dish of sea water, they moved back together to form a new sponge. If cells from the red sponge (*Microciona prolifera*) were mixed with cells from the purple sponge (*Haliclona occulata*), the red cells sought out other red cells and formed a red sponge, while the purple cells likewise formed a purple sponge. Red cells seldom stick to purple cells. Thus sponge cells could recognize their own specific species. This specificity served as a model for studying adhesive recognition in developing systems.

Moscona and Humphreys continued to investigate specific reaggregation in the sponge system. Using a rotary shaker method developed by Moscona, these workers found that when mixtures of red and purple sponge cells were rotated together, they still formed aggregates consisting mainly of only one type of cell. Red stuck to red, purple to purple (Figure 12–3). These results suggested that the major reason why the sponge cells formed species-specific aggregates was that each type of cell preferentially stuck to its own kind. The shaker method allowed cell aggregates to form as a result of adhesive stability only, without the directed movement (migration) that could occur in a stationary (nonrotated) system.

Moscona and Humphreys tried to determine the molecular nature of this species-specific adhesion. They found that sponges could also be disaggregated into single cells by immersing them in sea water free of calcium and magnesium. These cells, however, did not reaggregate if returned to sea water and rotated at low temperature (5° C). They did reaggregate when calcium was added to the disaggregation supernatant (the solution left over after disaggregated cells were removed), and the supernatant was added back to the cells. The supernatant from red sponges promoted the reaggregation of only red sponge cells. Similarly, the supernatant from the purple sponges promoted reaggregation of only purple sponge cells (Figure 12–4). It was concluded that a molecule responsible for species-specific sponge cell adhesion was released when the sponge disaggregated in sea water free of calcium

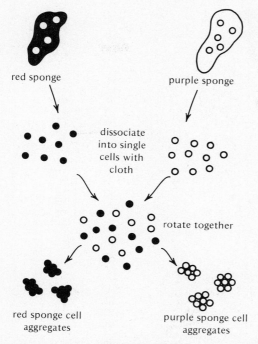

red sponge

purple sponge

dissociate
into single
cells with
cloth

rotate together

red sponge cell
aggregates

purple sponge cell
aggregates

Figure 12–3. Species-specific sponge cell reaggregation as demonstrated by Moscona and Humphreys.

and magnesium. These molecules could be added back to the cells from which they were removed, allowing them to stick back together. Apparently, cells could also resynthesize this molecule, because they reaggregated, to some extent, at higher temperatures (24° C) but not at low temperatures (5° C). At higher temperatures (24° C), cellular synthetic reaction could occur.

The molecules promoting adhesion in sponge cells were purified both by Moscona's group and by Humphrey's group. They appeared to be high-molecular-weight glycoproteins or proteoglycans (proteins associated with carbohydrate). Turner and Burger found the sugar D-glucuronic acid was on the adhesion-promoting molecule of *Microciona prolifera*. This sugar may be responsible for the binding with the receptor site on the sponge cell surface. The nature of the chemical bonds between readhering sponge cells is, however, not well understood. In any case, molecules have been isolated, purified, and characterized that appear to be responsible for allowing sponge cells to adhere to their own kind.

SELECTIVE ADHESION IN EMBRYONIC SYSTEMS

The methods used to study sponge cell adhesion have also been used to study embryonic cell adhesion. Townes and Holtfreter, Moscona, and Steinberg

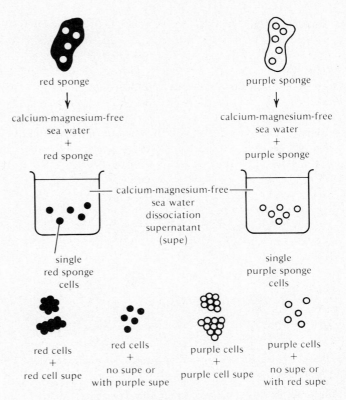

red sponge

↓

calcium-magnesium-free
sea water
+
red sponge

purple sponge

↓

calcium-magnesium-free
sea water
+
purple sponge

calcium-magnesium-free
sea water
dissociation
supernatant
(supe)

single
red sponge
cells

single
purple sponge
cells

red cells
+
red cell supe

red cells
+
no supe or
with purple supe

purple cells
+
purple cell supe

purple cells
+
no supe or
with red supe

rotation of sponge cells at 5°C with and
without dissociation supernatant (calcium added back)

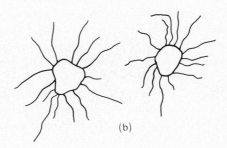

(b)

Figure 12-4. (a) Species-specific sponge aggregation factors. From experiments of Moscona and Humphreys.
(b) ''Sunburst'' Configuration of Microciona sponge aggregation factor. Based on an electron micrograph by S. Humphreys, in P. Henkart, *et al. Biochemistry* 12 (1973): 3045.

performed some pioneering adhesion experiments with embryonic cells. Embryos of all types can be disaggregated into single cells using alkaline solutions, solutions free of calcium and magnesium, or solutions containing proteolytic (protein degrading) enzymes such as trypsin. Embryonic cells disaggregated by such means can be rotated in the medium using a gyratory shaker (the Moscona shaker method). The cells from a whole embryo or from a variety of embryonic tissues recombine into aggregates of many cell types. After a day or two, however, cells tend to "sort out" into groups of only one specific cell type. The mixed cell aggregate gradually transforms into an aggregate containing several areas, each composed of only one cell type (Figures 12-5 and 12-6). For example, ectoderm cells stick to ectoderm cells, retina cells stick to retina cells, preheart cells stick to preheart cells. The nature of this "sorting out" phenomenon is unclear, but it appears to involve the relative adhesiveness between the different cell types. The most tightly adhesive cells squeeze together toward the center of the aggregate, pushing the less adhesive cells to the periphery. McGuire and coworkers recently

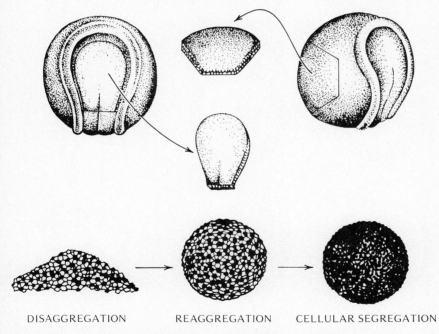

DISAGGREGATION   REAGGREGATION   CELLULAR SEGREGATION

Figure 12-5. Sorting out of reaggregated amphibian embryo cells. A piece of the medullary plate and a piece of prospective epidermis were excised and disaggregated by alkali. The free cells are intermingled (epidermal cells are indicated in black). Under readjusted conditions, the cells reaggregate and subsequently segregate so that the surface of the explant becomes entirely epidermal. From Townes and Holtfreter, *J. Exp. Zool.* 128 (1955): 53–120.

MEDULLARY PLATE +
EPIDERMIS

MEDULLARY PLATE +
EPID. + FOLD

MEDULLARY PLATE
ON ENDODERM

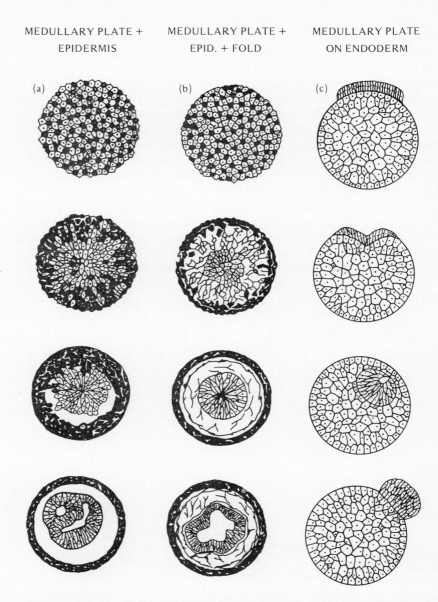

Figure 12-6. Sorting out of reaggregated combination of amphibian embryonic cells. Sections through successive stages of composite reaggregates. (a) Randomly arranged cells of epidermis (black) and medullary plate (white) move in opposite directions and reestablish homogeneous tissues. (b) Cells from the neural fold were added. The latter move to occupy the space between the neural tissue and the epidermal covering. (c) A piece of medullary plate or of larval forebrain first moves into, then out of a matrix of endoderm. From Townes and Holtfreter, *J. Exp. Zool.* 128 (1955): 53–120.

found that if embryonic cells were disaggregated from the embryos very gently, they behaved like the sponge cells. That is, embryonic liver cells mainly adhered immediately to other liver cells and not to cells from other embryonic tissue cell types. Thus, there is a tissue-specific adhesive recognition between like embryonic cells. This adhesive recognition may help govern embryonic cell rearrangements.

Let us now turn to some experimental evidence suggesting that specific adhesive recognition plays an important role in morphogenesis. Roth's group (Barbera, Marchase, and Roth) used the chick embryo retina-tectum system in their study. Nerves grow from the retina of the eye to specific regions of the optic tectum, the brain center that interprets visual impulses. Nerves from the dorsal portion of the retina always grow to the ventral portion of the optic tectum, not to the dorsal portion. Roth and coworkers sought to determine whether adhesive recognition plays a major role in these specific nerve connections. They first cut retinas into dorsal and ventral halves and tecta into dorsal and ventral halves. They then labeled retina cells from each half of the retina with radioactive precursors (molecules that become incorporated into cells) or by their own pigment. Single labeled dorsal or ventral retina cell suspensions were agitated with either dorsal tectal halves or ventral tectal halves. This technique of shaking chunks of tissue with more than one cell type was developed by Roth and Weston to measure the adhesion of like to like and unlike to unlike cell types. In general, the dorsal retina cells were picked up more often by ventral tectal halves than by dorsal tectal halves. Ventral retina cells, on the other hand, were picked up more often by dorsal tectal halves than by ventral halves (Figures 12–7, 12–8). Roth and coworkers concluded that there is a specific adhesive recognition between dorsal retina and ventral tectum and between ventral retina and dorsal tectum. This adhesive recognition may help determine where nerves from the retina become located on the tectum. In the embryo, after all, dorsal retina nerves innervate ventral tectum and ventral retina nerves innervate dorsal tectum. The nerve endings may recognize the surrounding area and get to their final resting place as a result of specific adhesion to that area.

The results of Barbera, Marchase, and Roth have been recently confirmed and extended using other approaches. Trisler, Schneider, and Nirenberg identified a monoclonal antibody that bound more strongly to dorsal retina than to ventral retina. They then cut 14-day-old chick embryo retinas into eight segments along a dorsal-ventral axis and determined how well each segment bound the monoclonal antibody. The most dorsal segment bound 35 times as much antibody as the most ventral segment. These results demonstrate that a molecular gradient of a specific antigen exists in the chick retina. The existence of this gradient supports the idea that specific interactions between certain cells of the retina and tectum could be governed by information contained in the gradient.

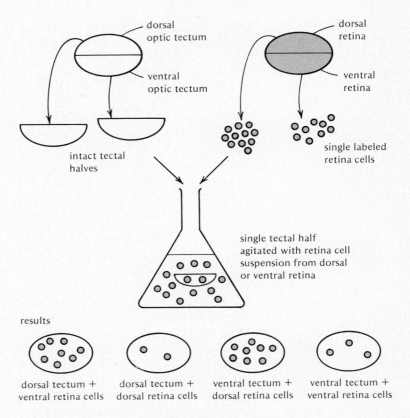

Figure 12-7. Retino-tectal adhesive recognition. Barbera, Marchase, and Roth found that dorsal retina cells adhered best to ventral tectum and ventral retina cells to dorsal tectum. The time needed for adherence varied in these experiments.

MOLECULAR MODELS OF ADHESIVE RECOGNITION IN EMBRYONIC SYSTEMS

In the previous sections, we saw that molecules could be isolated that promote the adhesiveness of sponge cells in a species-specific way. Other research indicated that molecules can also be isolated from embryonic cells that serve a similar function. Moscona and Lilien have isolated a molecule from a culture medium in which chick embryo neural retina cells were grown. This molecule promoted the adhesion of embryonic neural retina cells, but not that of other cells. When chick embryo retina cells were rotated with the adhesion-promoting factor, the cell aggregates formed over time were much larger than when the adhesion factor was absent. Similar adhesion-promoting factors have been isolated from mouse teratoma cells by Oppenheimer and Humphreys, from cellular slime molds by Rosen and Barondes, and from sea urchin sperm by Vacquier and colleagues (see Chapter 5).

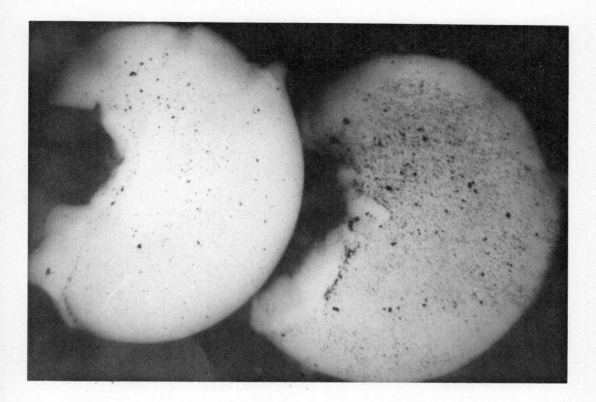

Figure 12-8. Adhesion of ventral retina cells to dorsal and ventral tectal halves.
Right: Dorsal tectum. Left: Ventral tectum. Retina cells are from pigmented retina and
appear as black dots on the tecta. From Barbera, Marchase, and Roth, *Proc. Nat. Acad. Sci.
U.S.*, 70 (1973): 2482–86. Courtesy of Stephen Roth.

Edelman and colleagues have identified a number of *cell-adhesion molecules*
(CAMs) in specific tissues of various vertebrates. One of these molecules
controls neuron–neuron and neuron–muscle adhesion. During embryonic
development, this neuron cell-adhesion molecule (N-CAM) loses much of its
sialic acid content. Such a molecular change could account for the changing
adhesive properties of developing cells and tissues. The N-CAMs do not
change in the cerebellum of mice with the staggerer mutation, a mutation in
which nerve-cell interactions are defective. These observations are consistent with the hypothesis that changing CAMs help to control cellular
interactions during embryonic development.

Oppenheimer and Meyer isolated an aggregation-enhancing protein
from blastulas of the sea urchin *Strongylocentrotus purpuratus* by disaggregating
them in sea water free of calcium and magnesium (Figure 12–9). The isolated
protein dramatically promoted the reaggregation of *Strongylocentrotus purpuratus* blastula cells, but not the corresponding gastrula cells. It has no effect

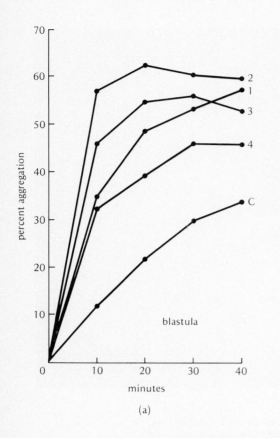

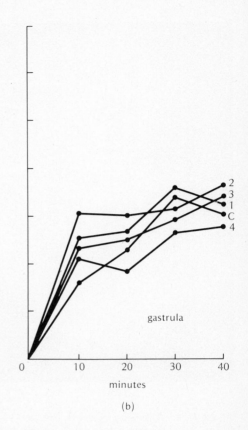

percent aggregation

(a)   (b)

Figure 12-9. (a-b) Developmental stage-specificity of *Strongylocentrotus purpuratus* blastula aggregation protein. *Strongylocentrotus purpuratus* blastula cells (a) or gastrula cells (b) were rotated in sea water alone (c) or with different batches of the supernatant obtained by disaggregating *S. purpuratus* blastulas in sea water, free of calcium and magnesium. (Numbers 1–4 refer to the specific batch of supernatant used.) As can be seen, *S. purpuratus* blastula dissociation supernatant substantially promoted the rotation-mediated reaggregation of *S. purpuratus* blastula cells, but not that of gastrula cells.
(c) Species specificity of *Strongylocentrotus purpuratus* blastula aggregation protein. Blastula cells from the sea urchin *Lytechinus pictus* were rotated with *Strongylocentrotus purpuratus* blastula dissociation supernatant (numbers 2, 4) or in sea water (c). As can be seen, *S. purpuratus* supernatant had no aggregation enhancing effect on *L. pictus* cells. In all experiments, aggregation was measured with an electronic particle counter by counting the number of single cells that formed aggregates over time.

whatsoever on blastula cells from *Lytechinus pictus*, a different species of sea urchin. Evidence suggests that this protein binds cells together at their surfaces, possibly by attaching to carbohydrates. This molecule, like the sponge aggregation factor, is clearly species specific. It is also stage specific in its action. Furthermore, this material was isolated from intact embryo cells

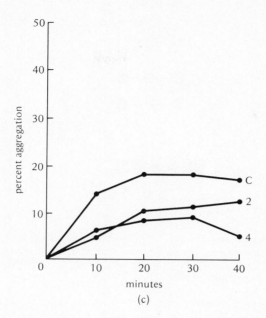

(c)

and promotes the adhesion of specific embryo cells in their natural medium, sea water. Taken together, these results support the contention that the molecule helps mediate cell adhesion in the embryo. Continued work with this material and with other factors isolated in different systems should illuminate the nature of specific cellular adhesiveness during embryonic development. Isolating, purifying, characterizing, and determining the functional groups of these adhesion factors may help elucidate the nature of the specific cell adhesion at the molecular level.

The mechanisms controlling selective cell association are probably not the same in all systems. A couple of models, however, have recently been proposed to explain cell adhesion in dynamic systems such as developing embryos and tumors. One model, the cell surface glycosyl transferase–carbohydrate acceptor model, was proposed by Roseman and developed further by Roth and others. This model suggests that embryonic cell adhesion and "de-adhesion" result from cell surface enzymes called glycosyl transferases, which attach to or let loose of the carbohydrate chains of glycoproteins and glycolipids at the cell surface (Figures 12–10, 12–11). Glycosyl transferases are enzymes that catalyze the transfer of single sugar residues from nucleoside diphosphate sugars to the ends of the carbohydrate chains of glycoproteins and glycolipids. Thus, these enzymes are responsible for synthesizing sugar chains. Are these enzymes present on the cell surface? Most experiments suggest that they are. Are the sugar chains present on the cell surface? Definitely yes. So it makes sense that a glycosyl transferase on one cell would grab a sugar chain on an adjacent cell. Many such enzyme-

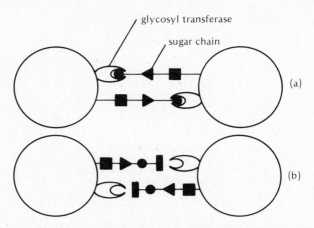

Figure 12–10. Hypothetical model of embryonic cell adhesion. (a) Adhesion takes place when the appropriate glycosyl transferases and sugar chains are present. (b) Adhesion fails because the glycosyl transferases and sugar chains do not match. After Roseman, *Chem. Phys. Lipids* 5 (1970): 270.

substrate complexes would cause the cells to adhere. Release of the sugar chains would allow the cells to separate.

The beauty of this hypothesis is that it can explain the complexity of specific cell associations and disassociations in embryos. Each transferase is specific for a given receptor chain and for the sugar it transfers. Thus, certain cells could stick together if the right transferases and the right sugar chains were present. It should be emphasized that this model only explains initial cell contacts. It does not deal with the more stable associations that result from the secretion of cementing substances in differentiated tissues. Experimental evidence is accumulating from several systems in support of the glycosyl transferase hypothesis.

GLYCOSYL TRANSFERASE AS A MEDIATOR OF SPERM BINDING TO EGGS

We discussed the hypothesis advanced by Roseman and colleagues that cell-surface glycosyl transferases mediate cell-cell adhesion by binding to cell surface molecules containing carbohydrate. A variety of more recent studies have also suggested that glycosyl transferase-acceptor interaction mediates cellular interactions in several developmental systems.

In one such study, Shur and Hall showed that mouse sperm bind to mouse eggs by means of a specific glycosyl transferase located on the surface of the sperm cell. The specific enzyme involved is a galactosyltransferase that binds to N-acetylglucosamine groups on the zona pellucida of mouse eggs. Findings that support this mechanism included the following:

1. Two reagents that interfere with surface galactosyltransferase activity inhibit sperm-egg binding. The milk protein alpha-lactalbumin

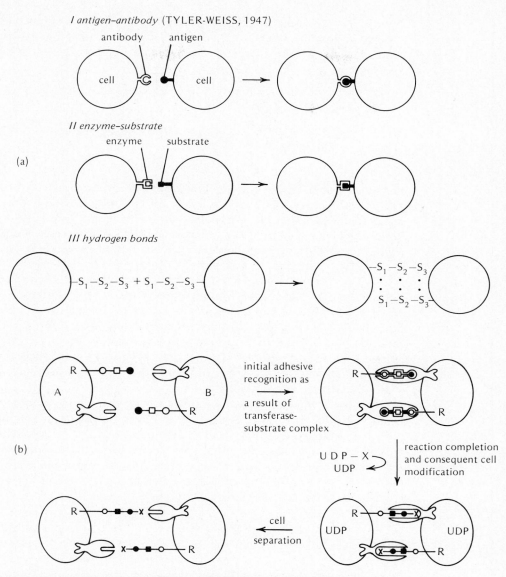

Figure 12–11. Hypotheses of cell–cell adhesion: antibody-antigen, enzyme-substrate, hydrogen bonds, modification of cell–cell adhesion by glycosyl transferase reaction. In the last hypothesis, the cells adhere as a result of the binding of a complex carbohydrate chain to the corresponding glycosyl transferase. Once bound, an internally generated sugar-nucleotide (UDP-X) provides X to the enzyme on the internal face of the membrane. The transferase uses X to complete the reaction, and in the third step (cell separation) the product of the enzymatic reaction dissociates from the enzyme. From: Roseman, The biosynthesis of complex carbohydrates and their potential role in intercellular adhesion, in A. A. Moscona, ed., *The Cell Surface in Development* (New York: John Wiley, 1974), pp. 255–72.

changes the ability of the transferase to bind to *N*-acetylglucosamine and also inhibits sperm-egg binding; UDP-dialdehyde inhibits sperm-egg binding and sperm surface galactosyltransferase to identical degrees.

2. Sperm-surface galactosyltransferase can add sugar to egg cell surfaces.

3. Enzymatic removal of egg-surface *N*-acetylglucosamine inhibits sperm binding by 86%.

The molecular basis of sperm-egg binding is indeed becoming clearer!

Substantial evidence supports a role for cell-surface carbohydrates in cell associations. Experiments from laboratories including those of Roseman, Roth, Oppenheimer, Steinberg, Moscona, Burger, Rosen, and Barondes suggest that cell-surface carbohydrates are involved in the adhesion process. If embryonic cells are treated with purified glycosidases (enzymes that catalyze the removal of sugar residues from carbohydrates), their adhesion and aggregation rates change. Molecules that promote aggregation are often inactivated when treated with glycosidases. In addition, specific carbohydrate-binding proteins have been isolated from slime mold and embryo cells. These proteins promote cell association when the cells are rotated with these molecules. The proteins apparently bind specific sugar residues on the cell surface. Thus, protein-carbohydrate interaction may be of key importance in controlling cell-cell associations.

In summary, the molecular nature of cell associations in embryos is not well understood. Cell adhesion, however, plays an important role in the cellular rearrangements that occur in developing embryos. Carbohydrate-protein interaction may control embryonic cell associations in some systems. Since the carbohydrate chains of cell surface glycoproteins and glycolipids reach farthest away from the cell, these chains must play a role in at least the initial contact with approaching cells and substrates.

OTHER MECHANISMS

Specific adhesive recognition among cells is clearly not the only mechanism that controls the development of form in embryos. We have treated cell adhesion in length because this area has been intensively investigated in recent years, and because specific testable models have been developed of the molecular nature of adhesive recognition among cells. Many other specific mechanisms are important in controlling embryonic morphogenesis. Let us briefly examine some of these.

CYTOSKELETAL ELEMENTS IN
MORPHOGENESIS

Throughout, we have mentioned that cytoskeletal elements such as rod-shaped proteins, microtubules, and microfilaments—the "microskeleton" of the cell—are involved in a variety of embryonic cell activities. For instance, cytoplasmic microtubules are associated with elongating cells. Cell elongation often occurs during embryonic cellular rearrangements and during morphogenesis of embryonic organs. When cell elongation takes place, microtubules become visible, lined up in the long axis of the cell. Microtubules are observed in elongating cells around the amphibian blastopore, in the neural plate cells, in cells moving into the primitive groove in bird embryos, and in outgrowing nerve cells. Cell elongation is prevented by colchicine, which specifically disrupts microtubules. Colchicine prevents nerve cell outgrowth and inhibits cell elongation in the chick primitive streak and in the neural plate. Colchicine is not known to affect anything other than microtubules. Microtubules thus play a role in embryonic cell elongation. They may directly force the cells to elongate, or they may serve as tracks for cytoplasmic flow in the direction of elongation.

Cytoplasmic microfilaments cause cell narrowing. Micofilaments serve as a simple cellular contractile system. The morphogenesis of tubular glands, for example, may be aided by bundles of microfilaments that contract and exert force on the sides of cells. This could narrow the cell at one end. If a sheet of cells narrows at one side, the sheet will bend. Other hypotheses, however, have been advanced to explain how certain cell layers (epithelia) bend or fold (Figure 12–12). For example, suppose that the outer surface cells of the layer were tightly attached to each other so that they cannot move laterally at all. If cell division takes place in the layer, the cell sheet will buckle inward. Microfilaments may also act here by causing the outer cells to become less elastic. Bundles of such microfilaments have been observed in the surface regions of embryonic pancreas, lens, thyroid, and lung.

EXTRACELLULAR MATERIALS AND
MORPHOGENESIS

Two groups of extracellular molecules play an important role in morphogenesis. These are glycosaminoglycans and collagen. Glycosaminoglycans are sugar polymers consisting of uronic acids and amino sugars. They are often linked to proteins, forming proteoglycans. (Recall that one of the sponge cell adhesion factors is a proteoglycan.) Collagen is a long structural protein that is fibrous and tough. Collagen is often associated with proteoglycan.

Collagen may exert a specific morphogenetic stimulation on certain cells. For example, Hauschka and Konigsberg showed that collagen stimulates prospective muscle cells to fuse together, forming multinucleated cells

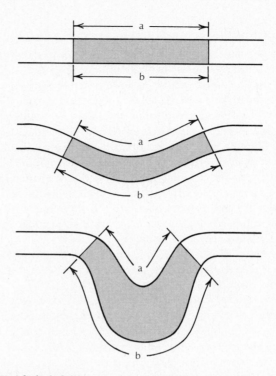

Figure 12–12. Epithelial folding: a hypothetical mechanism. Surface (a) (outer surface) is tightly adherent. Cell division takes place within the layer, and the sheet buckles. Modified from Zwann and Hendrix, *Amer. Zool.* 13 (1973): 1039.

that become adult muscle. Collagen is present in embryonic regions where skeletal muscle forms, and may play a role in its development and location. Also, an oriented collagen pattern is needed in feather development. If an oriented pattern is absent, normal, organized feather pattern development does not occur.

Glycosaminoglycans are also important in morphogenesis. These molecules apparently help maintain the shape of branching epithelial structures. For example, if embryonic salivary gland, lung, or kidney epithelia is treated with enzymes that remove surface glycosaminoglycans, normal morphogenesis is inhibited and the structures tend to round up and form a ball.

The branching or lobulation of epithelial layers in structures such as the salivary gland, mammary gland, lung, kidney, and thymus is a morphogenetic event that has been carefully studied in recent years. Although the mechanisms involved are not completely understood, a rather well-documented picture is emerging (Figure 12–13). Branching glands consist of an epithelium associated with mesenchyme. The entire organ rudiment can be removed from the organism and studied in culture. In this way, the roles of the different parts can be studied. Let us examine some of the interesting

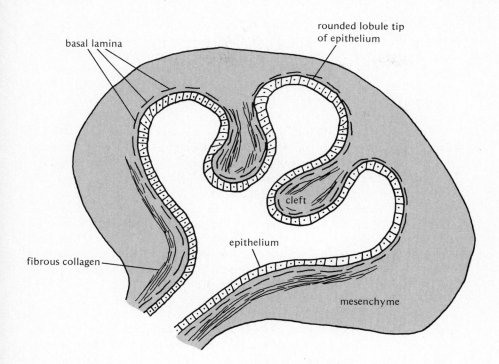

Figure 12–13. Branched gland showing relationships between the epithelium, the mesenchyme, the basal lamina, and the fibrous collagen. Based on work by Bernfield's group.

experiments that have helped us understand the nature of branching morphogenesis.

Branching begins when the epithelial layer folds inward, forming clefts in the epithelium. One of the mechanisms that control this infolding is the contraction of cytoplasmic microfilaments at one end of the cells in the folding layer. Also, the outer surface cells of the epithelial layer may be tightly attached to each other, so that when cell division occurs within the layer, the cell sheet will buckle inward. Cleft formation in epithelial layers is only one step in the formation of a lobulated gland.

Another component involved in epithelial lobulation is the mesenchyme that envelops the epithelium. The epithelium will not branch in the absence of mesenchyme. The influence of mesenchyme on epithelial branching is not well understood (we shall return to this question shortly). A great deal of attention, however, is being paid to extracellular material between the epithelium and mesenchyme. Some of these materials are synthesized by the epithelium, others by the mesenchyme.

In the well-studied mouse salivary gland, the extracellular material between the mesenchyme and epithelium is composed of two layers. One consists of collagen fibers together with glycoprotein and proteoglycan. The other layer, the *basal lamina*, is rich in collagens, glycoproteins, and polysaccharides such as glycosaminoglycans. It adheres tightly to the epithelium adjacent to the plasma membrane of the epithelial cells. It is important in maintaining the branched gland morphology (Figure 12–13). Bernfield and colleagues showed that if the basal lamina is removed, the epithelium making up the gland rounds up.

In one experiment, the epithelium of the gland was isolated and its basal lamina was removed enzymatically. When the epithelium was recombined with mesenchyme, the epithelium lost its lobular shape and rounded up. When the epithelium resynthesized basal lamina, the branching began again. Thus, the epithelium produces the basal lamina and the basal lamina maintains the branched morphology of the epithelium. The basal lamina acts like a tight-fitting glove that prevents the epithelium from losing its form. The basal lamina is thickest in the clefts and thinnest at the rounded lobule tips. This arrangement should keep the clefts stable, while allowing new clefts to form at the tips. Collagen fibers are deposited outside the basal lamina; these are also most densely packed in the clefts. Collagen fibers apparently help stabilize the indentations.

A major role of the mesenchyme is apparently to degrade the basal lamina near the rounded lobule tips so that clefts can form in these regions. Bernfield and colleagues labeled the glycosaminoglycans of the basal lamina with [³H] glucosamine. When mesenchyme was present, the label was lost from the basal lamina outside the rounded lobule tips. Very little label was lost from the basal lamina in the clefts. When mesenchyme was absent, no significant label was lost in any region of the basal lamina. These experiments suggest that the mesenchyme selectively degrades the basal lamina in the morphogenetically active regions of the branching epithelium (that is, the rounded tips). Once the basal lamina has been degraded, some sort of signal might stimulate epithelial cleft formation in that region. For instance, Ca^{2+} ions released during glycosaminoglycan degradation can trigger microfilament contraction in the adjoining epithelium. Alternatively, basal lamina degradation might stimulate cell division in the epithelium, which could also lead to cleft formation. The exact mechanism by which mesenchyme influences epithelial branching is not yet clear. Because of the active investigations in this area, however, more of the answers to these questions should soon be in hand.

In summary, in this section we examined some of the factors involved in morphogenesis. Cytoskeletal elements (microtubules and microfilaments), localized cell division, and extracellular substances such as collagen and glycosaminoglycans play important roles in developing epithelial structures.

MECHANISMS OF NEURAL CREST MIGRATION

We noted earlier that neural crest cells differentiate into many structures, some far from the original neural crest ectoderm. These cells must travel long distances to their final locations. Work from many laboratories, including those of Weston, Noden, Cohen, and Le Douarin, indicates that neural crest cells migrate along well-defined pathways. Neural crest cells of the trunk migrate in two streams. One moves ventrally into the mesenchyme located between the somites and the nerve tube. Some of the cells migrate into the somites, forming dorsal root ganglia; others form sympathetic ganglia and the adrenal medulla. Other crest cells move into the surface layers of the embryo, forming pigment cell precursors.

Experiments suggest that the environment in which neural crest cells find themselves plays a major role in controlling their differentiation. This conclusion is based on experiments in which neural crest cells transplanted from one trunk region could substitute for those of other trunk regions, and neural crest cells transplanted from one cranial region could substitute for those of other cranial regions. In one experiment, neural crest cells from the mesencephalon area, which normally give rise to the rostral parts of the skull, were transplanted to the more posterior metencephalon region. The resulting cells differentiated into the lower jaw, which is normally formed from crest cells in the metencephalon region. Thus, the final environment of the crest cells can change the fate of the cells to produce the stuctures that normally develop at that location. These transplantation experiments used chick and quail embryos, so that the donor cells in specific structures could be distinguished from the recipient cells.

While trunk crest can be interchanged with other levels of trunk crest and cranial crest can be interchanged with other levels of cranial crest, trunk crest cannot be completely interchanged with cranial crest, and *vice versa*. In other words, cranial crest cells are somewhat restricted or predetermined to form cartilage, while trunk crest cells cannot form cartilage no matter what their final environment is.

Finally we come to one of the most intriguing questions: What are the mechanisms that control the migration of neural crest cells to distant sites in the body? Recent work by Marianne Bronner-Fraser has led to a better understanding of this problem. We noted above that neural crest cells migrate along well-defined pathways. Bronner-Fraser injected neural crest cells, retinal pigment cells (non–neural crest cells), or latex polystyrene beads (uncoated or coated with bovine serum albumin) onto the ventral neural crest pathway of chick embryos. All of these, including the beads, translocated ventrally along the pathway in the same way as normal neural crest cells. If, however, the latex beads were coated with the glycoprotein fibronectin, the beads remained near the site of implantation and did not move ventrally. These results suggest that neural crest cell migration is influenced by some driving force imparted by the embryonic environment. Also, molecules such as fibronectin on the cell surface may serve as a recognition mechanism that prevents non–neural crest cells from entering the ventral neural crest migratory pathway. Cells of the somite, fibroblasts, and beads coated with fibronectin have fibronectin on their surfaces and do not move onto the ventral pathway. Neural crest cells, retinal pigment epithelial cells, and latex beads without fibronectin lack surface fibronectin and can move along the ventral pathway.

Exactly what forces control the translocation of cells along neural crest pathways is unclear. These experiments show, however, that the entrance to the ventral pathway is controlled by the presence of specific surface substances. Once able to enter the pathway, cellular forces (not necessarily originating in the translocating cells) may move the cells to their final location. These forces may involve movements in cells surrounding the translocating particles, allowing the particles to move in response to, for example, adhesiveness toward substances such as fibronectin. Fibronectin on the cells apparently prevents cells from entering the ventral pathway; whereas fibronectin concentrated at specific points along the pathway stops translocating cells at these points. This concept is supported by the observation that both beads and cells locate in areas where fibronectin is present. If fibronectin disappears from the substratum, cells adhering to the substratum may come loose, and the cells may form clusters with each other. This could account for the aggregation of some neural crest cells into ganglia. This is only speculation, but such speculations form good working hypotheses that surely will lead to additional work in this exciting area.

CELL DEATH IN MORPHOGENESIS

Before concluding our discussion of the mechanisms that influence the development of form in embryos, let us briefly mention another important factor that shapes certain structures. Some structures are molded by cell death that occurs in specific regions of the structures. For example, early in embryonic development our hands and feet resemble paddles or flippers. Our fingers and toes form as the result of cell death in the regions between the prospective digits, as shown in Figure 12–14. This cell death fails to occur in certain abnormalities. Webbed hands or feet result. Cell death occurs in many developing embryonic structures. We saw, for example, that certain structures in each sex degenerate after the indifferent stage of sexual development. The tadpole tail regresses during metamorphosis as the result of cell death.

Increased levels of lysosomal enzymes (for example, acid phosphatase, ribonuclease, cathepsins, etc.) are observed in many embryonic regions that are "programmed to die." In the urogenital system, sex hormones control death in certain structures. For example, androgens trigger cell death of Müllerian ducts in males by influencing lysosomal enzymes. In amphibians, on the other hand, the regression of the tail and gills is controlled by thyroid hormone. If the thyroid gland is removed from tadpoles, metamorphosis does not occur—the tadpole does not change to become a frog. Tails and gills remain. If thyroid hormone is given to the "thyroidless" tadpoles, meta-

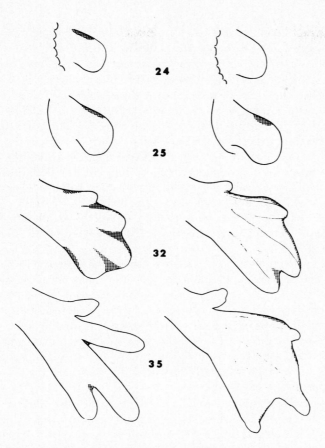

Figure 12-14. Cell death during digit development in chick and duck embryo hindlimb. Shaded areas die in forming limb. From Saunders and Fallon, in M. Locke, ed., *Major Problems in Developmental Biology* (New York: Academic Press, 1967), pp. 289–314. Courtesy of J. W. Saunders.

morphosis proceeds. Lysosomal enzymes increase in the areas that die. It should be stressed that specific hormones such as thyroid initiate cell death in only certain amphibian cells (tail, gills, and horny teeth) while in others (for example, the limb bud) it causes cell proliferation. The hormone may turn on certain genes in some cells, but other genes in other cells.

PATTERN MORPHOGENESIS

Pattern refers to the orderly arrangement of parts of the organism. It results from many types of morphogenetic and differentiative events. In other portions of this text, we discuss pattern formation in the context of such topics as nerve growth to end organs, limb morphogenesis and regeneration, finger and toe development, plant embryo polarity, mosaic cleavage patterns, and cytoplasmic segregations. In this section, we shall deal with only a few simple systems in which some fundamental findings have been made. How does an ordered arrangement of parts come about?

When we think about pattern formation, we usually think of several sorts of mechanisms. For instance, there must be gene action. That is, genes that control specific differentiations (hair bristle, pigment formation) must be turned on so that the relevant parts become differentiated.

The second important factor is the environment in which the differentiated cells reside. We have seen that cells often interact with other cells or with noncellular environments. This interaction initiates selective gene activation or aids functioning in responding cells.

A third set of events is cell migration and selective adhesion. In many systems, these events get the relevant cells to the future sites of pattern development.

In the fruit fly *Drosophila*, some of the mechanisms mentioned above come into play in forming bristle patterns in the epidermis. The presence or absence of a bristle at a specific site in the epidermis depends upon the genotype of that portion of the epidermis. If the tissue is mutant at a given locus, no bristle forms even if this section of tissue is surrounded by nonmutant, bristle-forming epidermis. If the tissue is wild type at that locus, a bristle forms even if the tissue is surrounded by mutant tissues. Such results indicate that specific genes control bristle formation in tissue. In other cases, however, surrounding tissue also plays a role.

Melanocytes are pigment-forming cells, most of them derived from the neural crest. Only the melanocytes in the pigmented retina of the eye are derived from the neural tube itself. The neural crest melanocyte precursors migrate from the neural crest to a variety of sites in the body, such as the skin and hair follicles. Pigment pattern in mammals can be studied in depth because the controlling genes are known in such mammals as mice. One can transplant a piece of skin from a black mouse onto the back of a yellow mouse, or vice versa. When yellow skin is placed on the back of a black mouse, the melanocytes from this skin move out into adjacent hair follicles in the host tissue and begin to produce black pigment despite their own genotype. In the reciprocal experiment, the melanocytes from the black transplant migrate into the hair follicles of the yellow host and produce yellow pigment. Thus, the formation of yellow or black pigment appears to be controlled by the hair follicle cells, not by the melanocytes. Sulfhydryl compounds, such as glutathione, may be the hair follicle factors that stimulate the formation of yellow pigment in the melanocytes. That is, the glutathione present in yellow hair follicles may inhibit black pigment synthesis while stimulating yellow pigment synthesis. This suggestion is supported by the finding that isolated, *in vitro* cultured yellow melanocytes form black pigment. If sulfhydryl compounds are added to the culture medium, however, these melanocytes revert to synthesizing yellow pigment.

The most dramatic examples of pigment patterns are the spotted coats of many mammals and birds. What about an animal with white spots on a dark background? How does such a pattern form? Cytological examination of hair follicles in the white spots of such an animal indicates that differentiated melanocytes are absent in these areas. Can the hair follicles in the white spots permit melanocytes to form pigment? Yes. Melanocytes from a black skin transplant can migrate into a white spot region, enter the hair follicles, and make pigment. Thus, the white-spotted areas can sustain the differentiation of melanocytes that are already mature. The white-spotting pattern, therefore, is probably caused by some factors present in prewhite spots that prevent melanocyte differentiation. Even if melanocyte precursors enter the white spot areas, they do not differentiate.

Pigment pattern formation, therefore, involves all three of the mechanisms mentioned earlier. These are

1. migration of premelanocytes from the neural crest to the skin and hair follicles;

2. interaction of the melanocytes with the environment (that is, with products of the hair follicle cells); and

3. specific gene activation in the melanocytes that results in the synthesis of pigment.

Pattern formation, therefore, is not a unique phenomenon but results from mechanisms that operate in many embryonic processes. We shall completely understand the formation of organized arrangements in organisms—that is, patterns—only when we understand the molecular nature of cell-cell interactions and differentiation.

MORPHOGENESIS IN A SIMPLE SYSTEM: CELLULAR SLIME MOLDS

Morphogenesis in vertebrates is complex and difficult to study. As yet, little is known about the mechanisms that control morphogenesis in vertebrate embryos. Slow progress is being made, but the major problem is that such studies must usually be performed with isolated embryo cells *in vitro*. Thus, for example, studies of cell-cell interactions are often done in culture, an environment that is not the same as that in the embryo. A few simpler systems, however, have been studied that have helped increase our understanding of the forces that shape the organism. One of these systems is the cellular slime mold.

What are cellular slime molds? At one stage in their life cycle, these organisms are amoebae. At another stage, the amoebae aggregate and form a multicellular fruiting body composed of a stalk and a mass of spores at the tip of the stalk (Figure 12–15). The spores give rise to amoebae. We have here a fairly simple system that goes through morphogenetic processes like some of those occurring in higher organisms. There is an aggregation phase in which the free amoebae congregate together to form an integrated migratory cellular mass (slug). This process resembles some of the cellular rearrangements that occur in vertebrate embryos. The cells in the slug differentiate, and morphogenesis occurs in which the slug is transformed into the fruiting body (sporocarp). Thus, the same sorts of events, namely morphogenesis and differentiation, occur in this rather simple system as occur in higher embryos. What is known about the mechanisms of slime mold morphogenesis?

The aggregation phase in several slime molds, such as *Dictyostelium discoideum*, has been examined by several groups. The free amoebae continue to divide and remain single as long as there is an adequate supply of the bacteria they feed on. When the bacteria are exhausted, the amoebae in

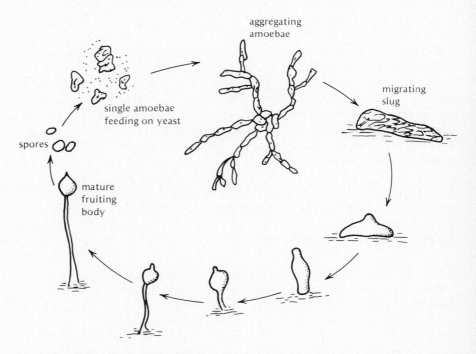

Figure 12–15. Vegetative growth and fruiting body formation in *Dictyostelium discoideum*.

dense cultures begin to aggregate, forming the multicellular slug. What causes this aggregation to occur? Certain cells secrete the nucleotide cyclic AMP (cyclic adenosine monophosphate). The amoebae move in the direction of increasing concentration of cyclic AMP. Such movement of cells is called chemotaxis. Thus, specific chemotaxis controls one phase in slime mold development, the aggregation phase.

Cyclic AMP causes amoebae to move together. Other work by Rosen's and Barondes's groups suggests that the amoebae stick together by protein-carbohydrate interaction. These workers have isolated carbohydrate-binding proteins (lectins) from slime mold cells that were competent to aggregate. The carbohydrate-binding proteins apparently occur on the cell surfaces of amoebae. These molecules bind specific sugar residues on carbohydrate chains that also apparently occur on the amoebae cell surfaces. Thus, cyclic AMP gets the amoebae together, and the carbohydrate-binding protein-sugar chain interactions stick the amoebae together (Figure 12–16).

What causes the mass of amoebae, the slug, to differentiate into the fruiting body? Cells at the leading end of the migrating slug become prestalk cells; the remainder become prespore cells. The position of the cells in the slug therefore plays a role in determining the fate of the cells. The fate of cells is reversible, because if the slug is divided into the prestalk cells and prespore cells, each part can produce a complete fruiting body. Thus some cells shift their fates to produce the missing cell types. Position with respect to other cells and the external environment probably plays a major role in the fates of cells in a cut slug, just as in an intact slug. Different, previously internal cells are now at the tip of the cut slug. New positions appear to determine new fates.

In summary, specific factors that control development in the cellular slime molds, such as cyclic AMP, have been investigated. Other factors such as carbohydrate-binding proteins also play a role, but these are still somewhat less well understood. Cell position also governs differentiation in slime molds, but no one yet knows how position "works" at the molecular level. Still, simple systems such as this help us understand the mechanisms that operate in the morphogenesis of higher organisms.

MORPHOGENESIS IN PLANTS

Throughout the text, we have stressed that cell-cell interactions play a key role in development. Animal embryos, for instance, take form as a result of cells and cell layers moving over and under one another. Such morphogenetic events are the result of intimate contact between the surfaces of the embryonic cells. Sometimes, these contacts result in stable associations; at other times, the contacts are only transitory and the cells move on to establish other, more permanent contacts. To many of us, plants are rather static structures whose cells are encased in thick cellulose walls. How do plants

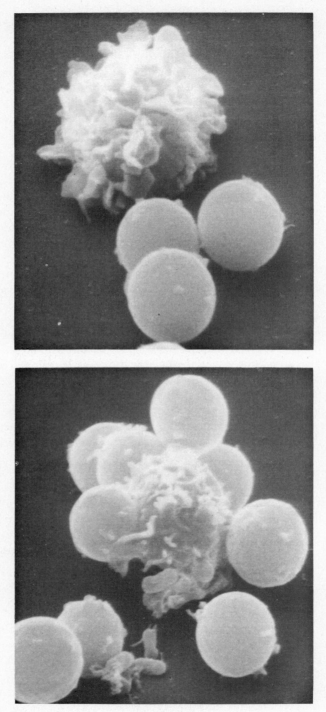

Figure 12–16. Scanning electron micrographs of slime mold cell (*Polysphondylium pallidum*) **and fixed sheep erythrocytes.** (above) In the prescence of D-galactose, cell-cell interaction is inhibited. (below) D-glucose does not inhibit cell-cell interaction. Courtesy of Steven Rosen.

develop? How does the plant embryo take shape? What effect do the cell walls have on development? Clearly, the full range of movements and contacts associated with animal embryos are not possible in plant embryos. As we will see, the plant embryo generates its form by cell formation and growth. Cell migration does not play a role in plant embryo morphogenesis.

Before we describe plant embryonic development, it might be useful to summarize some of the principal differences between animal and plant development. Plant cells develop where they are formed. There is no gastrulation, no process that shapes the embryo by cellular rearrangements. There is no real cleavage stage in plant embryos following fertilization; instead, immediate differentiation occurs.

DEVELOPMENT IN FUCUS

As in animals, some key aspects of plant development are best observed in plants whose embryos develop outside the parent. Among animals, sea urchins and amphibians are prime candidates for study because they are easily observable in the laboratory, a large number of gametes are available, and the gametes develop in simple media. *Fucus* is a brown seaweed that offers many of the same advantages. Large numbers of gametes can easily be observed in the laboratory. About 15 hours after fertilization, a striking event occurs: in the zygote. An outgrowth appears at one side of the zygote. This tubular outgrowth, the *rhizoid*, occurs only on one side of the zygote. Thus polarity is established early in the form of a specific differentiative event. About 9 hours later, a nuclear division occurs and a wall is formed perpendicular to the rhizoid outgrowth (Figures 12–17, 12–18).

In *Fucus*, we now have two cells that are clearly not the same in appearance. Each cell has a different developmental fate. Derivatives of the rhizoid cell produce a holdfast, a root-like anchor, that attaches the plant to the substrate. The derivatives of the other cell (the terminal cell) form a leaf-like body. What causes this polarity in the *Fucus* zygote? What causes a rhizoid to develop at one side of the zygote? It should be stressed that this kind of polarity is not limited to *Fucus*. Figure 12–19 shows that similar polar development patterns occur in many other plant embryos. The study of *Fucus* polarity, therefore, should be applicable to the development of form throughout the plant kingdom.

A rhizoid forms in zygotes in which nuclear division is blocked by colchicine. Colchicine, you will recall, blocks the assembly of the spindle fiber microtubules, the spindle components involved in separating the

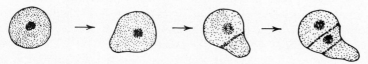

Figure 12–17. Early development of *Fucus* embryo. From Jaffe, *Adv. Morph.* 7 (1968): 295–328.

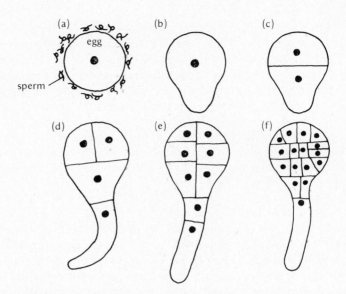

Figure 12-18. Early development in *Fucus*. From experiments of Nienburg (1928, 1933); Smith (1955).

chromosomes during cell division. The formation of the rhizoid, therefore, depends on differences in the cytoplasm rather than differences in early nuclei. What factors influence rhizoid development? The zygote establishes polarity in response to environmental factors. Rhizoids form on the warm side of a temperature gradient if one is set up in cultures. Rhizoids form on the more acid side of a pH gradient in cultures. Rhizoids form on the shaded side of a white-light gradient in cultures. These environmental factors apparently trigger cytoplasmic rearrangements that account for the development of polarity (Figure 12-20).

How do environmental factors influence polarity in the zygote? Studies of the ultrastructure of zygotes exposed to illumination from only one side show that the nuclear surface becomes polarized before any visible rhizoid appears. Finger-like projections radiate from the nucleus toward the site of rhizoid formation. These projections appear to be extensions of the nuclear membrane. Ribosomes, mitochondria, and fibrillar vesicles also concentrate in the region where the rhizoid will form. These changes occur about 12 hours after fertilization, before the rhizoid appears.

Within fifteen minutes after illumination with unilateral light, *Fucus* zygotes show some signs of polarity. For example, if one lyses (dissolves) the cells with hypotonic (dilute) solutions, lysis begins on the shaded side, on the side where the rhizoid develops 15 hours later! It seems that unilateral illumination produces a very early response.

Other changes in the *Fucus* zygote include the accumulation of the

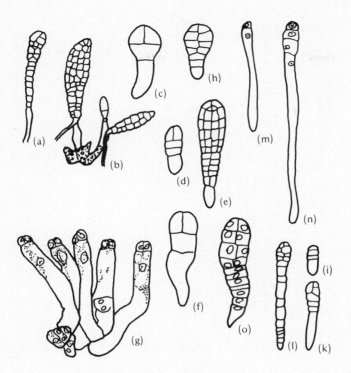

Figure 12-19. Early stages in the development of embryos in different major groups, illustrating: (1) the filamentous or polarized development of the young plant, and (2) Errera's principle of cell division by walls of minimal area. (a) *Fritschiella tuberosa* (a green alga; after M. O. P. Iyengar). (b) *Laminaria digitata* (a brown alga, young sporophytes still attached to oogonia; after F. Oltmanns). (c) *Fucus vesiculosus* (a brown alga; after G. Thuret and F. Oltmanns). (d), (e) *Radula complanata* (Hepaticae: Jungermanniales; after H. Leitgeb). (f) *Selaginella spinulosa* (Lycopodiales; after H. Bruchmann). (g) *Sequoia sempervirens* (a gymnosperm; after J. T. Buchholz): several embryos. (h) *Poa annua* (a monocotyledon; after R. Souèges). (i), (k) *Goodyera discolor.* (l) *Orchis latifolia* (Orchidaceae; after M. Treub). (m), (n) *Cardamine pratensis* (a dicotyledon; after A. Lebègue). (o) *Daucus carota* (a dicotyledon; after H. A. Borthwick). From C. W. Wardlaw, *Embryogenesis in Plants* (New York: John Wiley, 1955).

sulfated polysaccharide fucoidin near the cell wall at the rhizoidal end of the cell. Also, *Fucus* embryos generate electric currents in a capillary tube that is illuminated at one end. The end at which the rhizoids form becomes electronegative compared to the other end, suggesting that current moves through the zygotes as a result of sodium, calcium, and chloride ion movements.

Jaffe suggested that this current plays a major role in the development of polarity. He proposed that the current moves charged cytoplasmic macromolecules to specific points in the cell. For example, the current may cause fucoidin to build up at the rhizoidal end of the cell. These suggestions are

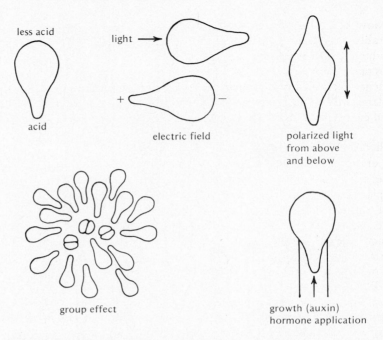

less acid

acid

light →

electric field

+   −

polarized light
from above
and below

group effect

growth (auxin)
hormone application

Figure 12–20. Polarity in *Fucus*. From data of Kniep (1907); Whitaker (1931, 1939); Olson (1937); Jaffe (1956).

supported by the finding that the polarity of *Fucus* zygotes can be determined by passing an artificial current through the cells.

Quatrano and others have suggested a working hypothesis on the development of polarity in *Fucus* zygotes. First, polarity is determined by the point of sperm entry unless one of the externally imposed factors or gradients is applied to freshly fertilized *Fucus* zygotes. The point of sperm entry (or, for example, the dark side of a light gradient) determines the area of the cell surface to which move the membrane components involved in transporting ions (such as Ca^{2+}) inward. Translocation of ion-transporting cell surface components is mediated by a process sensitive to cytochalasin B. This process may involve microfilaments, which move the components and also stabilize their accumulation at the specific site of rhizoid development. Cytochalasin B does reversibly block polar axis fixation in *Fucus*. Thus, one of the major events in the development of polarity probably is the movement of ion-transporting membrane patches to the prospective rhizoid area. Specific current fluxes are set up in this way, and these in turn help build up the concentration of cytoplasmic macromolecules, such as rhizoid cell wall polysaccharides, in the appropriate area. The microfilaments or other cytoskeletal components may also provide a "track" for accumulating the vesicles needed for extending the cell wall. The components of this hypothesis are now being tested. The *Fucus* system and others (such as *Pelvetia*) will probably provide answers to many basic questions regarding the mechanisms by which form develops in both plants and animals. Studies with the simple *Fucus* plant embryo have helped us understand a few of the many factors that influence early differentiation in plants. Let us now turn to a brief description of plant embryonic development. We should keep in mind the differences noted earlier between early plant and animal development.

EARLY PLANT EMBRYO

The embryo of *Capsella bursa-pastoris*, a flowering plant called shepherd's purse, has been extensively studied. We shall summarize the development of this plant embryo in this section and then move on to the interesting story of plant embryo culture. The *Capsella* embryo, like that of other flowering plants, develops in the flower, in the *ovule* (Figures 12–21 and 12–22). Thus, it is inaccessible to direct observation at early stages. Early development has been reconstructed by studying the sectioned embryos with light and electron microscopes (Figure 12–23).

The *Capsella* egg is metabolically inactive. It does not contain a great deal of aggregated ribosomes, endoplasmic reticulum, Golgi apparatus, or reserve nutrients. It is a small, pear-shaped cell contained in the *embryo sac* in the ovule. The haploid nucleus is situated at the broad end of the egg. The male gamete consists of a haploid nucleus in a cytoplasmic sheath.

Fertilization occurs after the male gamete is transported from the stigma

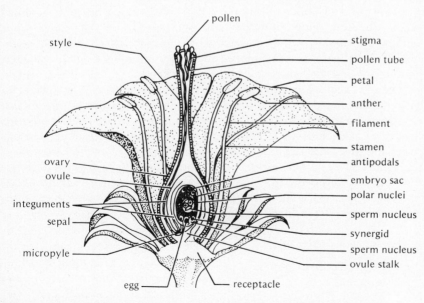

Figure 12–21. Parts of a flower. Fertilization has occurred in the ovule.

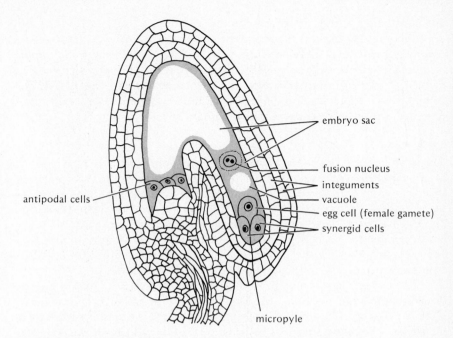

Figure 12–22. *Capsella* **ovule just prior to fertilization.** From Ebert and Sussex, *Interacting Systems in Development* (New York: Holt, Rinehart & Winston, 1970), p. 114.

of the flower by the pollen tube to the vicinity of the egg. The egg is activated after fertilization. New ribosomes are synthesized. The ribosomes come together to form polysomes, aggregates of ribosomes attached to messenger RNA. Golgi apparatus and endoplasmic reticulum increase in abundance. The thin cell wall that surrounds the egg thickens.

The first nuclear division separates the embryo into two unequal cells. One cell is small and located at the rounded end of the embryo; the other is vacuolated, large and elongated. Thus, as in *Fucus*, an early differentiation has occurred. Each of these cells has a different fate. The elongated cell divides and gives rise to a linear suspensor consisting of five to seven large vacuolated cells. This linear suspensor anchors the embryo. The small terminal cell gives rise to the globular part of the embryo, forming all of the embryonic organs. By about the 50-cell stage in the globular mass, three primary tissue systems differentiate:

1. a surface epidermis;

2. a central core of elongated procambium that gives rise to the vascular system; and

3. a cylinder of cells that will form the cortex, situated between the two other tissues.

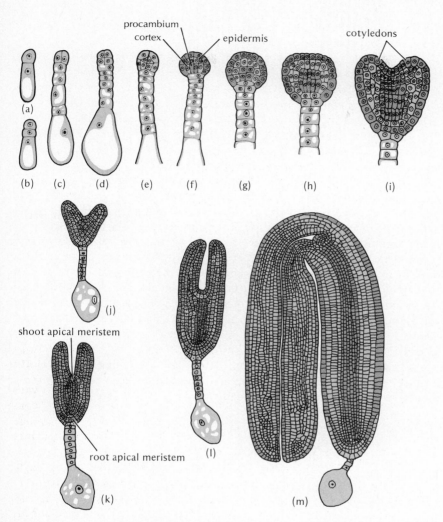

Figure 12-23. Development of *Capsella* embryo. From A. W. Haupt, *Plant Morphology* (New York: McGraw-Hill). (j)–(m) From M. Schaffner, *Ohio Naturalist* 7: 6.

The globular mass becomes heart-shaped (heart stage) as two hemispherical mounds form the cotyledons (embryonic leaves). At one pole, a group of cells between the *cotyledons* remains relatively undifferentiated. These cells form the shoot apical *meristem*. At the other pole, between the cotyledons and near the suspensor, another group of cells forms the root apical meristem. The meristems remain embryonic (Figure 12-23). Each meristem continues to divide, giving rise to the other tissues; after seed germination, they give rise to cells that differentiate into stem and root tissues. The embryo continues to enlarge in the ovule cavity. The rate of growth then declines and finally stops. The embryo is now part of a dormant seed that will not grow again until germination.

We see, therefore, that embryonic development in plants is different from that in animals. No fertilization membrane forms. No one knows what prevents polyspermy in plants such as *Capsella*. The plant embryo is shaped by the formation of cells and growth of these cells, not by cell migrations as in animals. No gastrulation occurs in plant embryos; the cells of plant embryos develop in place. The shoot and root complexes are represented only by the meristems, no vegetative leaves, shoots, or roots exist in the plant embryo. Such development occurs only after germination.

The type of embryonic development seen in *Capsella* is characteristic of most plant embryos. Some differences, however, exist. For example, gymnosperms are cone-bearing plants whose seeds are not enclosed in an ovary. The early nuclear divisions of gymnosperms are not followed by cell wall formation; thus, the young embryo is a mass of cytoplasm with many nuclei (coenocytic). A globular cell mass and a region of larger suspensor cells also develop in these embryos. In ferns and mosses, the embryo does not develop in an ovule, but in a free-living gametophyte stage. In mosses and ferns, early embryonic differentiation occurs. A region of large vacuolated cells called the foot develops, resembling the suspensor of *Capsella*. The fern and moss develop to the adult stage without going through an intervening dormant seed period.

Plant embryos are different from animal embryos in that they develop in a more static fashion, without cellular migration. The end is the same, however: an adult organism eventually forms.

HORMONAL MECHANISMS IN PLANT MORPHOGENESIS

The centers of cell division in plants, the meristems, play a role in the development of plant form. The location of these centers gives each type of plant its characteristic pattern of shoot and root growth and branching.

Meristems produce hormones that determine which buds will grow. In that way, they influence the overall shape and growth pattern of the plant. The hormones control the division and elongation of cells, the formation of vascular tissue, the breaking of seed dormancy, the growth of fruit, the abscission (release from plant) of flowers, fruits, and leaves, and the expression of sex. Morphogenesis and differentiation in plants, therefore, is under the control of specific hormones.

Shoot meristems produce the group of hormones called *auxins*, such as indole-3-acetic acid (IAA). Auxins promote stem elongation, mainly by causing vacuoles to enlarge, thus expanding the cells, and by synthesizing new cell walls. Auxin also controls the response of seedlings to light. Auxin accumulates on the shaded side of shoots and causes the cells on that side to elongate, thus bending the shoot toward the light.

Auxin controls many other activities in plants. Together with *gibberellin*, described in the next paragraph, it helps stimulate the growth and differentiation of cambium tissue into xylem and phloem (water and food carrying tissues, respectively). It is also involved in root and berry development.

Gibberellins, a group of hormones including gibberellic acid, promote stem elongation. These hormones are involved in the differences between short and tall plants. Dwarf plants, for example, will grow tall when treated with gibberellin. Gibberellins are also involved in breaking plant dormancy and in controlling the production of flowers. In some plants, the level of gibberellins and auxins influence the frequency of male or female flowers. Generally, high auxin levels stimulate the development of female flowers, while high gibberellin levels stimulate that of male flowers.

Other plant hormones include *ethylene*, which suppresses leaf growth until seedlings emerge from the ground, ripens fruit, stimulates flowering, and helps bring about abscission. *Abscisic acid* is a hormone that promotes the release of fruit and leaves and promotes dormancy. *Cytokinins* are a group of hormones that appear to be made in the roots and travel upward to the stems. In conjunction with auxins they maintain the balance between root and shoot growth. For example, excess cytokinin causes the extensive formation of lateral branches. Normally, auxin produced in the shoot tip inhibits the development of lateral branches near the tip. High levels of cytokinin can counteract this inhibitory effect, permitting lateral buds to develop. The shape of a plant therefore depends upon the interplay of these hormones.

How do plant hormones act? One mechanism is seen by studying the effect of gibberellic acid on the aleurone layer of barley seed endosperm. Gibberellic acid stimulates the synthesis or accumulation of α-amylase mRNA. α-Amylase is an enzyme that hydrolyzes the starch reserves of the seed, providing nutrients for the growing embryo. Like animal hormones, plant hormones can act by inducing the synthesis or accumulation of mRNA. The study of plant hormones will play an important role in unraveling the complex mechanisms that underlie the processes of morphogenesis and differentiation in higher organisms.

PLANT EMBRYO CULTURE AND TOTIPOTENCY

Exciting experiments performed by Steward and others have shed a great deal of light upon plant totipotency, the ability of plant cells to form tissues other than those that they have already formed. We have examined animal systems with respect to the factors that govern differentiation. In animal systems, cells in embryos are often not fixed in their fates. If such cells are transplanted to another part of the embryo, they differentiate into the type of tissue that now surrounds them. Position or environment, therefore, appears to govern differentiation. There is little evidence, however, that an adult animal tissue can redifferentiate into another type of tissue, although specific conditions may eventually be found that will allow such reversals. With some exceptions, all cells in the body possess the same genes. Since in most cases differentiation represents selective gene expression in specific types of cells, one can thus assume that differentiation would be reversible if certain genes could be turned off and others turned on.

Plants provide an important answer to the question of the reversibility of the differentiated state. Steward and others have cultured fragments of phloem tissue from adult carrot roots under artificial conditions. Phloem is the tissue that transports nutrients from the leaves to the stem and roots. The carrot fragments grow as undifferentiated tissue masses in a salt solution containing glucose and coconut milk, which contains growth-promoting factors. If the cell masses are dissociated into single cells and cultured as free cells in a suspension containing coconut milk, a remarkable process occurs. The cells divide and form embryo-like structures called *embryoids*. These embryoids develop shoots and roots like normal carrot embryos, and can develop into complete carrot plants that bear flowers and seeds.

Pollen grains and mature plant tissue (except for root phloem) have yielded embryoids in culture. These tissues include flower buds, leaves, and stems. It is possible that the cells that give rise to the embryoids are not really derived from the differentiated tissues, but instead from the actively proliferating meristem tissue often associated with differentiated tissues. There is evidence, however, that differentiated plant cells that are large and highly vacuolated actually alter their appearance and dedifferentiate. These cells lose their vacuoles, enlarge their nuclei and nucleoli, and increase their rate of DNA and ribosomal RNA synthesis. In tobacco plants, single differentiated cells will dedifferentiate into *callus* tissue. Under appropriate culture conditions, such callus can give rise to entire tobacco plants (Figure 12-24).

The growth factors that lead to dedifferentiation and differentiation of plant cells in culture have been investigated. Specific concentrations of the plant hormones auxin (indolacetic acid) and kinetin control root and shoot development from callus tissue in culture. Clearly, the propagation of new plants from cells of other plants has important implications for agriculture. Agriculture has been using such propagation methods for many years. This field will probably grow substantially in the years ahead.

One can isolate single plant cells in culture and remove their cell walls mechanically or with enzymes. These wall-less cells are called *protoplasts*. The cell will synthesize new cell walls, if the hormone kinin or auxin is present. Whole plants can arise from protoplasts derived from tobacco, carrot, petunia, belladonna, and asparagus.

The fact that wall-less plant cells can give rise to entire plants has very important implications for plant breeders. One wall-less plant cell can be fused with another in culture. Already, hybrid species of tobacco and petunia have been produced by fusing single protoplasts of the two species in culture. In addition, *chloroplasts* (cell organelles involved in photosynthesis) from one

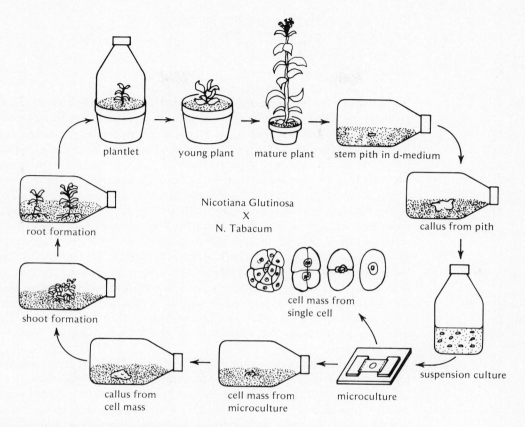

Figure 12-24. Culture of complete plants from single cells. From Vasil and Hilde-brandt, *Planta* 75 (1967): 139.

species have been introduced into the protoplasts of other species. Nuclear transplants have also been made possible by protoplast technology. Work is currently in progress to introduce nitrogen-fixing bacteria into the proto-plasts of plants that cannot normally fix nitrogen. If it were possible to transfer nitrogen-fixing ability to plants that usually require large amounts of nitrogen fertilizers, the need for such fertilizers could be lessened or eliminated!

In summary, plant and animal development differ in certain respects. Plant embryo cells generally develop in place; animal embryo cells rearrange to form organ rudiments. Like animal systems, plant systems offer unique qualities for the study of development. The question of the totipotency of cells, for example, has been studied in plant cell culture. New developments in protoplast plant culture are leading to important advances in the fields of plant breeding and agriculture.

LECTINS: TOOLS FOR STUDYING THE ROLE OF THE CELL SURFACE IN MORPHOGENESIS AND MALIGNANCY

We mentioned elsewhere in this text that lectins, carbohydrate-binding proteins or glycoproteins, have helped us understand certain cell surface phenomena. In the last decade or so, these molecules have become major tools for cell biologists in their study of the cell surface. Let us spend some time here to discuss these widely used molecules in more detail. We will sometimes stray a bit from the theme of morphogenesis in order to present the lectin story in an uninterrupted manner.

Lectins have been isolated from the seeds of plants and from extracts of sponges, snails, crabs, fish, slime molds, and vertebrate cells. A first question we might ask is: What is the function of lectins *in vivo*? We already mentioned that lectins are important in sperm-egg recognition. The sea urchin sperm bindin isolated by Vacquier is a lectin that recognizes carbohydrate groups on the cell surface of eggs. Lectins isolated from slime molds by Rosen's and Barondes's groups may play a role in the recognition and adhesiveness of slime mold amoebae as they aggregate to form a multicellular slug. Lectins, therefore, are important mediators of the cell recognition aspects of morphogenesis.

What about the function of lectins in plants, which are a major source of the lectins that have been studied to date? Lectins may function as plant antibodies to protect the plant against microbial attack. Lectins can agglutinate harmful microbes and can inhibit the hydrolysing enzymes used by fungi to attack the plant cell wall. Lectins may help plants store and transport sugars; they may also bind plant enzymes into multienzyme systems that function together in metabolic reactions. Lectins may control cell differentiation and mitosis, especially during seed germination. Lectins can induce animal cells, such as lymphocytes, to differentiate. The lectins bind and cross-link specific cell surface receptors that contain carbohydrate. Such cross-linking sets into motion a series of reactions that lead to differentiation and cell division.

An important function of lectins in plants is to bind nitrogen-fixing bacteria to the roots of legumes such as clover or beans. Legumes possess nodules of the nitrogen-fixing bacterium *Rhizobium* on their roots. These bacteria process gaseous nitrogen into a form usable by the plants and thus are of major importance to the plants. Lectins are secreted on the root surfaces of such plants. Specific lectins have been shown to bind only the strains of *Rhizobium* found on the roots of the plant from which the lectin is extracted. For example, soybean agglutinin, a lectin isolated from soybean plants, binds only strains of *Rhizobium* found on the roots of soybeans. It does not bind other *Rhizobium* strains that form nodules on the roots of other plants.

Some elegant work on the function of lectins in plants comes from the laboratory of Dazzo and his colleagues. Dazzo's group isolated a lectin from clover roots. They named the lectin trifoliin because it specifically binds to sugars on the nitrogen-fixing bacterium *Rhizobium trifolii*, which adheres to clover root hairs (Figure 12–25).

We can speculate that evolution produced a pair of organisms, clover and *Rhizobium trifolii*, that live together symbiotically, each deriving benefit from the association. The means by which they bind together probably evolved as a lectin-carbohydrate interaction at the surfaces of clover root hair and *Rhizobium* cells.

Before moving on to discuss how lectins are used to study the role of the cell surface in morphogenesis and malignancy, a brief description of the different lectins is in order. Lectins bind terminal sugars of the cell surface oligosaccharide chains or groups of sugars in the chains. The following is a list of lectins and the sugar(s) they preferentially bind.

Concanavalin A (Con A), isolated from the jackbean, binds *alpha*-D-mannose, D-glucose and *beta*-D-fructose groups.

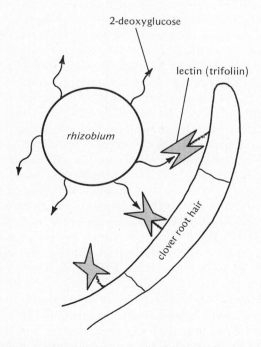

Figure 12–25. Lectin-mediated binding of nitrogen fixing bacterium (*Rhizobium*) to clover root hair. Hypothetical relationship based on results of Dazzo's group. Size relationships are exaggerated. From F. Dazzo, *J. Supramolec. Struct. and Cell Biochem.* (suppl. 5) (1981): 241.

Wheat germ agglutinin binds N-acetyl-D-glucosamine–like residues.

Ricinus communis agglutinin, isolated from the castor bean, binds D-galactose and N-acetyl-D-galactosamine–like groups.

Soybean agglutinin binds N-acetyl-D-galactosamine and D-galactose–like residues.

Garden pea lectin binds D-mannose groups.

Lentil lectin binds D-mannose and N-acetyl-D-glucosamine–like groups.

Lotus lectin binds L-fucose groups.

These are just a few of the many lectins that have been isolated. We should stress that these lectins preferentially bind the given sugars. The sugar specificity is seldom absolute, however, and most lectins will bind many sugars, though perhaps with less tenacity than the preferred sugars. The sugar specificity of lectin binding, however, has been a key reason for their successful use in the study of the cell surface.

How have lectins been used to study the cell surface in morphogenesis and malignancy? We presented one such experiment in the chapter on cleavage. Recall that Oppenheimer's group found that sea urchin embryo micromeres and mesomere-macromeres have a different reaction to the lectin concanavalin A. Although the micromeres, mesomeres, and macromeres all bound concanavalin A, only the micromeres showed a lectin-induced capping of concanavalin A cell surface receptor sites. In other words, the cell surface of the micromeres allowed the lectin to move the lectin receptor sites into a cap at one end of the cell. This suggested that the concanavalin A receptor sites are more mobile in the cell surface of the micromeres than in that of the macromeres or mesomeres. What does this have to do with morphogenesis? We know that the micromeres form the migratory primary mesenchyme cells that play a role in the gastrulation of the sea urchin embryo. The above-mentioned studies suggest that the cell surfaces of these premigratory cells are indeed different from the surfaces of the other cell types. The difference in lectin receptor site mobility in the cell surface may directly aid cell migration during morphogenesis. In any case, the use of lectin in this experiment showed that the cell surfaces of known populations of embryonic cells are very different in terms of the mobility of specific receptor sites.

There are many, many other studies using lectins. Let us just glance at some that relate to morphogenesis and malignancy. In 1888, Stillmark found that extracts from castor bean seeds can agglutinate (clump) human erythrocytes (red blood cells). Since then, lectins have been extracted from the seeds of over 800 species of flowering plants, especially the legumes. Lectins have been found not only in seeds but in the leaves, roots, and bark of plants and also in extracts from many animal cell types. Numerous studies have shown that, in general, malignant tumor cells, transformed cells, and early embryo cells are agglutinated with rather low concentrations of many lectins; whereas normal adult cells are seldom agglutinated at these concentrations. Why are embryonic cells and tumor cells agglutinated while normal cells are not?

Studies using lectins labeled with the fluorescent dye ferritin or with a radioactive isotope show that, in general, most cells bind lectin. That is, normal adult cells, tumor cells, and embryonic cells have lectin receptor sites in their cell surfaces. It has also been shown, however, that cells that are agglutinated with lectins (tumor and embryo cells, for example) tend to have mobile lectin receptor sites in their cell surfaces. Thus, capping and clustering of these sites will occur in the presence of lectin.

Lectins are multivalent molecules that can cross-link several surface receptor sites, thus clustering or capping them. This clustering or capping of the lectin receptor sites may explain how lectins agglutinate cells. A buildup of lectin-bound receptor sites in specific regions of the cell surface may facilitate lectin-bound cell agglutination. It has also been proposed that early embryo cells and malignant cells have altered cytoskeletal elements (microtubules and microfilaments) that fail to restrict the movement of cell surface lectin receptor sites. In normal adult cells, these cytoskeletal elements are attached to the inner membrane surface, restricting the movement of surface receptor sites. These suggestions are supported by the finding that drugs that disrupt cytoskeletal elements, such as colchicine, local anesthetics, and cytochalasin B, dramatically alter the mobility and distribution of cell-surface lectin receptor sites in a variety of cell types. Lectins, therefore, have been useful in identifying specific differences in the surfaces of tumor cells, normal cells, and embryonic cells.

Extensive studies with lectins have led to additional conclusions. Moscona's group showed that young chick embryo cells were agglutinated by concanavalin A, while older cells were not. Such studies were extended to other systems by several investigators. The findings in general suggest that young, motile embryo cells are agglutinable with lectins, while older embryo cells and those that do not move much are less so. In systems in which follow-up studies with labeled lectin were carried out, the results showed that agglutinable embryo cells displayed mobile surface lectin receptor sites like those in the sea urchin micromeres.

In summary, we can say that lectins have helped us identify cell-surface molecules containing sugar and have led to the conclusion that active cells in embryos and tumors possess mobile lectin receptor sites on their surfaces. It remains to be seen whether this characteristic plays a key role in facilitating the movements and interactions that occur during morphogenesis and the spread of cancer cells (Chapter 16).

In this chapter we have examined some of the mechanisms that play a role in morphogenesis. We have also looked at a variety of different systems and different techniques that have been used to investigate the problem of morphogenesis. With this information, let us now learn how specific parts of the embryo become different.

READINGS AND REFERENCES

Armstrong, P. B. 1971. Light and electron microscope studies of cell sorting in combinations of chick embryo neural retina and retinal pigment epithelium. *Wilhelm Roux's Archiv.* 168: 125–41.

Barbera, A. J., Marchase, R. B., and Roth, S. 1973. Adhesive recognition and retinotectal specificity. *Proc. Nat. Acad. Sci. U.S.* 70: 2482–86.

Barondes, S. H. 1981. Lectins: Their multiple endogenous cellular functions. *Ann. Rev. of Biochem.* 50:1.

Bernfield, M. R., David, G., Banerjee, S. D., Rapraeger, A. C., and Koda, J. E. 1981. Organization, assembly and remodeling of the epithelial basal lamina. *J. Supramolec. Struc. and Cell. Biochem.* (suppl. 5): 250.

Burnside, B. 1971. Microtubule and microfilaments in newt neurulation. *Develop. Biol.* 26: 416–41.

Dazzo, F. 1981. Lectin and its saccharide receptors in the *Rhizobium*-clover symbiosis, *J. Supramolec. Struc. and Cell. Biochem.* (suppl. 5): 241.

Edelman, G. M. 1983. Cell adhesion molecules. *Science* 219: 450–57.

Etkin, W. 1968. Hormonal control of amphibian metamorphosis. In W. Etkins and L. Gilbert, eds., *Metamorphosis*. New York: Appleton-Century-Crofts.

Frye, L. D., and Edidin, M. 1970. The rapid intermixing of cell surface antigens after formation of mouse-human heterokaryons. *J. Cell Sci.* 7: 319–35.

Hauschka, S. K., and Konigsberg, I. R. 1966. The influence of collagen on the development of muscle clones. *Proc. Natl. Acad. Sci. U.S.* 55: 119–26.

Humphreys, T. 1963. Chemical dissolution and *in vitro* reconstruction of sponge cell adhesions. *Develop. Biol.* 8: 27–47.

Karfunkel, P. 1971. The role of microtubules and microfilaments in neurulation in *Xenopus. Develop. Biol.* 25: 30–56.

Moscona, A. A. 1962. Studies on cell aggregation and demonstration of materials selective cell-binding activity. *Proc. Natl. Acad. Sci. U.S.* 49: 742–47.

Moscona, A. A., ed. 1974. *The Cell Surface in Development.* New York: John Wiley.

Oppenheimer, S. B. 1978. Cell surface carbohydrates in adhesion and migration. *Amer. Zool.* 18: 13–23.

Roseman, S. 1974. The biosynthesis of complex carbohydrates and their potential role in intercellular adhesion. In A. A. Moscona, ed., *The Cell Surface in Development*, pp. 255–72. New York: John Wiley.

Rosen, S. D., Barondes, S. H., and Haywood, P. L. 1976. Inhibition of intercellular adhesion in a cellular slime mold by univalent antibody against a cell surface lectin. *Nature* 263: 425.

Roth, S., and Weston, J. 1976. The measurement of intercellular adhesion. *Proc. Natl. Acad. Sci. U.S.* 58: 974–80.

Saunders, J. W., and Fallon, J. F. 1967. Cell death in morphogenesis. In M. Locke, ed., *Major Problems in Developmental Biology*, pp. 289–314. New York: Academic Press.

Schroeder, T. E. 1973. Cell constriction: Contractile role of microfilaments in division and development. *Amer. Zool.* 13: 949–60.

Shur, B. D., and Hall, N. G. 1982. A role for mouse sperm surface galactosyltransferase in sperm binding to the egg zona pellucida. *J. Cell Biol.* 95 : 574–79.

Singer, S. J., and Nicolson, G. L. 1972. The fluid mosaic model of the structure of the cell membrane. *Science* 175: 720–31.

Steinberg, M. S. 1970. Does differential adhesion govern self-assembly process in histogenesis? Equilibrium configurations and the emergence of a hierarchy among populations of embryonic cells. *J. Exp. Zool.* 173: 395–434.

Spooner, B. S., and Wessells, N. K. 1972. An analysis of salivary gland morphogenesis: Role of cytoplasmic microfilaments and microtubules. *Develop. Biol.* 27: 38–54.

Townes, P. L., and Holtfreter, J. 1955. Directed movement and selective adhesion of embryonic amphibian cells. *J. Exp. Zool.* 128: 53–120.

Trisler, G. D., Schneider, M. D., and Nierenberg, M. 1981. A topographic gradient of molecules in retina can be used to identify neuron position. *Proc. Natl. Acad. Sci. U.S.* 78: 2145–49.

Vacquier, V. D., and Moy, G. W. 1977. Isolation of bindin: The protein responsible for adhesion of sperm to sea urchin eggs. *Proc. Natl. Acad. Sci. U.S.* 74: 2456–2560.

Wessells, N. K., and Evans, J. 1968. The ultrastructure of oriented cells and extracellular materials between developing feathers. *Develop. Biol.* 18: 42–61.

Wessells, N. K. 1977. *Tissue Interactions and Development*. Menlo Park, CA: W. A. Benjamin.

Wilson, H. V. 1907. On some phenomena of coalescence and regeneration in sponges. *J. Exp. Zool.* 5: 245–58.

NEURAL CREST MIGRATION

Bronner-Fraser, M.E., and Cohen, A.M. 1980. The neural crest: What can it tell us about cell migration and determination? *Curr. Top. in Develop. Biol.* 15: 1–25.

Bronner-Fraser, M.E. 1982. Distribution of latex beads and retinal pigment epithelial cells along the ventral neural crest pathway. *Develop. Biol.* 91: 50–63.

LeLiever, C. S., Schweizer, G.G., Ziller, C.M., and LeDouarin, N.M. 1980. Restrictions of developmental capabilities in neural crest cell derivatives as tested by *in vivo* transplantation experiments. *Develop. Biol.* 77: 362–78.

Noden, D.M. 1975. An analysis of the migratory behavior of avian cephalic neural crest cells. *Develop. Biol.* 42: 106–30.

Weston, J.A. 1970. The migration and differentiation of neural crest cells. *Adv. in Morphogen.* 8: 41–114.

PATTERN MORPHOGENESIS

Crick, F. 1970. Diffusion in embryogenesis. *Nature* 225: 420–42.

Lock, M. 1967. The development of patterns in the integument of insects. *Adv. Morphogen.* 6: 33–88.

Markert, C. L., and Ursprung, H. 1971. *Developmental Genetics*. Englewood Cliffs, NJ: Prentice-Hall.

Poole, T. W. 1974. Dermal epidermal interactions and the site of action of the yellow (Ay) and non-agouti (a) coat color genes in the mouse. *Develop. Biol.* 36: 208–11.

Stern, C., and Tokunaga, C. 1967. Non-autonomy in differentiation of pattern-determining genes in *Drosophilia*. I. The sex-comb of eyeless-dominant. *Proc. Nat. Acad. Sci. U.S.* 57: 658–64.

Ursprung, H. 1966. The formation of patterns in development. In M. Locke, ed., *Major Problems in Developmental Biology*, pp. 177–216. New York: Academic Press.

Tokunaga, C. 1972. Autonomy or non-autonomy of gene effects in mosaics. *Proc. Nat. Acad. Sci. U.S.* 69: 3283–86.

Wolpert, L., Hicklin, J., and Hornbruch, A. 1971. Positional information and pattern regulation in regeneration of *Hydra*. *Symp. Soc. Exp. Biol.* 25: 391–415.

MORPHOGENESIS IN PLANTS

Ebert, J. D., and Sussex, I. 1970. *Interacting Systems in Development*, 2nd ed. New York: Holt, Rinehart & Winston.

Higgins, T. J. V., Zwar, J. A., and Jacobsen, J. V. 1976. Gibberellic acid enhances the level of translatable mRNA for α-amylase in barley aleurone layers. *Nature* 260: 166–69.

Jaffe, L. F. 1968. Localization in the developing *Fucus* egg and the general role of localizing currents. *Adv. Morphogen.* 7: 468.

Nuccitelli, R. 1978. Ooplasmic segregation and secretion in the *Pelvetia* egg is accompanied by a membrane-generated electrical current. *Develop. Biol.* 62: 13–33.

Pollock, E. G., and Jensen, W. A. 1964. Cell development during early embryogenesis in *Capsella* and *Gossypuim*. *Amer. J. Botany* 51: 915.

Quatrano, R. S. 1972. An ultrastructural study of the determined site of rhizoid formation in *Fucus* zygotes. *Exp. Cell Res.* 70: 1.

Quatrano, R. S. 1968. Rhizoid formation in *Fucus* zygotes: Dependence on protein and ribonucleic acid synthesis. *Science* 162: 468.

Quatrano, R. S. 1981. Cell wall formation in *Fucus* zygotes: A model system to study the assembly and localization of wall polymers. In R. M. Brown, ed., *Cellulose and Other Natural Polymer Systems: Biogenesis, Structure and Degradation*. Plenum.

Quatrano, R. S. 1978. Development of cell polarity. *Ann. Rev. Plant Physiol.* 29: 487–510.

Quatrano, R. S., Brawley, S. H., and Hogsett, W. E. 1979. The control of the polar deposition of a sulfated polysaccharide in *Fucus* zygotes. In S. Subtelny and I. R. Konigsberg, eds., *Determinants of Spatial Organization*, pp. 77–96. New York: Academic Press.

Sinnott, E. W. 1960. *Plant Morphogenesis*. New York: McGraw-Hill.

Skoog, F., and Miller, C. O. 1957. Chemical regulation of growth and organ formation in plant tissues cultured *in vitro*. *Symp. Soc. for Exp. Biol.* 11: 118.

Steward, F. C., Mapes, M. O., Kent, A. E., and Holsten, R. D. 1964. Growth and development of cultured plant cells. *Science* 143: 1.

Vasil, I. K. 1976. The progress, problems, and prospects of plant protoplast research. *Adv. Agron.* 28: 119.

Vasil, V., and Hildebrandt, A. C. 1965. Differentiation of tobacco plants from single isolated cells in microcultures. *Science* 150: 889.

Vasil, I. K. and Vasil, V. 1972. Totipotency and embryogenesis in plant cell and tissue cultures. *In Vitro* 8: 117.

Walbot, V. 1978. Control mechanisms for plant embryogeny. In M. Clutter, ed., *Dormancy and Developmental Arrest*, pp. 113–66. New York: Academic Press.

Wardlaw, C. W. 1955. *Embryogenesis in Plants*. New York: John Wiley.

Wardlaw, C. W. 1968. *Morphogenesis in Plants*. London: Methuen.

13

DIFFERENTIATION: NUCLEIC ACIDS

All embryo cells are derived from a single cell—the fertilized egg. How, then, do muscle cells, blood cells, nerve cells, pigment cells, and gland cells develop if they are all derived from the same original cell and presumably contain identical genes? We know that the differences between the cells are mainly in the specific proteins required for specialized function. This will be the topic of the next chapter. We also know that the structure of these proteins is coded by the sequence of nucleotides in the DNA of the cell's chromosomes. If cells with identical genes become different, surely only certain genes in certain cells have been activated. This would lead to the synthesis of specific proteins in the specific cell types.

In previous chapters, we saw that the environment in which the genes lie is responsible for activating or repressing certain genes. For example, the factors that tell cells to activate their genes for nerve-specific proteins are apparently passed from an inducing to a responding tissue. In this chapter, we shall examine differentiation at the nucleic acid level, since this level plays a major role in controlling protein synthesis. We shall examine evidence that specific genes are indeed activated at specific times during the differentiation of specific cells. In addition, we shall look at possible means by which these genes are repressed or activated. Let's briefly look at some of the techniques used to reveal differences in cellular nucleic acids, which reflect activation of specific genes.

RESEARCH TECHNIQUES

Techniques have been developed to determine whether the nucleotide sequences of two batches of DNA or of a batch of DNA and a batch of RNA are similar. This sort of information is needed to reveal the differences and similarities in nucleic acids during the differentiation process. These techniques are based upon the known pairing that occurs between the bases of nucleotides in nucleic acids. We know that adenine (A) on one DNA chain pairs specifically with a thymine (T) on the other chain, and that guanine (G) pairs with cytosine (C). We also know that RNA nucleotides contain uracil (U) instead of thymine. During RNA synthesis, uracil pairs with adenine on a DNA chain. These pairings occur by hydrogen bonds that form between the bases of the nucleotides (see Chapter 3 and Figure 13–1).

The strands of the DNA double helix can be separated by mild heating that breaks the hydrogen bonds between the nucleotides. The strands will reassociate (anneal) if the mixture is slowly cooled. The strands will pair up correctly so that A binds T and C binds G. This is the basis for the techniques we shall consider. Since one can easily separate double DNA helices by heating and then allow the strands to reassociate, DNA from different organisms or tissues can be compared for similarities and differences in base sequence. Since two strands of DNA will associate only if the sequence of bases are complementary, different DNAs can be compared by promoting

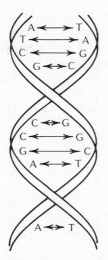

Figure 13-1. Section of DNA double helix showing nucleotide pairing. A pairs with T and C with G by specific hydrogen bonds.

reassociation between them under carefully controlled conditions. If reassociation is rapid and complete, the DNAs have similar or identical base sequences. If reassociation does not occur, the base sequences are not similar. Such experiments are called "hybridization experiments."

Many variations of the basic hybridization experiment have been developed. Some involve binding RNA to DNA in the presence of other types of RNA. These experiments help determine whether the two types of RNA have similar nucleotide sequences. If two types of cells have similar or different RNA sequences, we can determine whether the same or different genes have been activated in these cell types. These experiments will be described shortly.

The experimental protocols used in nucleic acid hybridization experiments are complex, and the conditions under which the nucleic acid strands reassociate must be carefully controlled. No attempt will be made to describe the details of these experiments. For these, the student is referred to the references at the end of the chapter. Let us, however, get a general idea of how these experiments are performed and what they can tell us about differentiation at the nucleic acid level.

In the basic DNA-DNA hybridization experiment, radioactive DNA is often produced by injecting a radioactive label, such as "hot" phosphate, into the organism. Nuclei with radioactive DNA are isolated by centrifuging the homogenized tissue, and the nucleic acids are chemically extracted from the nuclei. RNA is then removed enzymatically from the DNA preparation, isolating the radioactively labeled DNA. Nonradioactive DNA is also used in these experiments; it is extracted from other specimens in the same way. The nonradioactive DNA is heated to break the hydrogen bonds between the nucleotides. The DNA helices thus dissociate into single strands. These nonradioactive single pieces of DNA are attached to a filter or imbedded in a column of gel. The radioactive single stranded DNA segments are then passed through the filter or column. If the base sequence of the DNA on the filter is similar to that of the radioactive DNA, base pairing will occur and the radioactive DNA will anneal with the DNA on the filter. The amount of radioactive DNA bound to the DNA on the filter can easily be measured with a scintillation counter (Figure 13–2). If the two DNAs are not similar, the radioactive DNA will pass through the filter without binding. In this way, one can determine if two DNAs have similar or different base sequences.

The results of such DNA-DNA hybridization experiments suggest that there is species-specific differentiation of DNA. That is, DNA base sequences differ between species. Evolutionary relationships between different species have been confirmed using the DNA-DNA hybridization technique. In general, closely related species have more DNA base sequences in common than distantly related species do. In addition, such experiments suggest that there is no tissue-specific differentiation of the DNA base sequence in cells of the same organism. This makes sense because all cells in an organism are derived from the fertilized egg and should possess identical genes.

Major advances in nucleic acid technology have recently been made. These include the technique of gene cloning. In this technique, purified messenger RNA is used to make many copies of the gene (DNA) that codes

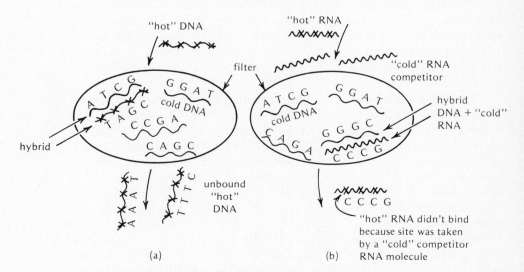

Figure 13–2.(a) DNA-DNA hybridization experiment. (b) RNA-DNA competition hybridization experiment.

for the messenger RNA. This is accomplished with the enzyme reverse transcriptase (RNA-dependent DNA polymerase). This enzyme copies RNA templates to make complementary DNA strands (cDNA). Many copies of a given gene can be produced in this manner. Such gene clones are extremely useful in determining whether a given species of messenger RNA is present in a specific cell at a specific time. This is done by using DNA-RNA hybridization techniques to measure how much of the RNA under study is complementary to the specific cDNA. In this way cDNA is used directly to determine the presence or absence of a particular species of RNA. The study of differentiation at the messenger RNA level, therefore, has become far more direct through the use of cDNA probes.

TISSUE-SPECIFIC DIFFERENTIATION OF RNA

Are different RNA molecules present in different tissues? In other words, is there tissue-specific differentiation of RNA? Intuition says yes, since different tissues possess different tissue-specific proteins, and these proteins must be coded for by specific RNA molecules. We would guess further that, although all cells in an organism possess identical genes, only some genes are active in given tissues. Thus, tissue cells can differentiate by synthesizing the specific proteins required for their specialized functions. Let us briefly describe an experiment that shows that different tissues do indeed contain different RNAs. Note again that it is the base sequence in messenger RNA (and DNA) that determines the amino acid composition of specific proteins. Differences and similarities in the base sequence of nucleic acids can be estimated using a modification of the previously described hybridization techniques. If pieces of DNA or a segment of DNA and a segment of RNA hybridize with each other, the base sequences are complementary. If they do not, the base sequences are not.

A modification of the hybridization technique is used to determine if tissue-specific differentiation of RNA exists. Segments of DNA are heated or chemically treated to separate them into single strands. These strands are adsorbed onto a nitrocellulose membrane filter. Radioactive RNA is then produced by injecting radioactive RNA precursors into the organism and extracting the RNA from different tissues, such as the spleen, liver, and kidney. Radioactively labeled RNA from a specific tissue is then incubated with the DNA on the filter in the presence of increasing concentrations of unlabeled RNA from different tissues.

If unlabeled RNA from a given tissue significantly inhibits the labeled RNA from binding to the DNA, this indicates that the base sequence in this unlabeled RNA is similar to that of the radioactive RNA. This is so because the unlabeled RNA competes for the hybridization sites on the DNA template. Fewer labeled RNA molecules can bind to the DNA because the sites have been taken by the similar unlabeled RNA molecules (Figure 13–2). Likewise, if unlabeled RNA from a given tissue does not inhibit the radioactive RNA from binding to the DNA on the filter, the base sequence of the unlabeled RNA is not similar to the base sequence of the radioactive RNA. This is so because the unlabeled RNA binds to DNA strands different from those to which the labeled RNA binds. The student is referred to the references at the end of this chapter for a more in-depth analysis of the many parameters and problems associated with these experiments.

Figure 13–3 gives an example of this type of competition-hybridization experiment. Labeled RNA from mouse kidney is incubated on the mouse DNA filter, together with unlabeled RNA from mouse kidney, spleen, or liver. As can be seen, kidney RNA is the most effective competitor for the hybridization reaction between kidney RNA and DNA. Spleen RNA is less effective, and liver RNA is the least effective competitor. These results suggest that the three tissues do not contain identical sets of RNA, but that different genes are active in the different tissues. Taken together with the known protein differences of specialized tissues (next chapter), these experiments provide strong evidence that differential gene activation is a key part of the differentiation process.

STAGE-SPECIFIC DIFFERENTIATION OF RNA

Are different RNA molecules present at different developmental stages? In other words, is there stage-specific differentiation of RNA? To answer this

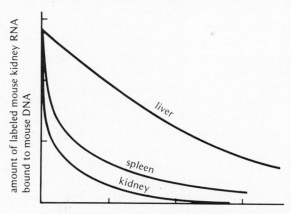

Figure 13–3. Tissue-specific differentiation of RNA. Labeled mouse kidney RNA is inhibited from binding to mouse DNA if it is incubated in the DNA filters in the presence of unlabeled RNA from different mouse tissues. The best inhibitor (competitor) is kidney RNA, followed by spleen RNA. Liver RNA is the least effective competitor. From B. J. McCarthy, and B. H. Hoyer *Proc. Natl. Acad. Sci. U.S.* 52 (1964) : 915–22.

question, we shall again use the competition-hybridization method. Recall that the base sequence in RNA (and DNA) is the genetic code that controls the synthesis of each specific protein. So, if we talk about similarities or differences in RNA during different developmental stages, we are concerned with base sequences. Similarities and differences in base sequence between nucleic acids can be determined using hybridization techniques. If segments of DNA or a segment of DNA and a segment of RNA hybridize with each other, the base sequences are complementary. If they do not, the base sequences are not similar.

The competition-hybridization technique is used as follows to determine if there is stage-specific differentiation of RNA during early embryonic development. Segments of DNA double helix are heated, or chemically treated, so that the DNA separates into single strands. The strands are adsorbed on a filter. Radioactively labeled RNA from a specific developmental stage of an embryo of the same species is then incubated with the DNA on the filter. These incubations are carried out in the presence of increasing concentrations of unlabeled RNA from the same or a different developmental stage of the organism. If a batch of unlabeled RNA from a given developmental stage significantly inhibits the labeled RNA from binding to the DNA, then the base sequence in this unlabeled RNA is similar to that of the labeled RNA (Figure 13-2). If, on the other hand, the unlabeled RNA from a given developmental stage does not inhibit the binding of the labeled RNA, then the base sequence of the unlabeled RNA is not similar to the base sequence of the radioactive RNA.

Figure 13-4 gives an example of this type of competition-hybridization experiment. Labeled RNA from the prism (postgastrula) stage of the sea urchin embryo is incubated on the sea urchin DNA filter together with unlabeled RNA from different sea urchin stages. As can be seen, unlabeled prism RNA is the best competitor of the labeled prism RNA binding to the DNA on the filter. The next best competitor is unlabeled gastrula RNA; unlabeled blastula and unfertilized egg RNA are significantly poorer competitors. Unlabeled *P. morganii* (starfish) RNA does not compete at all. From this experiment, we can conclude the following: (1) The sea urchin gastrula has many RNA sequences in common with the sea urchin prism stage. (2) The sea urchin blastula and unfertilized egg have fewer RNA sequences in common with the prism stage. (3) Starfish RNA is not similar to sea urchin prism RNA in terms of base sequence. This work suggests that there is indeed stage-specific differentiation of RNA. This makes sense, because surely different genes will be activated at different developmental stages to form the RNA needed to code for the specific and sometimes different proteins required for development to proceed at each stage.

Before leaving the discussion of nucleic acid hybridization, we should mention some important information that has been discovered using this technique about the genome of the cells of higher organisms (eukaryotic cells). Eukaryotic cells—all cells above the bacteria and blue-green algae—

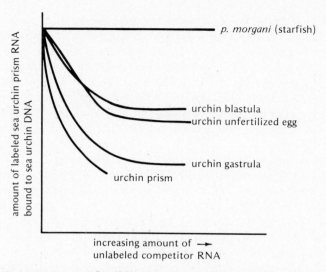

Figure 13–4. Stage-specific differentiation of RNA. The graph shows that labeled sea urchin prism RNA is inhibited from binding to sea urchin DNA when the DNA filters are incubated in the presence of a variety of unlabeled RNAs from different developmental stages of the sea urchin. The best inhibitors (competitors) are urchin prism and urchin gastrula RNA. Starfish RNA does not show concentration-dependent inhibition of binding. From A. H. Whitely, B. J. McCarthy, and H. R. Whitely *Proc. Natl. Acad. Sci. U.S.* 55 (1966) : 519–25.

have a well-defined nucleus. Prokaryotic cells (bacteria and blue-green algae) generally contain a single naked DNA double helix. Most eukaryotic cells contain thousands of times the amount of DNA found in prokaryotic cells. Much, but not all of this DNA is required to code for and regulate the tremendous genetic diversity expressed during the differentiation process in higher organisms.

Using the DNA-DNA hybridization technique, it has been shown that some single strands of eukaryotic DNA reanneal with each other rather quickly at low DNA concentrations. Other pieces, however, don't reanneal until the DNA concentration and length of time increase (Figure 13–5). This suggests that the strands that reanneal quickly at low concentrations have many base sequences in common. The more slowly annealing segments represent DNA strands present in only a few or single copies. The more quickly DNA strands hybridize at low DNA concentrations, the greater the number of identical DNA copies present. This is so because it is not difficult for a strand to find and bind with a complementary strand if many copies of the sequence are present. The DNA strands that recombine only under high DNA concentrations and after long incubation times are present in fewer copies and thus have difficulty in colliding with a complementary sequence.

Thus, some of the DNA of an eukaryotic cell is present in single copies,

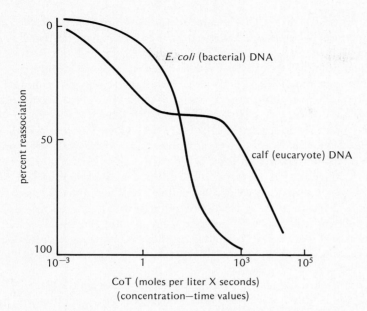

Figure 13-5. Reassociation of prokaryotic (bacterial) DNA and of eukaryotic (calf) DNA. The calf DNA curve shows early annealing sequences at low concentrations of DNA and late annealing sequences at high concentrations of DNA. The *E. coli* curve is S shaped, suggesting that only single copies of each gene exist. From R. J. Britten and D. E. Kohne, Repeated Sequences in DNA, *Science* 161 (9 August 1968) : 529–40, Fig. 3. Copyright 1968 by the American Academy for the Advancement of Science.

while multiple copies of other sequences exist. Prokaryotic cells apparently possess mostly single copies of given genes (Figure 13-5).

In summary, the eukaryotic genome contains unique and repetitive DNA sequences. Eukaryotic cells probably require regulatory genes and other types of genes in multiple copies in order to control the complex sequence of developmental events that characterize the differentiation process.

OTHER APPROACHES TO STUDYING NUCLEIC ACID DIFFERENTIATION

In the next chapter, we will examine differentiation at the protein level. Here we are looking at nucleic acids, the molecules that contain the genetic code and thus represent an earlier stage than protein synthesis. Thus, before we examine proteins, we should understand what happens in a cell at the nucleic acid level. Recent advances have been made that allow us to examine nucleic acids in developing systems. We have already considered one of these methods, nucleic acid hybridization. We have seen using this method that there is tissue-specific and stage-specific differentiation of RNA during development. This technique has also been used to home in specifically on the messenger RNA molecule. Heretofore, we have examined only total cellular RNA. It is also possible to identify specific messenger RNA molecules. This is done by taking purified molecules of messenger RNA and making DNA complementary to this message by using the enzyme reverse transcriptase. We now have DNA that is complementary to specific messenger RNA molecules. This complementary DNA (cDNA) is a very potent tool that can be used to determine whether its complementary messenger RNA is present in cells. Single strands of the DNA are adsorbed to a filter, and batches of RNA are passed through the filter. Any messenger RNA that is complementary to the cDNA will hybridize with the cDNA. These hybrids can be detected by any of the variety of methods previously described. Thus, we can determine if a specific message for a specific protein is present in a population of cells. We can also determine how many times this message was duplicated and from that learn a lot about specific gene activation at the nucleic acid level.

Figure 13-6 shows that if only cytoplasmic messenger RNA molecules are used in hybridization experiments, complex results are obtained. In this experiment, mRNA populations from different sea urchin developmental stages were hybridized with DNA made using gastrula stage mRNA as template. The experimental results indicate that the sea urchin oocyte possesses not only the same mRNA sequences as the gastrula, but also many mRNA sequences not present in the gastrula. Adult sea urchin tissues (tubefoot, intestine, and coelomocyte) possess only a small number of messages, some shared with the gastrula and others not. The unshared messages presumably code for specific proteins required in these specialized differentiated tissues. Like those presented previously, these results indicate that the characteristics of cellular RNA vary during embryonic development and differentiation.

These results have been extended to other systems, including plant embryos. Using the hybridization technology, Goldberg and colleagues found that there is a large overlap in the mRNA species of different developmental stages of the soybean embryo. However, they also identified a unique set of superabundant mRNAs that occur in different amounts at different developmental stages. This set of mRNAs probably codes for a spectrum of seed proteins that are synthesized during maturation. Some of these are probably metabolized by the rapidly growing seedling.

The issue is made more complex by recent results by Rosenthal, Hunt, Ruderman, and others. These results indicate that different mRNA molecules need not be present at different developmental stages. For example, oocytes and embryos of the surf clam Spisula solidissima contain identical sets of mRNA. Different groups of these message sets, however, are associated with

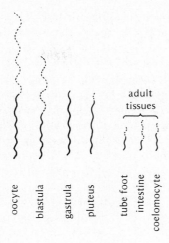

Figure 13-6. Comparison of messenger RNA from sea urchin gastrula with messenger RNA from other sea urchin stages and tissues. Solid strands show the relative amount of message sequences shared with the gastrula. Upper, dashed strands show the relative amount of message sequences not shared with gastrula. This work suggests that some messenger RNA sequences are the same at different stages while others are different. Based on experiments by Galau, *et al., Cell* 7 (1976) : 487–505.

ribosomes at the different stages. Thus, the surf clam embryo employs stage-specific utilization of different subsets of mRNA from a set present at both stages (oocyte and embryo). Stage-specific differences in protein types at the different developmental stages probably result from such stage-specific utilization of specific messages from the large population of messages contained in oocytes and later embryos. Some of the mRNA molecules are probably masked at certain stages, preventing them from associating with ribosomes and, hence, being translated. Future work should tell us more about mRNA masking and about how a message can be unmasked, making it functional at a specific developmental stage.

A second method for examining nucleic acid changes during differentiation uses the RNA synthesis inhibitor actinomycin D. Differentiating cells are incubated with this drug at various times. If actinomycin D prevents the synthesis of a specific protein, this suggests that the drug blocked the transcription of the messenger RNA for that protein. If, on the other hand, actinomycin D treatment does not block synthesis, the message for that protein was probably already present in the cell. In this way, we can ascertain whether a specific gene becomes activated shortly before protein synthesis occurs, or whether stable messenger RNA molecules were produced at some time prior to incubation with actinomycin D. Determining when a specific gene is activated is an important part of understanding differentiation at the nucleic acid level.

A third approach to investigating the changes in nucleic acids during differentiation is to identify specific messenger RNA molecules by their ability to direct the synthesis of specific proteins. This can be done by incubating the RNA with the necessary components of the protein synthesis system. These components include ribosomes, amino acids, transfer RNAs, and a variety of factors and enzymes. They can be prepared from bacterial cells. We can determine if messenger RNA molecules specific for a given protein are present by synthesizing that protein with such an in vitro protein synthesizing system.

These techniques have been used in many developing systems. For example, there comes a point in the development of red blood cells, muscle cells, and pancreas cells when actinomycin D no longer prevents specific protein synthesis. This suggests that stable messenger RNA molecules accumulate in these cells in preparation for the extensive synthesis of cell specfic proteins characteristic of the differentiated state.

In the mouse, actinomycin blocks the synthesis of hemoglobin in prospective red blood cells before day 10 of gestation. After day 10, actinomycin D does not do so. Thus, any hemoglobin message made before day 10 is stable and can synthesize the protein after day 10.

We might now ask, How does messenger RNA become stabilized? One possibility is that the rate of message synthesis is increased before day 10 so that there are too many message copies to be totally degraded. Also, messages may be synthesized and additional nucleotides, such as polyadenylic acid, may be attached that may combine with some protein to protect them from degradation. It has been proposed along the same lines that most or all genes in the genome are transcribed in eukaryotic cells, but that only some are protected from degradation. This sort of regulation might play a major role in determining which genes form messages that are active in synthesizing protein—many messages may be transcribed, but only some stay around long enough to reach the cytoplasm to synthesize protein. This concept (proposed by Wilt and others) may be one important piece of the differentiation puzzle.

The chromosomes of higher cells are complex. Before we move on to differentiation at the protein level, let's look at another interesting proposal that may help account for the observation that during differentiation certain genes are turned on while others are turned off.

CONTROL OF GENE TRANSCRIPTION DURING DIFFERENTIATION

Chromosomes in the cells of higher organisms contain a great deal of protein associated with the DNA. This protein may play an important role in differentiation by repressing certain genes while allowing others to become active. As discussed in Chapter 3, there are two major types of chromosomal proteins, histone proteins and nonhistone proteins. Histones are low-molecular-weight proteins rich in basic amino acids, such as lysine and arginine. Histones from different organisms and different cell types are structurally similar. Nonhistone proteins associated with DNA are larger and much more diverse than the histones. Most are phosphorylated and are acidic in nature rather than basic. The relationships between histone and nonhistone proteins were described in Chapter 3. In addition, experiments using DNA-RNA hybridization techniques indicated that estrogen-stimulated oviduct chromatin produces eight times as much oviduct-specific ovalbumin messenger RNA as chromatin from oviducts in which estrogen treatment was halted. The nonhistone protein fraction from estrogen-treated oviduct chromatin can be added back to DNA with histones from oviducts in which estrogen treatment was halted. This chromatin can now transcribe specific ovalbumin messenger RNA in amounts characteristic of estrogen-treated oviduct tissue. In the reverse experiment, nonhistone proteins from estrogen-halted chromatin lowered the amount of ovalbumin messenger RNA produced by DNA with histones from estrogen-treated oviducts.

To summarize, inducer complexes, hormone complexes, and other molecules that control differentiation may activate genes by binding to nonhistone proteins on chromosomes. This absorbing area of research is moving rapidly. The exact means by which gene action is regulated during the differentiation of eukaryotic cells may soon be completely understood.

RIBOSOMAL RNA GENES IN THE AMPHIBIAN EMBRYO

We have mainly considered the differences and similarities in messenger RNA during development because mRNA directly codes for the tissue-specific proteins that are important in differentiation. Before moving on to these proteins, we shall briefly consider a type of RNA that is required for assembling ribosomes, the seats of protein synthesis.

The synthesis of ribosomal RNA is controlled by specific genes that become activated at specific developmental times. This is nicely shown by the anucleolate mutant of the amphibian *Xenopus*. Homozygous mutant individuals, obtained by mating two heterozygous individuals, die after hatching. These homozygous individuals do not make ribosomes and have no nucleoli. Such mutants can be produced by mating two 1-nu individuals (each 1-nu individual has only 1 nucleolus in its cells). They produce offspring of the following types: homozygous wild type (2-nu—two nucleoli in their cells), heterozygous (like the parents—1-nu), and homozygous mutant (0-nu—no nucleoli). The anucleolate mutants (0-nu) do, however, possess ribosomes that were made by the heterozygous mother frog. Using these maternal ribosomes, the mutants can synthesize proteins for some time during early development. New ribosomes, however, are apparently required by the time the embryo hatches. Since the mutants can't make new ribosomes, they die by the early swimming tadpole stage.

Let's look at some specific evidence indicating that ribosomal RNA synthesis is controlled by specific genes active at specific developmental stages. Figure 13-7 shows that ribosomal RNA is synthesized in normal *Xenopus* from the neurula to the tail-bud stages. The figure also shows that ribosomal RNA is not synthesized in the mutant, though other RNA is still made. These results were obtained by incubating neurula-stage embryos with a radioactive label incorporated into RNA. Some of the different types of RNA synthesized were separated on the basis of their sedimentation (S, based on the size of the molecule) in a gradient of a substance such as sucrose. Ribosomal RNA includes 28S, 18S and 5S size classes. 4S RNA, also shown in the figure, is soluble, or transfer, RNA. 5S ribosomal RNA is not shown in the figure.

Does the anucleolate mutant possess the genes for ribosomal RNA but not express them? Or are the ribosomal RNA genes absent entirely? The

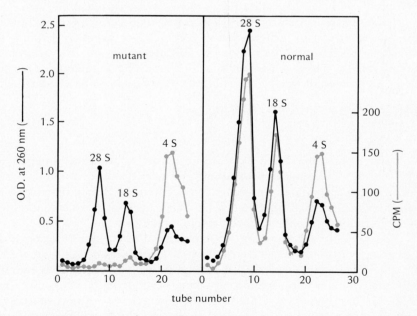

Figure 13-7. RNA synthesis in normal and mutant *Xenopus* embryos at neurula-tailbud stage. Light lines indicate radioactivity incorporated into newly synthesized RNA. Dark lines show where the specific types of RNA appear in fractions from a density-gradient centrifugation experiment. The anucleolate mutant does not synthesize ribosomal RNA (28S + 18S RNA), while the normal embryo does. From D. D. Brown, *J. Exp. Zool.* 157 (1966) : 101–14.

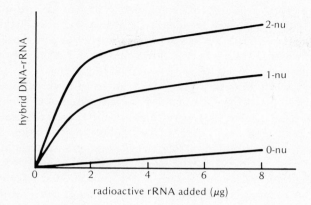

Figure 13-8. Hybridization of ribosomal RNA with DNA. 2-nu DNA from normal *Xenopus* + ribosomal RNA. 1-nu DNA from heterozygous *Xenopus* + ribosomal RNA. 0-nu DNA from a nucleolate mutant *Xenopus* + ribosomal RNA. From C. L. Markert and H. Ursprung, *Developmental Genetics* (Englewood Cliffs, NJ: Prentice-Hall, 1971), p. 12. Reprinted by permission of Prentice-Hall, Inc., Englewood Cliffs, New Jersey.

DNA-RNA hybridization technique was used to answer this question. Nonmutant *Xenopus* DNA was dissociated into single strands and incubated with purified radioactive nonmutant ribosomal RNA. The ribosomal RNA hybridized with the ribosomal RNA genes of the DNA. The extent of hybridization was determined by measuring the radioactivity bound to the DNA, as previously described. When DNA from the anucleolate mutant was used, however, no radioactive ribosomal RNA bound to the DNA (Figure 13-8). Thus, anucleolate mutant DNA does not contain ribosomal RNA genes. This can be visually confirmed by noting that the anucleolate mutant lacks certain secondary constrictions that are present on two chromosomes of nonmutant cells. These constrictions were found to be the sites from which nucleoli originate and where ribosomal RNA is produced (Figure 13-9).

Throughout this text we have discussed how cytoplasmic or environmental factors and inducers activate specific genes at specific times. Let us conclude this chapter by looking at this sort of activation in the frog embryo system.

Figure 13-9. Drawing of mitotic chromosomes of normal *Xenopus* larva. n is the nucleolar organizer region.

Nuclei from a variety of embryonic cells can be transplanted into enucleated amphibian eggs. In this way, scientists can answer the questions: Can a nucleus from a later developmental stage still support the development of the early embryo? What factors turn genes on and off at various developmental stages?

Gurdon and others transplanted nuclei from differentiated intestinal epithelium cells, tail-fin epithelial cells, or cultured skin cells of *Xenopus* tadpoles into enucleated frog eggs. Some of these nuclei could completely support normal frog development (Figure 13–10). A gut cell nucleus, for example, synthesized ribosomal RNA until it was transplanted into the egg. It then stopped. At gastrulation, when normal ribosomal RNA synthesis increases rapidly, nuclei derived from the transplanted intestinal nucleus resumed normal ribosomal RNA synthesis. This suggests that RNA synthesis is regulated by the type of cytoplasm surrounding the nucleus. Nuclei from many types of differentiated cells can still return to the embryonic state by turning on "old" genes that were active only in the early embryo, if the cytoplasmic conditions that control such a change should reappear. More will be said about this notion when we discuss cancer and embryology in Chapter 16.

GENES CODING FOR SPECIFIC MESSAGES

Specific mRNA molecules have been isolated from cells committed predominantly to synthesizing a particular protein. Such cells possess numerous copies of the message, making its isolation feasible. For example, precursor red blood cells (erythroblasts) synthesize large quantities of hemoglobin and possess large amounts of relatively pure globin mRNA. Using hybridization technology, it has been shown that globin message—and most other purified messages—bind to cellular DNA at a slow rate, suggesting that there is only one copy of a given message-coding gene in the genome. One exception is the set of genes that code for histone proteins. Histone mRNA binds to DNA faster than would be expected for single copy genes, suggesting that several hundred copies of histone genes exist per genome.

INTERVENING SEQUENCES

As noted in Chapter 3, some intriguing findings have recently been made with respect to gene structure and message differentiation or maturation. It has been shown that certain genes that code for mRNA molecules contain *intervening DNA base sequences*, or introns, that are not present in the messenger RNA coded by such genes. Introns have been found in the mouse and rabbit β-globin gene, in the mouse α-globin gene, in the mouse immunoglobulin light chain gene, and in the chick ovalbumin and chick lysozyme genes. Figure 13–11 shows the intervening sequences in the mouse β-globin gene. As

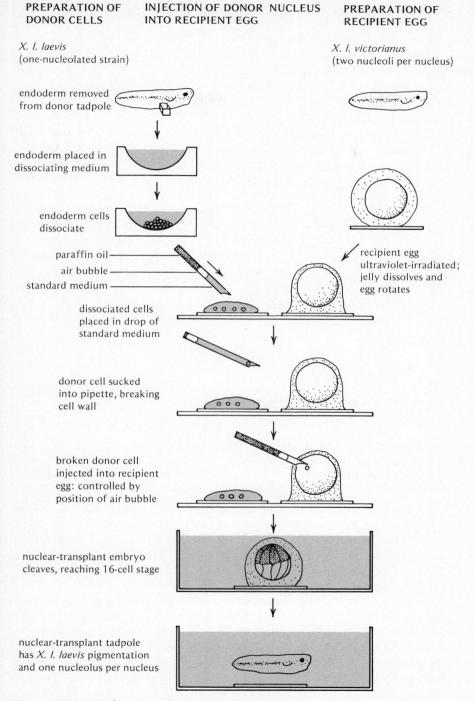

PREPARATION OF DONOR CELLS

INJECTION OF DONOR NUCLEUS INTO RECIPIENT EGG

PREPARATION OF RECIPIENT EGG

X. l. laevis
(one-nucleolated strain)

X. l. victorianus
(two nucleoli per nucleus)

endoderm removed from donor tadpole

endoderm placed in dissociating medium

endoderm cells dissociate

paraffin oil
air bubble
standard medium

dissociated cells placed in drop of standard medium

recipient egg ultraviolet-irradiated; jelly dissolves and egg rotates

donor cell sucked into pipette, breaking cell wall

broken donor cell injected into recipient egg: controlled by position of air bubble

nuclear-transplant embryo cleaves, reaching 16-cell stage

nuclear-transplant tadpole has *X. l. laevis* pigmentation and one nucleolus per nucleus

Figure 13–10. Nuclear transplantation in *Xenopus*. Donor nuclei from one strain are transplanted to host eggs of another strain. The tadpoles resulting from such transplants have characteristics of the nuclear parent, showing that the nuclei of the tadpoles were derived from the transplanted nucleus. From Gurdon, *Endeavour* 25 (1966) : 96.

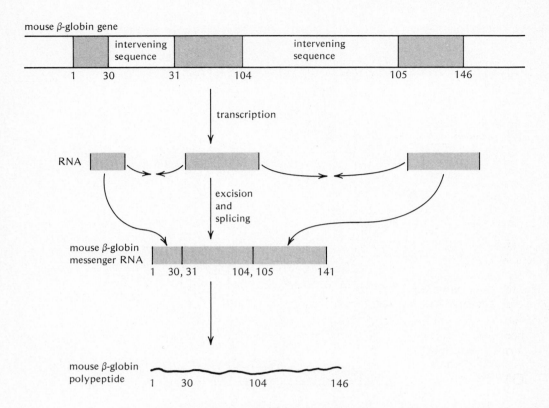

Figure 13-11. Mouse β-globin gene, showing intervening sequences and their elimination from the RNA. Numbers 1, 30, 31, 104, 105, and 146 indicate bases coding for the corresponding amino acids in β-globin polypeptide. Based on experiments by Konkel et al. (1978) and Tiemeier et al. (1978).

can be seen, the sequence coding for the β-globin polypeptide is interrupted by an intervening sequence between the bases coding for amino acids 30 and 31 and between those coding for amino acids 104 and 105. Intervening sequences are actually transcribed into RNA, but they are enzymatically excised from the newly transcribed RNA. The cut ends are then spliced to form the active mRNA.

Let us take a moment to describe how the intervening sequences were discovered. β-globin messenger RNA was used as a template to make complementary DNA using the enzyme reverse transcriptase. Both the β-globin RNA and the complementary DNA hybridized to only a portion of the β-globin gene. This was visible under the electron microscope as loops of unhybridized nucleic acid hanging from the β-globin gene DNA (Figure 13–12).

Many other findings also suggest that intervening sequences exist. For example, a direct analysis was made of the nucleotide sequences of the actual

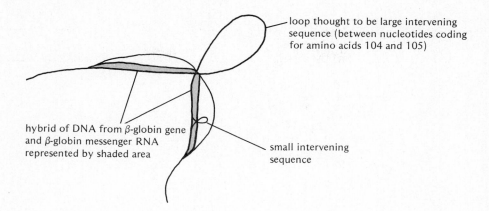

loop thought to be large intervening
sequence (between nucleotides coding
for amino acids 104 and 105)

hybrid of DNA from β-globin gene
and β-globin messenger RNA
represented by shaded area

small intervening
sequence

Figure 13–12. Intervening sequences of mouse β-globin gene. Diagram based on electron micrograph of Tilghman, et al. (1978).

genes, the messenger RNA, and the DNA complementary to the message. The results indicate that the intact genes contain many additional sequences not found in the message or in the complementary DNA. Intervening sequences are widespread among eukaryotic genes. Their function is a subject of current research.

RECOMBINANT DNA TECHNOLOGY

Before ending this section, it is appropriate to discuss how specific intact genes can be copied for use in experiments such as those dealing with intervening sequences.

Genes can sometimes be isolated by denaturing cellular DNA to the single strand form and separating it into fractions that bind labeled specific complementary DNA or the specific mRNA used to produce complementary DNA. Genes can also be chemically synthesized from nucleotides and other required building blocks. To do this, one must know the exact amino acid sequence in a given protein or know the actual nucleotide sequence of the gene. Among others, the human insulin gene has been chemically synthesized.

Once the gene is available, many copies of it must be made in order to provide enough for further experimentation. Genes are copied by using enzymes called *restriction endonucleases* to insert them into small circular DNA molecules (*plasmids*) that replicate in bacteria such as *E. coli*. Restriction enzymes make the ends of the gene "sticky," and break the plasmid circle so that its ends are also sticky. The sticky ends of the gene combine with the sticky ends of the plasmid, reforming the circle (Figures 13–13, 13–14). In this way, the gene is incorporated into the plasmid. This set of procedures is

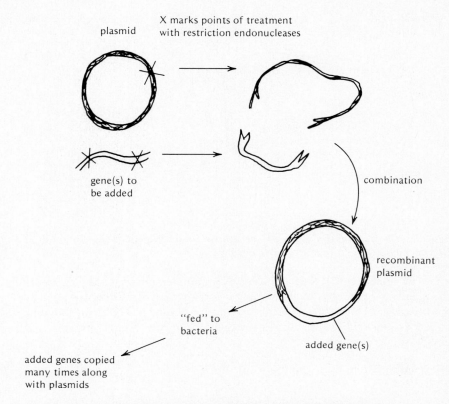

Figure 13-13. **Recombinant DNA technology, simplified drawing.**

called *recombinant DNA technology*. It has revolutionized the study of genes of higher organisms.

Once the recombinant plasmid has been constructed, it can be "fed" to cultures of bacteria such as *E. coli.* The plasmids are so constructed as to give bacteria a selective advantage. For example, a gene for antibiotic resistance can be included in the plasmid. If an antibiotic is added to the bacterial culture, only bacteria that took up the plasmid will grow. Thus, cultures of bacteria containing plasmids can be developed (Figure 13-14). The plasmid DNA replicates in the bacteria, producing clones of specific genes. These genes can be enzymatically recovered from the plasmids.

The potential of recombinant DNA technology is mind-boggling. Already, the human insulin gene has been cloned. *E. coli* possessing this gene can produce large quantities of life-saving human insulin. This technology is also revolutionizing the study of eukaryotic genes, and probably will lead to a better understanding of differentiation in developing systems.

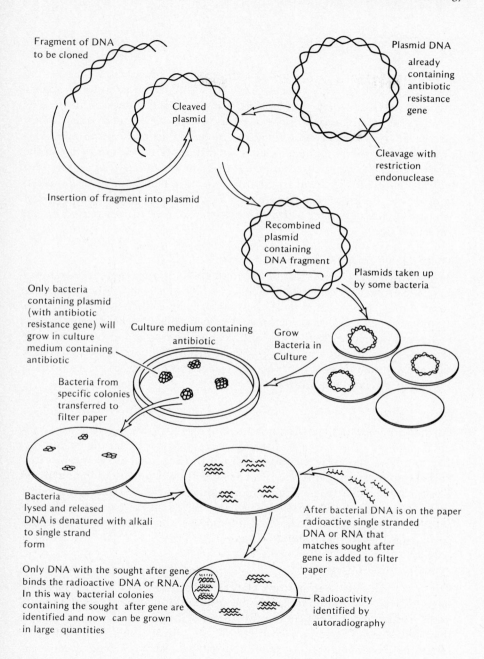

Figure 13-14. Recombinant DNA technology and gene cloning. From S. B. Oppenheimer, *Cancer: A Biological and Clinical Introduction* (Boston, Mass.: Allyn and Bacon, 1982).

DIFFERENTIAL GENE EXPRESSION

It is beyond the scope of this introductory text to go deeply into the exciting new work dealing with gene expression during differentiation in higher organisms. It is, however, important that we understand some of the key recent concepts in this field. Gene expression in eukaryotic cells is apparently influenced by a number of mechanisms, including gene loss, gene amplification, and gene rearrangement. Genes are differentially transcribed, and the RNA transcripts are treated in different manners and used in a variety of ways (Figure 13–15). Families of neighboring genes (multigene families) regulate the timing of gene expression as well as its amount and diversity. We shall briefly examine a few of these mechanisms in the context of cell differentiation.

MULTIGENE FAMILIES
As development proceeds from stage to stage, some genes that are neighbors on chromosomes are expressed in sequence. The clustering of such genes into families may permit sequential activation to occur. In other cases, a cell turns groups of genes on or off at one time. Whole families of genes may be needed at a given developmental stage, and gene clustering could facilitate their activation or repression in a stage-specific manner (Figure 13–16). Although clustering of related genes does occur in prokaryotes, little evidence for such clustering is found in eukaryotes.

GENE AMPLIFICATION
During development, massive amounts of a specific gene product are sometimes required. This happens, for instance, in oocytes of certain organisms. The requirement can be met by gene amplification, or multiple replication. For example, ribosomal RNA genes are specifically amplified in oocytes of *Xenopus* so that thousands of times more ribosomes can be synthesized than a single body cell usually produces. As another example, genes for chorion protein in *Drosophila* are amplified in ovarian follicle cells, which require large quantities of these proteins (Figure 13–17).

GENE REARRANGEMENT
Genes can change their position on chromosomes—that is, genes can be excised from one region of DNA and inserted into another. For example, widely separated regions coding for immunoglobulin can be rearranged so that they are adjacent to each other on the chromosome. In this way, a functional immunoglobulin gene is constructed. Such rearrangements may commit the cell to synthesizing a specific immunoglobulin. It is not yet known whether gene rearrangements are of general importance in the differentiation of other cell types (Figure 13–18).

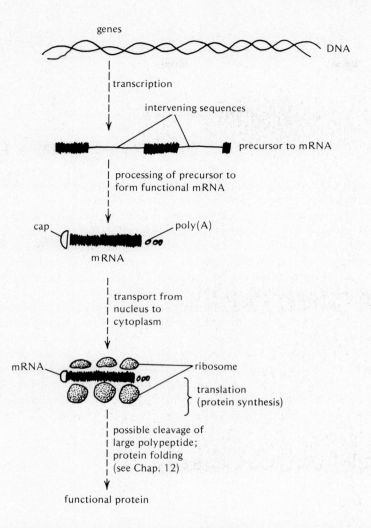

genes

DNA

transcription

intervening sequences

precursor to mRNA

processing of precursor to
form functional mRNA

cap   poly(A)

mRNA

transport from
nucleus to
cytoplasm

mRNA   ribosome

translation
(protein synthesis)

possible cleavage of
large polypeptide;
protein folding
(see Chap. 12)

functional protein

Figure 13–15. Some levels of control that occur as genes express themselves.

OTHER MECHANISMS

Many other mechanisms play a role in the expression of specific genes in
differentiating cells (Figure 13–15). For example, an inverted GTP residue is
often added to the 5′ end of mRNA molecules. Similarly, poly (A) (polyaden-
ylate) residues are often added to their 3′ ends. GTP caps may help bind the
mRNA to ribosomes and enhance its stability. Poly (A) "tails" also stabilize
mRNA, allowing it to stay active longer. Such alterations of mRNA may be

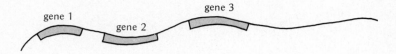

Figure 13–16. Multigene families in development. Gene 1, 2, and 3 in a family, for example, may all be expressed at one time at a certain developmental stage. In other multigene families, gene 1 may be expressed at one stage, followed by genes 2 and 3 in other stages.

important in controlling gene expression. The activity of a given mRNA may, in part, depend on such post-transcriptional events.

Gene expression during differentiation and during the entire life of a cell may also be controlled at the level of translating mRNA transcripts. Factors may mask stored mRNA until it is needed. Such a mechanism may explain how some cells can suddenly begin to synthesize specific proteins, even though they are not actively synthesizing mRNA at that time. How messages are masked and unmasked is not well understood.

Although this chapter deals primarily with differentiation at the nucleic acid level, it is appropriate to mention that proteins are sometimes first synthesized as part of much larger proteins that are cleaved to form the functional protein. ACTH (adrenocorticotropic hormone) is formed from a larger precursor that contains other biologically active polypeptides as well. Insulin A and B chains are cleaved from a single polypeptide precursor. Thus, the cleavage of large polypeptides into more than one active protein may be a mechanism for controlling gene expression in certain systems.

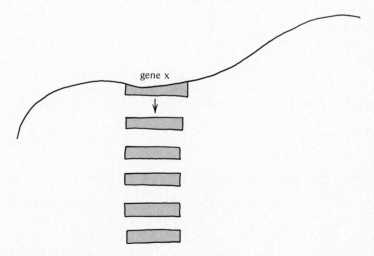

Figure 13–17. Gene amplification. Many copies of a single gene are sometimes made.

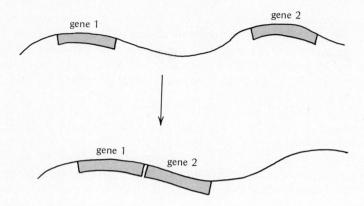

Figure 13-18. Gene rearrangement. Some genes can be excised from one region of DNA and inserted into another region.

LEVELS OF GENE CONTROL

Clearly, differential gene expression is controlled at many levels and by many mechanisms that are just beginning to be revealed. Future work will clear up mysteries that still cloud the picture of gene expression in eukaryotic cells. Progress will probably be very rapid in this area because of recent technological advances.

Several recent results show that the control of mRNA synthesis (transcriptional control) is essential to cellular differentiation. One experiment used the fact that estrogen stimulates the production of egg-white proteins in the chicken oviduct. Complementary DNA (cDNA) was formed from the mRNA molecules coding for the egg-white proteins ovalbumin, conalbumin, and ovomucoid. Labeled nuclear RNA was then isolated from chick oviduct epithelial cells that had either been treated with hormone (estrogen) or left untreated. RNA from the treated cells hybridized 10–50 times as much cDNA as that from untreated cells. This suggests that the production of egg-white protein mRNA in chick oviduct cells is controlled by hormone and that mRNA synthesis is a key step in the synthesis of these proteins. Similar results have been obtained for the estrogen-induced synthesis of mRNA for the proteins vitellogenin and very-low-density lipoprotein in the liver cells of roosters.

Additional evidence comes from studies on globin mRNA in the early chick. For the first few days after fertilization, chick embryos synthesize embryonic forms of alpha and beta globin. Later, two adult forms are produced. It was shown that erythrocyte stem cells form one set of cells that synthesize only the embryonic globins and another set that synthesize only the adult globins. Specific mRNA complementary to the cDNA for the embryonic globin gene was found in 5-day-old chick embryos; whereas message for adult globin was found in 12-day-old embryos. These results support the notion that, during differentiation, specific cells produce specific proteins because specific mRNA molecules have been synthesized. The level of transcription, therefore, is an important control point for differential gene expression during differentiation.

In another study, it was shown that only adult mouse liver cells contained specific mRNA for a major protein excreted in the urine of adult mice (major urinary protein). Newborn mice do not secrete this protein and do not synthesize much mRNA for major urinary protein. Again, the presence of a specific differentiation product—in this case, major urinary protein—depends upon the synthesis of the specific message coding for that protein.

Specific destruction or degradation of messenger RNA is another control mechanism in gene expression. During the differentiation of mammalian erythroblasts, the early developing cells synthesize many different mRNAs, including a small amount of globin mRNA. During the final four or so cell divisions, however, over 90% of the protein synthesized is globin. This suggests that globin mRNAs are preserved, while many other mRNAs are destroyed. Destruction of specific mRNA has also been observed during the development of the slime mold *Dictyostelium discoideum*. If the developmental cycle is interrupted at a specific point, the slime mold cells destroy the new mRNAs that would have been used in synthesizing the specific proteins required for later developmental stages.

Differentiation is also controlled at the translation level. For example, differential translation of mRNA molecules occurs in the cell cytoplasm of sea urchin eggs. Fertilized eggs synthesize protein much more actively than unfertilized eggs. The same amount of mRNA is present in the cytoplasm before and after fertilization. In the clam, two-dimensional gel electrophoresis reveals that the proteins formed before fertilization are different from those formed after, although the same mRNAs are present at both times. Translation, therefore, is another point at which gene expression during differentiation is controlled. More specific information on protein differentiation will be discussed in the chapter on proteins and differentiation.

What about post-translation? Several specific control mechanisms may come into play after proteins are synthesized to influence their final concentration in the cell. For example, proteases present in the cytoplasm can cleave specific polypeptides into smaller polypeptides. This apparently occurs in the posterior pituitary gland, where cells can produce either adrenocorticotropin or beta lipotropin, from the same primary translation product. Additional studies on the differential degradation of specific proteins will be described in detail in the protein chapter.

THE COMMITMENT TO DIFFERENTIATE

How do cells "decide" to differentiate? We have raised this question often because it is central to the understanding of embryonic development. Some new work will be cited here that brings us closer to answering this question, although we are still far from a total understanding of this puzzle.

In certain systems, DNA replication is not required for a cell to begin differentiating. This was shown by blocking DNA synthesis in certain systems with either ara C or FUdR, two inhibitors of DNA replication. Rat myoblasts are converted to differentiated myotubes whether or not DNA synthesis is blocked by these inhibitors. In another study, erythroid differentiation was unaffected by the inhibition of DNA synthesis.

Messenger RNA synthesis, followed by the synthesis of a specific protein, initiates the commitment to differentiate in certain systems. Inhibitors of mRNA (such as cordycepin or alpha amanitin) or inhibitors of protein synthesis (anisomycin or cycloheximide) block the differentiation of concanavalin-A-stimulated splenic lymphocytes and Friend cells into erythroid cells.

An influx of calcium ions appears to be necessary for committing some cells to differentiate. For example, exposing lymphocytes to calcium ionophore A23187 (recall the use of this agent to activate eggs) commits them to differentiate. When concanavalin A is used to initiate so-called lymphocyte blast transformation, an influx of Ca^{2+} is observed prior to the transformation. If EGTA is used to remove Ca^{2+} from the medium, lymphocytes no longer respond to concanavalin A by differentiating. Similar results have been obtained in the Friend cell erythroid differentiation system.

The puzzle of differentiation is becoming a little better understood. Intracellular calcium levels and messenger RNA synthesis (followed by the synthesis of specific new protein), are important in moving a cell toward the differentiated state.

GENETIC ENGINEERING: THE TRANSFER OF GENES TO MAMMALS

Genes have been injected into fertilized mouse eggs that were then implanted in the uteri of foster mother mice and allowed to develop. Some of the mice born possessed the foreign genes. Three big questions remained: (1) Can such new genes integrate into the recipient's genome? (2) Can they be transmitted to future generations? (3) Can they function in the recipient? The last question is critical to developmental biologists because such studies could produce a far better understanding of how specific genes are turned on and off at various stages and in specific tissues.

The answers to some of these questions are beginning to emerge. Work from many laboratories, including those of Mintz, Graham, Costantini, and Brinster, shows that 20–30% of the mice that develop from injected eggs carry the injected genes in their genomes and transmit them to their off-spring. Thus, questions (1) and (2) have been answered in the affirmative.

The answer to question (3) also appears to be affirmative, at least in some recent experiments. Brinster and Palmiter used a hybrid gene consisting of the viral thymidine kinase gene (*tk* gene) and control sequences from a mouse gene coding for the protein metallothionein. About 50% of the mice carrying the hybrid gene produced the viral thymidine kinase enzyme. Results have been mixed in studies by other investigators, using different genes. In some cases, the genes seem to be expressed; in others, they were not. The key point is, however, that we have already shown that it is *possible* for cells to express transferred genes.

The importance of such work cannot be overemphasized. Not only will it lead to a better understanding of the mechanisms of differential gene expression in developing systems, but it may lead to cures for many specific diseases that result from known genetic defects.

GIANT MICE FROM EGGS INJECTED WITH GROWTH HORMONE GENES

Recent work of Palmiter, Brinster, Hammer, Trumbauer, Rosenfeld, Birnberg, and Evans has exciting implications for accelerating animal growth and correcting genetic disease. A DNA fragment containing the promoter of the mouse metallothionein-I gene was fused to the structural gene of rat growth hormone. The resulting fusion gene was microinjected into the pronuclei of fertilized mouse eggs. Of 21 mice that developed from these eggs, 7 carried the fusion gene. Six of these grew significantly larger than their littermates. Several of these gene-injected mice had an extraordinarily high level of fusion mRNA in their liver and a similar level of growth hormone in their serum. This is the first time that genetic engineering has altered the pheno-type of an animal in such a profound manner.

This technology may lead to stimulating rapid growth in commercially important animals and to many other important practical applications. The age of helping humankind by introducing useful genes into the genome has finally arrived.

READINGS AND REFERENCES

Britten, R. J., and Kohne, D. E. 1968. Repeated sequences in DNA. *Science* 161: 529–40.
Braun, D. D. 1964. RNA synthesis during amphibian development. *J. Exp. Zool.* 159: 101–13.

Brown, D. D. 1981. Gene expression in eucaryotes. *Science* 211: 667–74. (*A particularly lucid and well-organized presentation.*)

Callan, H. G. 1966. Chromosomes and nucleoli of the axolotl, *Ambystoma mexicanum. J. Cell Sci.* 1: 85–108.

Chambon, P. 1981. Split Genes, *Scientific American* 244(5): 60–71.

Darnell, Jr., J. E. 1982. Variety in the level of gene control in eucaryotic cells. *Nature* 297: 365–71.

Davidson, E. H., and Britten, R. J. 1973. Organization, transcription and regulation in the animal genome. *Quart. Rev. Biol.* 48: 565–613.

Galau, G., Klein, W. H., Davis, M. M., Wold, B. J., Britten, R. J., and Davidson, E. H. 1976. Structural gene sets active in embryos and adult tissues of the sea urchin. *Cell* 7: 487–505.

Goeddel, D. V., Kleid, D. G., Bolivar, F., Heyneker, H. L., Yansura, D. G., Crea, R., Hirose, T., Kraszewski, A., Itakura, K., and Riggs, A. D. 1979. Expression in *Escherichia coli* of chemically synthesized genes for human insulin. *Proc. Natl. Acad. Sci. U.S.* 76: 106–10.

Goldberg, R. B., Hoschek, G., Tam, S. H., Ditta, G. S., and Breidenbach, R. W. 1981. Abundance, diversity and regulation of mRNA sequence sets in soybean embryogenesis. *Develop. Biol.* 83: 201–17.

Goldberg, R. S., Hoschek, G., Ditta, G. S., and Breidenbach, R. W. 1981. Developmental regulation of cloned superabundant embryo mRNAs in soybean. *Develop. Biol.* 83: 218–31.

Gurdon, J. B. 1974. *The Control of Gene Expression in Animal Development*. Oxford: Clarendon Press.

Konkel, D. A., Tilghman, S. M., and Leder, P. 1978. The sequence of the chromosomal mouse β-globin major gene: Homologies in capping, splicing, and poly (A) sites. *Cell* 15: 1125–32.

Levenson, R., and Housman, D. 1981. Commitment: How do cells make the decision to differentiate. *Cell* 25: 5–6.

McCarthy, B. J., and Hoyer, B. H. 1964. Identity of DNA and diversity of mRNA molecules in normal mouse tissues. *Proc. Natl. Acad. Sci. U.S.* 52: 915–22.

Markert, C. L., and Ursprung, H. 1971. *Developmental Genetics*. Englewood Cliffs, NJ: Prentice-Hall.

Marx, J. L. 1982. Still more about gene transfer. *Science* 218: 459–60.

Miller, O. R., Jr., and Beatty, R. R. 1969. Visualization of nucleolar genes. *Science* 164: 955–57.

O'Malley, B. W., and Means, A. R. 1976. The mechanism of steroid hormone regulation of transcription of specific eucaryotic genes. *Prog. Nucleic Acid Res. Mol. Miol.* 19: 403–19.

Palmiter, R. D., Brinster, R. L., Hammer, R. E., Trumbauer, M. E., Rosenfeld, M. G., Birnberg, N. C., and Evans, R. M. 1982. Dramatic growth of mice that develop from eggs microinjected with metallothionein-growth hormone fusion genes. *Nature* 300: 611–15.

Rosenthal, E. T., Hunt, T., and Ruderman, J. V. 1980. Selective translation of mRNA controls the pattern of protein synthesis during early development of the surf clam, *Spisula solidissima. Cell* 20: 487.

Stein, G. S., Spelsberg, T. C., and Kleinsmith, L. J. 1974. Nonhistone chromosomal proteins and gene regulation. *Science* 183: 817–24.

Tiemeier, D. C., Tilghman, S. M., Polsky, F. I., Seidman, J. G., Leder, A., Edgell, M. H., and Leder, P. 1978. A comparison of two cloned mouse β-globin genes and their surrounding and intervening sequences. *Cell* 14: 237–45.

Tilghman, S. M., Tiemeier, D. C., Seidman, J. G., Peterlin, B. M., Sullivan, M., Maizel,

J. V., and Leder, P. 1978. Intervening sequence of DNA identified in the structural portion of a mouse β-globin gene. *Proc. Natl. Acad. Sci. U.S.* 75: 725–29.

Tsai, S. Y., Harris, S. E., Tsai, M. J., and O'Malley, B. W. 1976. Effects of estrogen on gene expression in chick oviduct. *J. Biol. Chem.* 251: 4721.

Wessells, N. K., and Rutter, W. J. 1969. Phases in cell differentiation. *Sci. Amer.* 269: 43 (March 1969).

Whiteley, A. H., McCarthy, B. J., and Whiteley, H. R. 1966. Changing populations of messenger RNA during sea urchin development. *Proc. Natl. Acad. Sci. U.S.* 55: 519–25.

Wilt, F. W. 1974. The beginnings of erythropoiesis in the yolk sac of the chick embryo. *Ann. N.Y. Acad. Sci.* 241: 99–112.

14

DIFFERENTIATION: PROTEINS

In the last chapter, we saw that cellular RNA changes during development. We also noted that it is extremely difficult to interpret these changes in nucleic acids because it is uncertain which species of RNA are functional. Since proteins are the important products of nucleic acid messages, and since their synthesis or lack of synthesis can be definitively determined, many investigators study proteins as markers of development and differentiation. In this chapter we shall glance at proteins in early development, methods of analysis in the protein field, and some of the specific proteins that differentiate cells from one another.

PROTEIN SYNTHESIS IN EARLY DEVELOPMENT

As mentioned earlier, the sea urchin embryo is ideal for studying development because a vast number of embryos can easily be obtained and because normal development occurs in sea water medium. We can add a radioactive amino acid (such as radioactive methionine) to the sea water medium and measure the amount of this amino acid that is incorporated into newly synthesized protein. This is done by incubating the embryos with the "hot" amino acid for a short time, then placing the embryos in sea water without the radioactive amino acid. If we do this at various development stages, we can determine how much protein was synthesized at each specific stage by measuring how much radioactive protein was formed after administering each pulse of "hot" amino acid. Using such experiments, we can measure the amount of new protein synthesized with reasonable accuracy if we make sure that the radioactive amino acid is being taken up by the cells at all stages. We must also make sure that the intracellular supply (pool) of available amino acid remains constant at all stages. These and other factors must be considered if we are to reliably determine the relative amount of protein being synthesized at different developmental stages.

Figure 14–1 shows an example of the results obtained by investigators examining protein synthesis in the sea urchin embryo. Protein synthesis rapidly increases for the first 3 or 4 hours after fertilization and then declines, only to increase again at about the gastrula stage. Radioactive methionine was used in this experiment and was shown to be taken up by unfertilized eggs and other stages. Amino acid uptake was thus not a problem in these experiments. Also, the supply (pool) of intracellular methionine was shown not to change appreciably in the different stages examined. Thus the pool of methionine was also not a problem. These experiments suggest that different amounts of total protein are synthesized at various stages of embryonic development.

Figure 14–1 also shows that if embryos are treated with the RNA synthesis inhibitor actinomycin D, protein synthesis increases after fertilization but not at gastrulation. These results suggest that the proteins synthe-

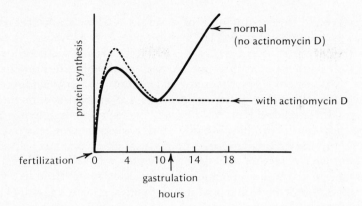

Figure 14-1. Protein synthesis at various stages in sea urchin embryo development. From H. Ursprung and K. D. Smith, *Brookhaven Symp. Biol.* 18 (1965): 1–13.

sized during early development use messenger RNA formed in the egg before fertilization. At gastrulation, however, new species of mRNA may be transcribed, and this would be blocked by actinomycin D. More recent work has, however, shown that some specific proteins, such as histones, are synthesized in the early hours of sea urchin embryonic development using newly formed mRNA, as well as that stored in the egg before fertilization.

We have described the synthesis of total protein at various developmental stages. What about specific proteins? Do all proteins in the cell follow the same pattern as that for total protein? Figure 14–2 shows that if individual marked proteins are separated chromatographically at different develop-

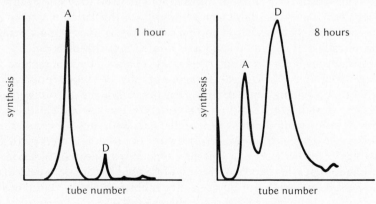

Figure 14-2. Specific proteins synthesized at different stages of sea urchin embryo development. Protein synthesis is measured as counts per minute of radioactive amino acid incorporated into the protein in each column fraction. After C. H. Ellis, *J. Exp. Zool.* 163 (1966): 1–22.

mental stages, different amounts of a single protein are synthesized at different developmental stages. A lot of protein A, for instance, is synthesized at the first hour after fertilization in the sea urchin, while only a little protein D is synthesized at this time. At the eighth hour after fertilization, however, less protein A is synthesized and a lot of protein D is synthesized. These results show that different proteins are synthesized in different amounts at different developmental stages. The experiment shown in Figure 14–2 does not identify proteins A and D. In fact, these proteins may be similar and may represent different numbers of identical subunits or an association of the protein with cofactors. Later in the chapter we shall examine specific proteins and reaffirm that given proteins are often synthesized in different amounts at different developmental stages. In this section we have looked at a few experiments in one model system to get an idea of how some of the early experiments in this area were performed. We have looked at protein synthesis in whole embryos. This introduces us to specific proteins in specific cells during differentiation.

MEASUREMENT OF SPECIFIC PROTEIN SYNTHESIS

Protein synthesis is generally measured by incubating embryo cells in a solution containing a radioactive amino acid such as ^{14}C methionine or ^{14}C leucine. The cells are then homogenized and large molecules are separated from the amino acids and other small molecules by acid precipitation and chromatography on gel columns that separate molecules according to size or charge. The amount of radioactive amino acid incorporated into each separated large molecule is easily determined with a scintillation counter. Thus, the amount of newly synthesized protein can be determined by measuring the amount of radioactive amino acid incorporated into this material.

We mentioned that specific proteins are separated by such techniques as gel chromatography. Using gel chromatography in conjunction with radioactive amino acid incorporation, we can determine the rate of synthesis of the specific proteins separated on the gels. We can often identify specific proteins by comparing their profiles on the gels with those of characterized protein standards. Electrophoresis (separation of proteins in electric fields) is also used to compare a protein sample with a known protein standard.

Immunological methods are also widely used to identify specific proteins. The specific protein is injected into rabbits or goats, which produce antibodies against the protein that can then be isolated from the blood of the animal. When the antibody is added to a specific protein in a mixture of many proteins, it complexes with the specific protein, forming a precipitate.

Newly synthesized specific protein can be identified using the immunological technique coupled with the radioactive amino acid method. Cells are incubated with radioactive amino acid for a short time and then homogenized. The homogenate (or an extract thereof) is then incubated with anti-bodies against a single specific protein, precipitating it out of solution. The amount of radioactivity in the precipitate is then measured. This tells us how much radioactive amino acid was incorporated into the specific protein precipitated by the antibody. In this way we can determine how much of the specific protein was synthesized while the cells were being incubated with radioactive amino acid.

This technique is very useful in identifying the pattern synthesis of specific proteins during differentiation. We should remember that just meas-uring the activity of a given enzyme or the presence or absence of a specific protein at a given developmental stage does not determine synthesis per se. Synthesis must be measured by a method such as that described above in which *newly formed* protein is determined. The simple presence or absence of a specific protein or the presence or absence of specific enzyme activity could result from the activation or inactivation of an enzyme by specific cofactors or from the differential degradation of the protein by degradative enzymes present in the cells.

Before moving on to look at some specific proteins formed during differentiation, let us see how the immunological technique was used to determine specific protein synthesis in one model system. The system is the synthesis of the enzyme tryptophan pyrrolase by the rat liver. This enzyme opens the indole ring of tryptophan, a precursor of the important vitamin, nicotinamide. The activity of this enzyme rises rapidly at birth. Its activity is also increased by the hormone hydrocortisone and by the amino acid trypto-phan (Figure 14-3). The immunological technique described previously was used to determine how these substances increase tryptophan pyrrolase activ-

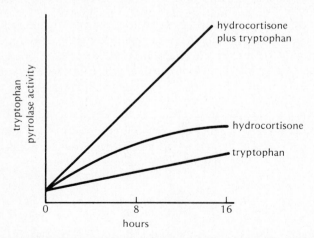

Figure 14-3. Stimulation of tryptophan pyrrolase activity by hydrocortisone, tryptophan, or both. Redrawn and modified from R. T. Schimke, *Natl. Cancer Inst. Monograph* 27 (1967): 301-14.

ity. These experiments are interesting because they not only show how the technique is used, but also help us understand how protein synthesis and degradation are regulated.

Purified tryptophan pyrrolase was injected into rabbits, which produced specific antibody against this enzyme. Rats were inoculated with hydrocortisone, tryptophan, or saline and, after a few hours, with radioactive amino acid. Their livers were then removed and homogenized. Tryptophan pyrrolase antibody was added to the homogenate to precipitate out tryptophan pyrrolase, and the amount of newly synthesized enzyme was determined by measuring the radioactivity of the precipitate.

Figure 14–4 shows that hydrocortisone increases the synthesis of tryptophan pyrrolase. This conclusion is based upon the observed increase in radioactive tryptophan pyrrolase in the hydrocortisone-treated livers over that of the controls. There was, however, no increase in tryptophan pyrrolase synthesis in the tryptophan treated livers. How does tryptophan produce the observed increase in tryptophan pyrrolase activity (Figure 14–3)? This question was resolved in another experiment. Liver proteins were labeled as before with radioactive amino acid. Later, either saline or tryptophan was injected. Still later, the livers were homogenized and the amount of radioactive tryptophan pyrrolase present was determined after antibody precipitation. Figure 14–5 shows that the amount of radioactively labeled tryptophan pyrrolase declined steadily after saline injection. Tryptophan, however, prevented this decline. These results suggest that tryptophan keeps tryptophan pyrrolase enzyme active by protecting it from degradative enzymes.

In summary, we can tell whether a specific protein is being synthesized by measuring how much radioactive amino acid is incorporated into the protein precipitated by a specific antibody. We saw that the activity of an enzyme is regulated by stimulating its synthesis or by preventing its degradation. Other work has shown that enzyme activity is also regulated by inhibitor or activator molecules. Now that we know how one goes about measuring protein synthesis and know some of the factors that regulate the

Injected material	Counts per minute of radioactive amino acid in protein precipitated by antibody against tryptophan pyrrolase.
NaCl	1406
Hydrocortisone	9466
Trytophan	1954

Figure 14–4. Stimulation of tryptophan pyrrolase synthesis by hydrocortisone but not by tryptophan. Modified from Schimke et al., *J. Biol. Chemistry* 240 (1965): 322–31.

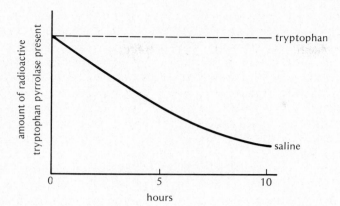

Figure 14-5. Prevention of tryptophan pyrrolase degradation by tryptophan.
Tryptophan pyrrolase was labeled with radioactive amino acid. At 0 time and at 4 and 8
hours the animals were injected with saline or tryptophan. The amount of radioactive
tryptophan pyrrolase was determined at 3, 6, and 9 hours. From R. T. Schimke, *Natl.
Cancer Inst. Monograph* 27 (1967): 301-14.

activity of specific enzymes, let us examine some of the proteins synthesized
in differentiating cells.

SPECIFIC PROTEINS AND DIFFERENTIATION

HEMOGLOBIN

Hemoglobin is a *tetramer*, a molecule composed of four folded polypeptide
chains. Each polypeptide chain carries a nonprotein heme group (Figure
14-6). The heme groups contain iron and bind oxygen. The most important
function of hemoglobin is to carry oxygen to different parts of the body.
Hemoglobin is the major protein contained in red blood cells and thus is a
distinct differentiation marker in these highly specialized cell types.

There are 4 different types of hemoglobin, embryonic hemoglobins,
fetal hemoglobins, and normal and abnormal adult hemoglobins. Stage-
specific differences in hemoglobin probably evolved as a result of the different respiratory needs of the embryo, fetus, and adult. For example, mammalian fetuses obtain oxygen by diffusion across the placenta from the maternal
blood. The hemoglobin of the fetal red blood cells must have a higher oxygen
affinity than that of the mother if such an exchange is to occur.

What is the nature of these stage-specific differences? As mentioned
previously, adult mammalian hemoglobin consists of four polypeptide chains.
Two of these chains are coded for by one gene; the other two chains are
coded for by another. Adult hemoglobin can thus be designated as $\alpha\alpha\ \beta\beta$,
indicating that two polypeptide chains are coded by gene A and two are by

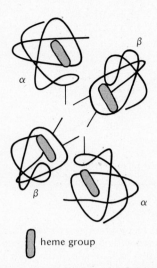

heme group

Figure 14–6. Diagram of hemoglobin showing 4 subunits. A heme group is associated with each subunit.

gene B. In the early mouse embryo, other genes code for hemoglobin chains. The A gene is active in the embryo, as are the X, Y, and Z genes. Embryonic hemoglobin does not contain β chains because the B gene is not active until late in fetal development. Thus, important changes in hemoglobin take place when the B gene is activated, making β polypeptide chains part of the new hemoglobin molecules.

Abnormal types of adult hemoglobin include hemoglobin S and hemoglobin C. Hemoglobin S is found in individuals who have the disease sickle-cell anemia. In this disease, the red blood cells are sickle shaped. Hemoglobin S differs from normal hemoglobin in only the β subunits; the α subunits are identical to those in normal adult hemoglobin. The normal β subunit consists of 146 amino acids. The only difference between the β subunits of normal and sickle-cell hemoglobin is a single valine residue that replaces the number 6 amino acid, glutamic acid, of the normal β chain. This single amino acid change can be caused by a single base change in the mRNA coding for the β subunit. Glutamic acid is coded for by the base sequence GAA; whereas, valine is coded for by GUA. Thus, a difference of a single base can produce a disease that can be fatal. This change probably occurred as a result of a single point mutation in the DNA of the β subunit gene.

Hemoglobin C, another abnormal hemoglobin, also results from a single amino acid change. The same β chain amino acid replaced in hemoglobin S is also replaced in hemoglobin C. Instead of valine, however, a lysine substitutes for the glutamic acid residue at position 6. The bases AAA code for lysine; whereas the bases GAA code for glutamic acid. Again, a single base change, an A for a G, leads to an amino acid change in the β subunit.

Hemoglobin C, however, does not cause a severe anemia in people. Thus, if one specific amino acid out of 146 amino acids in the β subunit is changed to a valine, a major disease results; whereas a change to a lysine causes a minor problem. Now that we know a little about hemoglobin, let us look at the differentiation of red blood cells.

Red blood cells in the embryos of humans and mice originate first in the yolk sac, then (later in development) in the liver. After birth, red blood cells are produced by the bone marrow.

Important questions we may now ask are: When do the cells that give rise to red blood cells begin to differentiate? What events mark the beginning of differentiation? Recall also that embryonic hemoglobin differs from adult hemoglobin. Do the same cells, or cells derived from the same parent cell, give rise to the different types of hemoglobin? Or do different cell lines give rise to the different hemoglobins?

In the mouse, hemoglobin begins to appear in red blood cell precursors in the blood islands of the yolk sac by the eighth day of gestation. On the ninth day, these precursor cells enter circulation. These cells synthesize embryonic hemoglobin. They divide four times, then their nuclei condense and they become inactive. On the twelfth day of gestation, the liver begins to produce red blood cells. This new population synthesizes adult hemoglobin. The yolk sac continues to produce red blood cells through the fourteenth day, so both types of red blood cells (yolk sac and liver) are found in fetal circulation. Spleen and bone marrow take over the production of red blood cells between day 15 and birth, while liver production declines. Thus, there are at least two populations of red blood cells, one that synthesizes embryonic hemoglobin and another that synthesizes adult hemoglobin. No one knows whether a common ancestor cell originally seeded the yolk sac and the liver, spleen, and bone marrow.

In the yolk sac, hemoglobin (globin) mRNA is synthesized before day 10 of gestation. After day 10, hemoglobin synthesis is not inhibited by actinomycin D. This suggests that the mRNA made before day 10 is stable and used later on for synthesizing the protein chains of hemoglobin. Thus, the main event that triggers the formation of red blood cells is probably the activation of the genes that code for the polypeptide chains of hemoglobin (globin).

Messenger RNA that codes for the polypeptide chains of hemoglobin has been isolated. Such mRNA is identified by its ability to direct the synthesis of globin in a cell-less system and by hybridization experiments such as those described in the last chapter. Early fetal mouse liver cells do not contain globin messenger RNA. The glycoprotein hormone, erythropoietin, causes these cells to synthesize globin mRNA. Hemoglobin synthesis begins shortly thereafter. This hormone is produced by the kidney in response to decreases in tissue oxygen content. It is a key factor in inducing prospective erythrocytes to differentiate into mature red blood cells. It probably acts by turning on globin genes. Exactly how it works is unknown, but it may act in a way similar to that proposed for estrogen (estradiol) in the previous chapter. Its final action may be on specific chromosomal proteins that function to regulate gene action. Exactly how certain cells commit themselves to forming red blood cells is unknown. However, the events that immediately lead to differentiation—globin message transcription followed by hemoglobin synthesis—are becoming better understood.

MYOSIN

The protein myosin is probably present in many cell types. Muscle cells, however, contain a large amount of this protein. Myosin is thus a major marker for muscle cell differentiation. Myosin is a part of the contractile networks present in all eukaryote cells. Since the main function of muscle cells is to contract, it makes sense that these cell types should be enriched in myosin. Myosin makes up ten to twelve percent of the fresh weight and nearly fifty percent of the dry weight of adult muscle.

Myosin is a fibrous protein composed of two identical subunits wound around each other (Figure 14-7). In muscle, myosin molecules aggregate into filaments. This assembly phenomenon will be examined in the next chapter. Myosin combines with other proteins, such as actin. Units composed of myosin, actin and certain other proteins are contractile and serve as a cellular

Figure 14-7. Myosin consists of two subunits wound around each other.

contractile apparatus. Proteins associated with actin filaments include α-actinin, β-actinin, tropomyosin, and troponin. These proteins serve a variety of functions to keep the contractile apparatus intact and allow it to work smoothly at appropriate times.

Let us look first at developing muscle cells and then zero in on the differentiation of these cell types. In this chapter, we shall stress myosin synthesis. In the next, we shall gain insight into how muscle proteins assemble into functional contractile units.

It has been shown using fluorescent labeled myosin antibody, that muscle cell contractile proteins are present in many prospective muscle cells before any organized contractile apparatus is visible there. Prospective muscle cells (myoblasts) are derived from the myotome region of the somite and hypomere mesoderm (Chapter 9). The prospective muscle cells divide, forming many additional myoblasts. When they stop dividing, they fuse together, forming long muscle tubes (myotubes) containing many nuclei in a common cytoplasm.

In two-day-old chick embryos, somite myotome cells are actively dividing. Some of these cells form spindle shapes containing detectable myosin filaments. By the fourth day, most of the myoblasts contain muscle contractile fibrils (myofibrils). All this occurs before the cells fuse into the multinucleated myotubes. In limb muscle, however, myofibrils appear only in the multinucleated myotubes and not in the prefusion myoblasts.

We have said that elongated-myotubes arise as a result of fusion of single cells rather than by continued nuclear divisions without cell division. This was elegantly shown by an experiment in which cells from two different mouse embryos were aggregated. Each embryo strain differed in the nature of one enzyme, isocitrate dehydrogenase, which was made up of subunits. The mosaic embryos were implanted in the uterus of a foster mother mouse, and developed into normal mice consisting of cells from the two different strains of mice. The skeletal muscle of these mice contained some enzyme like that of each parent, but also some hybrid enzyme. This hybrid enzyme consisted of subunits from one parent's enzyme and from subunits with the other parent's enzyme. This type of hybrid enzyme could have been produced only if the myotubes contained nuclei from cells of each parent in a common cytoplasm (Figure 14-8). The nuclei from each will produce a message that codes for its own enzyme subunits. These subunits, when synthesized, then polymerize into the hybrid enzyme observed. Such enzymes are easily detected using electrophoresis, which separates proteins on the basis of size and charge. Thus, skeletal muscle forms when single cells fuse into elongated myotubes.

Myoblast fusion has also been studied *in vitro* to learn about the specific factors that influence this phenomenon. Myoblasts in culture do not fuse until six to eight hours after cell division has stopped. Presumably these cells use this time to acquire specific cell surface properties that facilitate the fusion

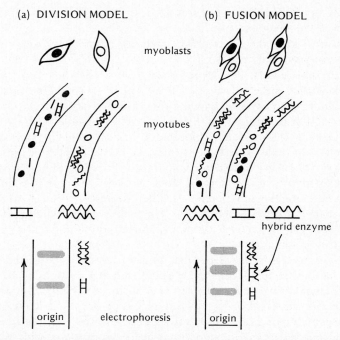

Figure 14-8. Isocitrate dehydrogenase in myotubes. Model (b) (the fusion model) is correct because hybrid enzyme was indeed found during the electrophoresis of muscle from allophenic mice (mice produced by aggregating cells from different mouse strains). From Mintz and Baker, *Proc. Natl. Acad. Sci. U.S.* 58 (1967): 593, Fig. 1.

process. Limb myoblasts in culture will synthesize myosin and assemble myofibrils even if fusion is prevented by chelating agents that bind divalent cations. Thus, even in limb muscle, fusion is not absolutely necessary for muscle differentiation to occur.

Yaffe and others have shown in culture that much of the mRNA that codes for myosin is probably synthesized hours before the myoblasts fuse. Actinomycin D inhibits the synthesis of muscle proteins if applied more than six hours before fusion. If applied within six hours of fusion, actinomycin D has little effect. Thus, stable mRNA molecules that code for muscle proteins are synthesized long before fusion occurs.

Cell culture experiments have also shown that specific factors may be required for muscle cell fusion and differentiation. Hauschka and Konigsberg found that collagen can stimulate the fusion of myoblasts to form myotubes. Collagen is present in the somite and hypomere where skeletal muscle forms. This substance may play a role in controlling myoblast fusion and differentiation *in vivo*. Exactly what factors cause certain cells to become muscle cells is, as yet, unknown.

PANCREATIC PROTEINS AND DIFFERENTIATION

Before we examine the differentiating pancreas, let us say a few words about one of the pancreatic proteins—the hormone insulin. This hormone is extremely important in metabolizing carbohydrates. Without it, glucose cannot be tranformed to glycogen. In its absence, blood sugar builds up, eventually killing the organism. The disease diabetes results from a lack of insulin. It can be controlled with insulin injections. Recent work has shown that the human insulin gene can be incorporated into the genome of the bacteria *E. coli*. Bacteria so altered begin to synthesize human insulin. Nobel prizes were recently awarded to Arber, Smith, and Nathans for their work in discovering enzymes that cut DNA, allowing the incorporation of new genes such as the insulin gene. Human insulin should be very useful in treating many difficult cases of diabetes that do not respond well to animal insulin. Virtually unlimited supplies of this hormone may become available. Such work is leading to major breakthroughs in medicine. This genetic engineering is clearly a new frontier in biology today.

Insulin is a small protein hormone with a molecular weight of 6000 daltons. It consists of two polypeptide chains (Figure 14-9). It is one of many important proteins produced by the pancreas. In the mouse embryo, insulin is first detected in the pancreas rudiment 9–10 days after fertilization. The cells that produce insulin pinch themselves off from other pancreas tissue, forming separate masses called Islets of Langerhans (Figure 14-10). The islet cells differentiate into two distinct cell populations: A-cells, which synthesize the hormone glucagon, and B-cells, which produce insulin. B-cells are easily distinguished from A-cells by their dense cytoplasm and beta granules. Both

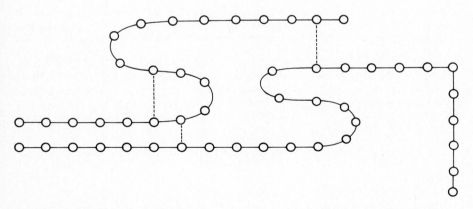

Figure 14-9. Insulin. The hormone insulin is composed of two polypeptide chains. Disulfide bridges form between the sulfhydryl groups of specific amino acids (cystein residues).

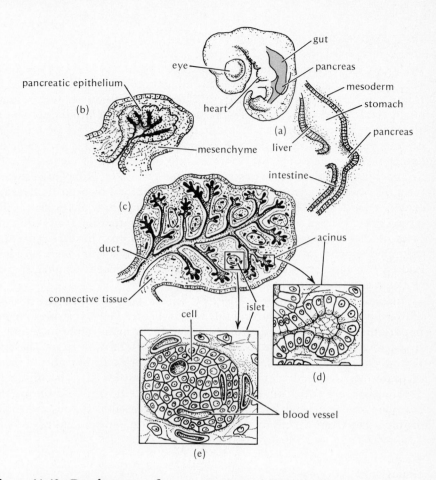

Figure 14–10. Development of mouse pancreas. (a) Initial elevation of pancreas epithelium—would have low levels of specific proteins (about 12-day embryo). (b) Acini and islets are now present. The acini are composed of exocrine cells with zymogen granules. The islets are endocrine in function (c). Secretory acini begin to form (about 12-day embryo). (d), (e) Enlargement of an acinus, containing secretory (zymogen) granules and an islet, respectively. From W. Rutter and N. Wessells, "Phases in cell differentiation," *Scientific American* (March 1969). Copyright © 1979 by Scientific American, Inc. All rights reserved. Used with permission.

hormones are involved in carbohydrate metabolism. The cells that produce digestive enzymes are located at the ends of a branched duct system in pouches called acini. The digestive enzymes produced in the acini include lipase, amylase, trypsin, and chymotrypsin. Acinar cells have extensive endoplasmic reticulum, Golgi apparatus, and storage granules (zymogen).

The mouse pancreas rudiment appears as an evagination from the gut tube 9–10 days after fertilization. Pancreatic enzymes and hormones are detectable, but in very small amounts. A period of rapid cell division follows in which the enzyme and hormone levels remain low. By about day 15, cell division stops and the synthesis of specific enzymes and hormones rapidly increases. During this time, the cells of the pancreas also differentiate cytologically. Endoplasmic reticulum, Golgi apparatus, and zymogen granules become increasingly abundant in the nondividing acinar cells.

The differentiation of the endodermal epithelium of the mouse pancreas is apparently induced by the mesenchyme that overlies the gut. The mesenchyme merges with the endodermal pancreatic rudiment as the rudiment evaginates from the gut tube at day 9 or 10. If an 11-day pancreas rudiment is separated from its mesenchyme and cultured on a porous filter, the endodermal epithelium does not proliferate and differentiate. If the mesenchyme is placed on the other side of the porous filter, however, cell division and differentiation occur. Specific proteins are synthesized and characteristic tubules develop. Mesenchyme from other tissues—and even from other species—also induces differentiation. The mesenchyme apparently produces a factor that can pass through the filter pores to stimulate cell division in the epithelium.

A cell-free extract of whole chick embryos or of mesenchymal tissues can be added to the medium in which pancreas epithelium is cultured. These extracts cause the epithelium to divide and differentiate in the absence of mesenchyme. Cell division is needed for differentiation, because if DNA synthesis is inhibited in predividing cells by FUdR (5-fluorodeoxyuridine), differentiation does not occur. BUdR (5-bromodeoxyuridine) is a thymidine analog that replaces thymidine in the DNA of dividing cells but does not prevent DNA synthesis or cell division. The DNA produced, however, is abnormal. This substance also prevents the differentiation of pancreas cells treated during the division period. In the presence of BUdR, mitosis continues and some common cell proteins are synthesized, but increased levels of pancreas-specific proteins and differentiation do not occur. BUdR may prevent the production of mRNA that codes for cell-specific proteins. The RNA synthesis inhibitor actinomycin D also inhibits the increase in protein synthesis if the inhibitor is applied during the cell division period or before the increase in specific protein synthesis begins. It has no effect once protein synthesis accelerates. This implies that stable messages are produced before the increase in specific protein synthesis. To sum up, a period of normal DNA synthesis must occur before a high rate of specific protein synthesis can begin.

The mesenchymal factor (MF) that stimulates cell division and differentiation in pancreas epithelium has been isolated by Rutter and colleagues. This factor was covalently bound to inert beads, and the beads were cultured with pancreas epithelium cells. The cells attached to the beads and continued DNA synthesis and cell division. Zymogen granules formed in the cells, indicating that differentiation had occurred (Figure 14-11). Since it is unlikely that MF attached to the beads entered the cells, these results suggest

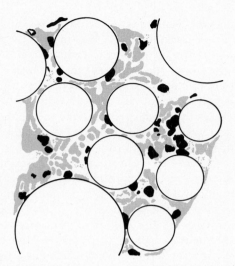

Figure 14-11. Pancreas epithelial cells attached to sepharose beads coated with mesenchymal factor (MF). These cells are dividing. Cell division seldom occurs if cells are added to beads without MF. After R. Pictet and W. J. Rutter, *Nature New Biology* 246 (1973): 49.

that MF stimulates mitosis and differentiation by some action at the surface of the pancreas epithelial cells.

In summary, the pancreas develops as an evagination of the gut tube. The pancreatic rudiment joins with the mesenchyme surrounding the gut tube. In the presence of mesenchyme, morphogenesis begins and specific proteins are synthesized at a low rate. Cell division and morphogenesis continue. Finally, specific protein synthesis increases rapidly and cytological differentiation finishes. Stable mRNA coding for the specific proteins is apparently synthesized before the synthesis of specific protein accelerates. Mesenchyme apparently induces cell division in the pancreas epithelium. Wessels and others have proposed that increased cell mass in the developing pancreas (and in other systems) may be a key factor in entering the state of increased specific protein synthesis and cytodifferentiation. Large masses of cells could alter their immediate environment, creating conditions that accelerate specific protein synthesis. This idea may suggest additional experiments that could uncover some of the key control mechanisms of differentiating systems such as the pancreas.

TISSUE- AND STAGE-SPECIFIC ISOZYME PATTERNS

We have seen that proteins often consist of subunits coded for by separate genes. Hemoglobin subunits, for instance, are coded for by several genes. The relative activity of these genes varies at different developmental stages, so fetal hemoglobins and adult hemoglobins have clearly different subunit compositions. These different hemoglobins are presumably those best adapted to the different environments of the embryo and the adult.

Isozymes are enzyme molecules that are functionally similar but that behave differently during electrophoresis because of differences in the charge of the enzyme molecules. Some forms are more negatively charged than others. These charge differences occur because the enzymes are composed of subunits and, as in hemoglobin, different subunits can combine to form the isozymes. Each different subunit is coded for by a different gene. Thus, given genes may be more active than others at a given stage or in a given tissue, and the resulting proteins will also be different.

Lactate dehydrogenase (LDH) is an enzyme that exists in multiple isozymes. Different isozymes of LDH predominate in different species, in different tissues, and at different developmental stages. LDH catalyzes the interconversion of pyruvate and lactate. Pyruvate is a key intermediate when glucose is oxidized to carbon dioxide and water, releasing energy that is essential for various metabolic activities. LDH makes it possible for cells to function effectively when oxygen is temporarily unavailable. As LDH converts pyruvate to lactate, it generates NAD from NADH. NAD is essential for completing an early step in a set of metabolic reactions called glycolysis. Like hemoglobin, different LDH isozymes presumably function best under different metabolic conditions. Thus, one isozyme might predominate at a specific stage or in a specific tissue because it is best adapted to function under those specific cytoplasmic conditions. A brief consideration of LDH isozymes during development is appropriate now that we know a little about isozymes. First, however, we will see how it was determined that LDH exists as isozymes in the first place.

When crude homogenates of tissues such as mouse skeletal muscle are subjected to electrophoresis, the various proteins present in the homogenate separate on the basis of their net electrical charge. After a time, the gel block is immersed in a staining solution that contains lactic acid (an LDH substrate), the required cofactor NAD, and a tetrazolium salt. If LDH is present in the gel, the lactic acid is oxidized and the tetrazolium salt is reduced to colored formazan that precipitates at the site of LDH activity. Five of these colored zones appear after a mouse muscle homogenate is subjected to electrophoresis (Figure 14–12). Each colored zone identifies an LDH isozyme. Five isozymes of LDH have been found in most of the mammal and bird tissues examined. Why do five LDH isozymes form?

Markert and Ursprung purified LDH by several methods. The pure LDH also electrophoresed as five isozyme bands. By analytical ultracentrifugation in sucrose density gradients, they determined that the molecular weight of each LDH isozyme is about 140,000 daltons. When the pure LDH isozymes were subjected to agents or conditions that separated proteins into

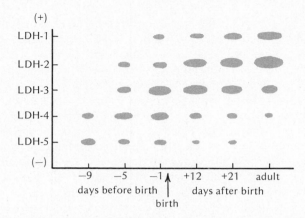

Figure 14–12. Changing LDH isozyme pattern in heart muscle at different stages of mouse development. LDH-1 migrates toward the anode (+) and LDH-5 migrates to the cathode (−) during electrophoresis. Diagram based upon photo from C. L. Markert and H. Ursprung, *Developmental Genetics* (Englewood Cliffs, NJ: Prentice-Hall, 1971), Fig. 4–7, p. 44. © 1971. Reprinted by permission of Prentice-Hall, Inc., Englewood Cliffs, New Jersey.

subunits, only two, not five, electrophoretic bands appeared that stained for protein. Each band represented a subunit of LDH. Each subunit had a molecular weight of about 35,000 daltons. The two bands were designated LDH subunit A and LDH subunit B.

The separated LDH subunits recombine to form the five LDH isozymes as follows:

LDH–1   BBBB
LDH–2   ABBB
LDH–3   AABB
LDH–4   AAAB
LDH–5   AAAA

Such combinations of subunits probably occur in cells and in this way form the five LDH isozymes.

Different genes code for LDH subunit A and LDH subunit B. Thus, the amount of each isozyme that appears in cells of different tissues or at different developmental stages reflects, in part, the ratio of the A subunits to the B subunits produced at that time in that tissue (Figure 14–12, 14–13). For example, mouse eggs produce mostly LDH-1 (the B tetramer). Thus, the B gene is the most active subunit during oogenesis. As development proceeds, the A gene becomes activated and the B gene is suppressed. LDH-5 (the A tetramer) begins to predominate. By birth, the B gene has been activated in

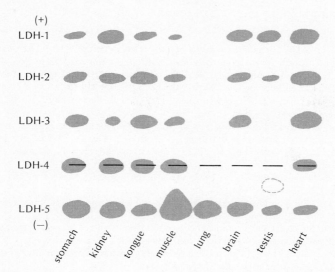

Figure 14-13. LDH isozyme patterns in 8 adult rat tissues. Electrophoretic patterns show tissue-specific differences in LDH isozymes. Dotted circle in testis pattern represents an isozyme that may be testis-specific. Diagram based upon photo from C. L. Markert and H. Ursprung, *Developmental Genetics* (Englewood Cliffs, NJ: Prentice-Hall, 1971), Fig. 4-6, p. 43. © 1971. Reprinted by permission of Prentice-Hall, Inc., Englewood Cliffs, New Jersey.

many tissues of the body, so the isozyme pattern shifts again toward LDH-1. Thus, shifts in LDH patterns reflect the shifting relative activities of the A and B genes.

Adult rat tissues also differ with respect to LDH pattern (Figure 14-13). For example, skeletal muscle shows a lot of LDH-5, whereas heart muscle has more LDH-1, LDH-2, and LDH-3. Here, too, shifting relative activities of the A and B genes occur. Specific cytoplasmic factors such as pH and ionic species probably control the rates of subunit synthesis. Such conditions may also affect subunit accumulation and assembly. We must also remember (recall the work on tryptophan pyrrolase) that the amount of a given protein present is not only a function of synthesis of the protein, but also of its rate of degradation. Thus, the presence of specific LDH isozymes represents a balance between the synthesis and degradation of those isozymes.

To sum up, both stage-specific and tissue-specific patterns of proteins appear during the differentiation process. Such protein patterns result from differential synthesis, degradation, and activation of specific proteins. Now that we have been introduced to differentiation at the nucleic acid and protein levels, we can move on to look at how larger components of cells are assembled during differentiation.

READINGS AND REFERENCES

Ellis, C. H., Jr. 1966. The genetic control of sea-urchin development: A chromatographic study of protein synthesis in the *Arbacia punctulata* embryo. *J. Exp. Zool.* 164: 1–22.

Hauschka, S. D., and Konigsberg, I. R. 1966. The influence of collagen on the development of muscle clones. *Proc. Natl. Acad. Sci. U.S.* 55: 119–26.

Ingram, V. M. 1963. *The Hemoglobins in Genetics and Evolution.* New York: Columbia University Press.

Konigsberg, I. R. 1963. Clonal analysis of myogenesis. *Science* 140: 1273–84.

Markert, C. L. 1968. The molecular basis for isozymes. *Ann. N.Y. Acad. Sci.* 151: 14–40.

Markert, C. L., and Ursprung, H. 1971. *Developmental Genetics.* Englewood Cliffs, NJ: Prentice-Hall.

Marks, P. A., and Rifkind, R. A. 1972. Protein synthesis: Its control in erythropoiesis. *Science* 175: 955–61.

Paul, J. *et al.* 1973. The globin gene: Structure and expression. *Cold Spring Harbor Symp. Quant. Biol.* 38: 885–90.

Rutter, W. J., Wessells, N. K., and Grobstein, C. 1964. Control of specific synthesis in the developing pancreas. *Nat. Cancer Inst. Monogr.* no. 13, p. 51.

Schimke, R. T. 1967. Protein turnover and the regulation of enzyme levels in rat liver. *Nat. Cancer Inst. Monogr.* no. 27, pp. 301–14.

Schimke, R. T., Sweeney, E. W., and Berlin, C. M. 1965. The roles of synthesis and degradation in the control of rat liver tryptophan pyrrolase. *J. Biol. Chem.* 240: 322–31.

Ursprung, H., and Smith, K. D. 1965. Differential gene activity at the biochemical level. *Brookhaven Symp. Biol.* 18: 1–13.

Wessells, N. K. 1964. DNA synthesis, mitosis and differentiation of pancreatic acinar cells *in vitro. J. Cell Biol.* 20: 415.

Wessells, N. D. 1977. *Tissue Interactions and Development.* Menlo Park, CA: W. A. Benjamin.

Wilt, F. W. 1974. The beginnings of erythropoiesis in the yolk sac of the chick embryo. *Ann. N.Y. Acad. Sci.* 241: 99–112.

Yaffe, D., and Dym, H. 1972. Synthesis and assembly of myofibrils in embryonic muscle. *Curr. Topics in Develop. Biol.* 5: 235–80.

15

DIFFERENTIATION: HIGHER ORDERS OF STRUCTURE

In the last two chapters, we have examined some of the changes in nucleic acids and proteins that occur in differentiating cell systems. In addition to these molecular changes, one often sees major changes in cellular characteristics in differentiating cells. For example, differentiating cells sometimes exhibit increases in endoplasmic reticulum, ribosomes, microtubules, and microfilaments (some dramatic views of cellular components are given in Figure 15-1). In this chapter, we shall look at how some of these subcellular structures are formed. We shall stress that many of these cellular components assemble themselves from preformed subunits when the chemical conditions in the cell are conducive to such assembly. To get a feeling for what is going on, we shall not restrict our study to differentiating cells. Instead, we shall focus on the assembly process per se to better understand the specific factors that lead to the differentiation of higher orders of structure in developing systems.

PROTEIN ASSEMBLY

In the last two chapters, we stressed processes by which DNA is transcribed to form messenger RNA, and those by which this message translates into proteins. We did not focus on the next level—that is, how the amino acid chains fold and assemble themselves into functional entities that often consist of several protein subunits. Here, we'll briefly examine the assembly process to better understand how proteins become important structural and functional units that support the life of a cell.

Polypeptide chains are synthesized on polysomes, or groups of ribosomes. These chains of amino acids seldom remain as straight chains. Instead, the chains twist into a so-called secondary structure (Figure 15-2). The twisted structure is the functionally active form, as will be described later. What are the forces responsible for such "protein morphogenesis"? Hydrogen bonds, hydrophobic associations, electrostatic interactions, and covalent bonds are formed between certain amino acids in the polypeptide chains. It has been shown in the laboratory that folded proteins will unfold if conditions such as ionic strength, temperature, or pH are changed, and return to their normal state if the original conditions are returned. Most proteins carry out their functions in the folded state. These functions include, among others, catalyzing reactions (enzymes); maintaining cell and organelle structure (ribosomal proteins, membrane proteins, and microtubule protein); regulating cell activities (hormones and genetic repressors); binding and transporting molecules (hemoglobin and hormone receptors); and binding foreign materials (antibodies). Recall that a single amino acid substitution resulting from a single gene mutation in the hemoglobin gene can cause sickle cell hemoglobin to form. The mutation changes the three-dimensional folding of normal hemoglobin.

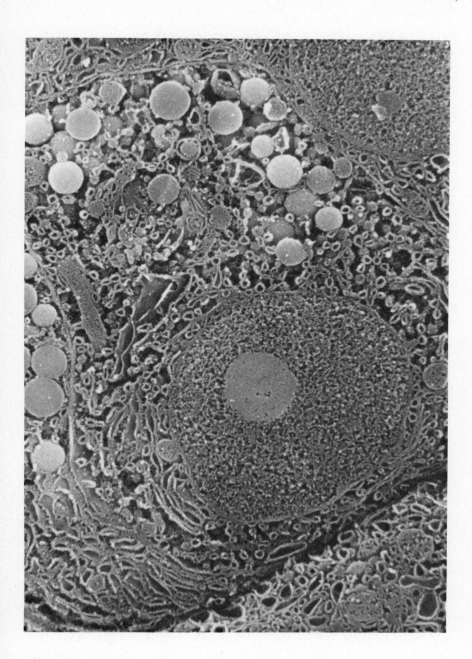

Figure 15-1.
(a) Cellular organelles of a eukaryotic cell. Scanning electron micrograph courtesy of
K. Tanaka, M.D.

(b) Golgi apparatus.

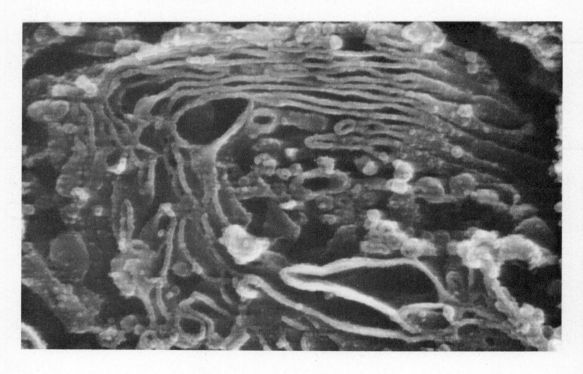

(c) Enlarged view of (b). Scanning electron micrographs courtesy of K. Tanaka, M.D.

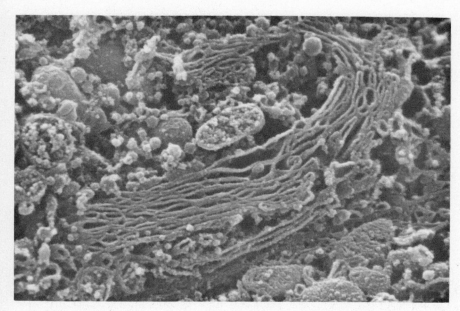

(d) Endoplasmic reticulum with attached ribosomes (rough endoplasmic reticulum). Scanning electron micrographs courtesy of K. Tanaka, M.D.

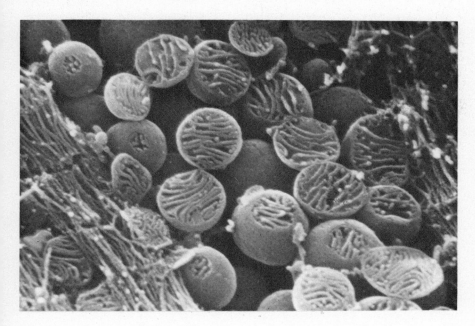

(e) Mitochondria.

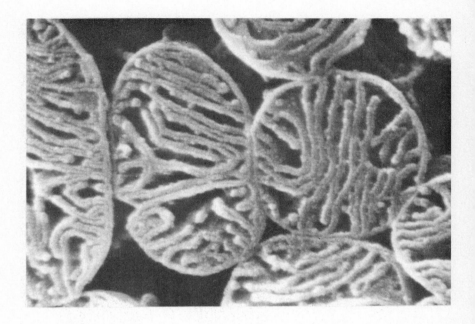

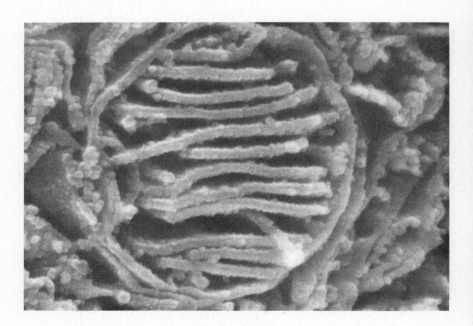

(f)–(g) Mitochondria, progressively enlarged views. Scanning electron micrographs courtesy of K. Tanaka, M.D.

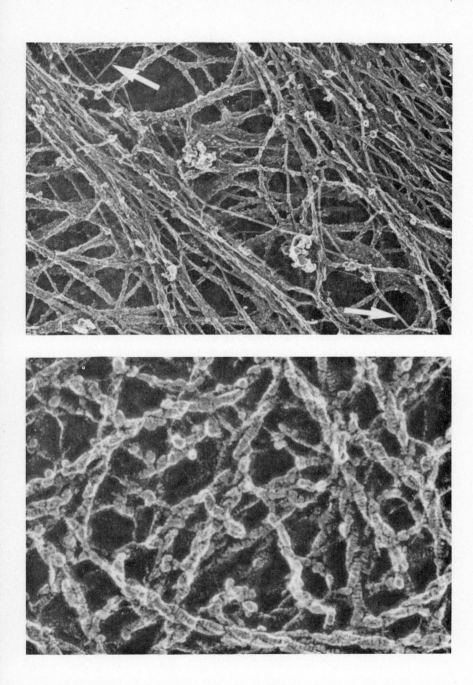

(h), (i) Replica of a freeze-dried cytoskeleton. Micrographs produced by Dr. John E. Heuser of Washington University School of Medicine, St. Louis, Mo.

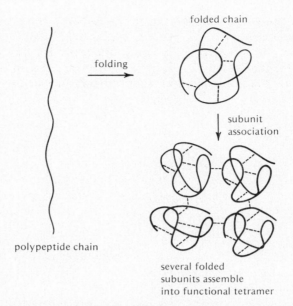

Figure 15-2. The formation of a folded protein subunit from a polypeptide chain, followed by the assembly of four subunits into a tetramer. The tetramer is the functional unit of this protein.

Thus, the amino acid sequence in a protein chain is directly responsible for the specific interactions between chain "links" that determine the folded structure of the chain. The assembly of a functional folded protein from a nonfunctional linear amino acid chain is determined by the specific groups present in the chain and by the conditions of ionic strength and pH in which the chain resides.

Note also that many functional proteins are composed of multiple units (Figure 15-2). The same interactions that fold single protein chains also assemble protein subunits into larger complexes. By altering the conditions of ionic strength and pH or by exposing the protein to agents such as urea or detergents that break some of the bonds mentioned above, the protein aggregate will dissociate into its component subunits. If the denaturing agents are removed and the original conditions returned, the subunits will often fold and reaggregate into the active complexes. An example of such a complex is hemoglobin, which consists of four subunits. Hemoglobin does not actively bind oxygen unless all four folded protein chains are assembled into the tetramer with the iron-containing heme groups in their proper positions in each subunit.

To summarize, folding proteins and aggregating these folded chains depends upon the sequence of amino acids in the single protein chains. The sequence folds into the most stable configuration as the result of hydrogen bonds, hydrophobic associations, electrostatic interactions, and covalent bonds. In many proteins, the folded subunits interact similarly with each other. Conditions in the cell such as pH and ionic strength play an important role in controlling the assembly of specific cellular proteins. Now that we have a general idea of how functionally active proteins are formed, let us look at some examples of the assembly of cellular components that are above the protein level of structure.

MICROTUBULES

We mentioned elsewhere that cytoskeletal elements such as microtubules and microfilaments are important components in developing systems. How do microtubules and microfilaments come about? Let us begin with microtubules. Microtubules make up cilia, flagella, and mitotic spindles. They apparently cause some of the changes in cell shape and movements of organelles that occur in morphogenesis and in the everyday life of a cell. In Chapter 12 we noted that cytoplasmic microtubules are associated with elongating cells. These cytoskeletal elements line up along the long axis of the cell. Cell elongation is prevented if the microtubule-disrupting agent colchicine is present. Microtubules have been observed in elongating cells around the amphibian blastopore, in the neural plate, in cells moving into the primitive groove of bird embryos, and in outgrowing nerve cells. Colchicine prevents the outgrowth of nerve cells and inhibits cell elongation in the chick primitive streak and in the neural plate. Microtubules may directly force the cells to elongate, or they may direct cytoplasmic flow in the direction of elongation. These structures, therefore, are of major interest to developmental biologists. How are microtubules assembled?

Microtubules are cylinders about 250 A in diameter, are composed of globular subunits of the protein tubulin (Figure 15–3). Each subunit, in turn, consists of two proteins, α tubulin and β tubulin. α and β tubulin have different amino acid sequences and can be distinguished from each other by electrophoresis. Each monomer has a molecular weight of about 54,000 daltons.

The assembly of microtubules can be studied *in vitro* to determine the factors involved in assembly. The conditions of neutral pH, low Ca^{2+}, low Mg^{2+}, presence of GTP, and temperature of 37° C favor the assembly of tubulin into discs that, in turn, assemble into elongating microtubules. Thus, tubulin molecules assemble into microtubules when conditions are right. The exact nature of *in vivo* microtubule assembly is not well understood. The same factors, however, are probably required. Other factors that may play a role in the arrangement of polymerizing microtubules in cells include the presence of initiation sites where microtubule assembly begins. Such sites include basal granules in cilia and centrioles in spindles. Centrioles may also serve as microtubule initiation sites in nondividing cells. Key factors that may control cell elongation, as described previously, could be conditions that set up

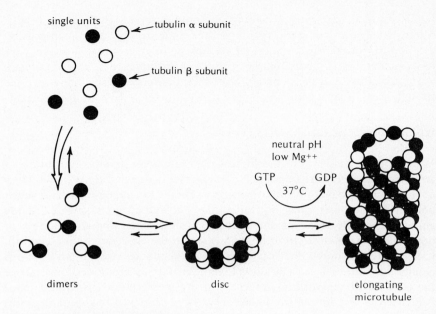

single units

tubulin α subunit

tubulin β subunit

neutral pH
low Mg++

GTP   GDP

37°C

dimers   disc   elongating
                                                  microtubule

Figure 15-3. Microtubule assembly.

microtubule initiation sites that serve to orient microtubule assembly in a specific axis of the cell.

MICROFILAMENTS

In Chapter 12 and elsewhere we mentioned that cytoplasmic microfilaments are involved in cell narrowing, an important factor in morphogenesis and motility. The morphogenesis of tubular glands, for example, may be aided by bundles of microfilaments that contract and exert force on the sides of cells, making them narrow at one end. If a sheet of cells narrows at one side, the sheet will bend. Microfilaments may also play a role in controlling cytoplasmic division, nerve outgrowth, cytoplasmic streaming, invagination during gastrulation, smooth muscle contraction, cardiac muscle contraction, and cortical contraction in eggs. These processes are inhibited by the drug cytochalasin B, which disrupts microfilaments. This drug, like colchicine, has been used to suggest a role for microfilaments in many processes sensitive to the drug. Colchicine, however, binds specifically to microtubule subunits and is not known to affect other cell processes. Cytochalasin B, on the other hand, also affects many cellular processes, including respiratory metabolism, membrane permeability, and protein and glycoprotein synthesis. Cytochalasin B, therefore, provides useful preliminary evidence for a role of microfilaments in cellular processes, but cannot be used to definitively establish such a role because of the multiple effects it has on cells.

How are microfilaments assembled in cells? Microfilaments are commonly found in cells as rods 30–80 Å in diameter. Thicker (100–120 Å) filaments have also been observed attached to plasma membrane specializations called desmosomes. Very thick (up to 160 A) filaments have been observed in nerve axons. The 30–80 Å filaments are similar to F-actin, a protein contained in thin filaments of muscle. Actin filaments (F-actin) consist of globular subunits (G-actin) (Figure 15–4). These subunits polymerize into F-actin filaments *in vitro* with increasing ionic strength. Small "seed" fragments of F-actin stimulate the polymerization of G-actin into elongating F-actin filaments. Here, as well, we see that microfilament assembly is dependent upon the presence of G-actin subunits and upon specific conditions of ionic strength. When the preformed subunits are present and when conditions are right, microfilaments assemble and become involved in the cellular processes and morphogenetic events described above.

Studies have been made using a fluorescent antibody that specifically binds to given proteins associated with the cellular contractile apparatus. The results suggest that α-actinin serves as sites for the initiating actin polymerization and attachment. Also, the protein tropomyosin may assist in actin polymerization and stabilization. Thus, as with microtubules, microfilaments assemble from preformed subunits under specific ionic conditions. Initiation sites (possibly α-actinin) may govern where in the cell the microfilaments polymerize.

FLAGELLA

The bacterial flagellum is another model system that illustrates the self-assembly of structure from preformed subunits. Flagella are composed of a globular protein called flagellin. Thousands of these flagellin subunits assem-

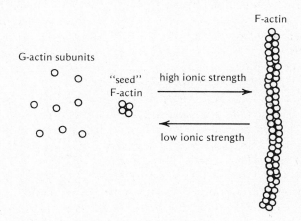

Figure 15–4. Self-assembly of F-actin microfilament.

ble to produce each flagellum (Figure 15-5). Flagella dissociate into the subunits if the pH is lowered. Returning the pH to neutrality repolymerizes the subunits into the flagellum. Here, again, we see that when subunits are present and conditions are right, polymerization occurs. As with microtubules and microfilaments, flagellum assembly is speeded up by the presence of a small cluster of aggregated subunits. This shows again how specific proteins contain within themselves all the information needed to assemble themselves into higher orders of structure.

RIBOSOMES

Throughout this text we have looked at differentiating cell systems. We saw that differentiation often brings an increase in the number of cytoplasmic ribosomes in cells. These are apparently required for synthesizing the many new proteins that appear during differentiation. It is therefore of interest to briefly examine the assembly of ribosomes and to see if these key cellular organelles also form spontaneously, as did the other components discussed in this chapter.

The ribosome is a magnificent complex of nucleic acid and protein that functions in synthesizing protein. Let us focus on the well-studied bacterial ribosome, which consists of ribosomal RNA combined with many (20–40) separate proteins. The two subunits dissociate at low Mg^{2+} concentrations. When the Mg^{2+} concentration is raised, the subunits reassemble into an active ribosome that can synthesize protein *in vitro*. The two subunits cannot synthesize proteins unless they are assembled into the intact ribosome.

Let us now look at one of the subunits, the 30S component. The 30S subunit consists of one ribosomal RNA molecule and about 21 different proteins. If the specific proteins and the ribosomal RNA molecules are combined, 30S particles assemble. When combined with the other subunits,

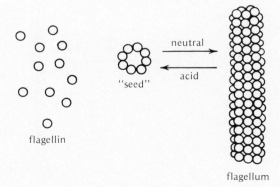

flagellin

"seed"

neutral

acid

flagellum

Figure 15-5. Self-assembly of the bacterial flagellum.

the 50S components, functional ribosomes assemble. Thus, ribosomes can assemble themselves from RNA and proteins *in vitro* under relatively simple conditions. Specific ribosomal RNA and specific ribosomal proteins probably contain in themselves all the information needed for ribosome assembly when relatively simple environmental conditions are met. It has been shown that some of the specific ribosomal proteins are required to assemble the ribosome, while other proteins are required to maintain the structural integrity or function of the ribosome. Most of the ribosomal proteins must be present if stable, functional ribosomes are to be assembled. An important control mechanism in ribosome assembly may be the synthesis of one or more specific ribosomal proteins at specific times in the development and differentiation of cells.

SUMMARY

In summary, many higher orders of structure in cells can assemble themselves. The information needed for such assembly or polymerization is carried in the subunits of these structures. If relatively simple conditions are met, these subunits polymerize into functional cellular components such as specific enzymes, microtubules, microfilaments, flagella, and ribosomes. It should be stressed that self-polymerization from preformed subunits is not the only way in which cellular components are assembled. In many cellular organelles, specific components are added to already existing structures. For example, membrane systems in mitochondria and chloroplasts are formed by adding specific components to preexisting units. Some of these membranes differentiate by incorporating new enzymes or proteins into the existing lipid bilayer. The assembly of many basic cellular components, however, does appear to be controlled by the amino acid sequence of the subunit proteins. Basic chemical interactions such as hydrogen bonding, hydrophobic interactions, electrostatic interactions, and covalent bonding take place between amino acids, within subunits and between subunits. These interactions spontaneously form stable units. Such self-assembly probably plays a major role in at least the beginning stages of the morphogenesis and differentiation of many cellular organelles.

READINGS AND REFERENCES

Anfinsen, C. B. 1973. Principles that govern the folding of protein chains. *Science* 181: 223–30.

Caspar, D. D. 1966. Design and assembly of organized biological structures. In K. Hayashi and A. G. Szent-Gyorgi, eds., *Molecular Architecture in Cell Physiology*, pp. 191–207. Englewood Cliffs, NJ: Prentice-Hall.

Grant, P. 1978. *Biology of Developing Systems*. New York: Holt, Rinehart & Winston.

Johnson, K. A., and Borisy, G. G. 1975. The equilibrium assembly of microtubules *in vitro*. In S. Inoue and R. E. Stephens, eds., *Molecules and Cell Movements*, pp. 119–39. New York: Raven Press.

Kushner, D. J. 1969. Self assembly of biological structure. *Bacteriol. Rev.* 33: 302–45.

Lazarides, E. 1976. Actin, α-actinin and tropomyosin interaction in the structural organization of actin filaments in non-muscle cells. *J. Cell Biol.* 68: 202.

Nomura, M. 1973. Assembly of bacterial ribosomes. *Science* 179: 864–73.

Olmsted, J. B., and Borisy, G. G. 1973. Microtubules. *Ann. Rev. Genet.* 8: 411–70.

Oosawa, F., Kasai, M., Hatano, S., and Asakura, S. 1966. Polymerization of actin and flagellin. In G. E. W. Wolstenhome and M. O'Connor, eds., *Principles of Biomolecular Organization*, pp. 273–303. London: Churchill.

Osborn, M., and Weber, K. 1976. Cytoplasmic microtubules in tissue culture cells appear to grow from an organizing structure towards the plasma membrane. *Proc. Nat. Acad. Sci. U.S.* 173: 867–71.

Raff, R. A., and Makler, H. R. 1972. The non-symbiotic origin of mitochondria. *Science* 177: 575–82.

Reinert, J., and Ursprung, H., eds. 1971. *Origin and Continuity of Cell Organelles*. New York: Springer-Verlag.

Tilney, L. G., and Goddard, J. 1970. Nucleating sites for the assembly of cytoplasmic microtubules in the ectodermal cells of blastulae of *Arbacia punctulata. J. Cell Biol.* 46: 564–75.

Wessells, N. K., Spooner, B. S., Ash, J. F., Bradley, M. D., Luduena, M. A., Taylor, E. L., Wrenn, J. T., and Yamada, K. M. 1971. Microfilaments in cellular and developmental processes. *Science* 171: 135–43.

16

CANCER AND EMBRYOLOGY

Why are we discussing cancer in a book on embryology? As will be seen, cancer cells and certain embryonic cells are similar in many ways. A study of cancer and embryology will help us tie together many of the concepts developed in this text. We shall show how our understanding of embryonic cells is leading to important insights into a major disease, cancer.

The story of cancer is multi-faceted. A complete discussion of the topic includes the causes of cancer, the types of tumors, the diagnosis and treatment, and the cellular aspects of cancer. The relation of cancer cells to embryonic cells appears in the last category. Thus, we shall begin with the causes of cancer and then unravel the entire story, ending with cellular aspects. In this way, we shall develop an understanding of the cancer problem as a whole. Our discussion of the clinical aspects of cancer will be brief.

CAUSES OF CANCER

Cancer is a disease in which cells grow in an unregulated manner. These cells may eventually kill the individual by growing into blood vessels and causing hemorrhage, or by interfering with some vital function. Most of us realize that many things can cause cancer. These causative factors include radiation, viruses, and chemicals. Many of the agents that cause cancer may act on the chromosomes of the cells. Certain genes may code for proteins able to transform cells into a cancerous condition. These genes may be activated by the cancer-causing factors listed above. These factors may also act by damaging DNA, thus altering normal genes enough to transform the cell into a cancerous state. Cancer may also be caused by the direct insertion of a viral genome into the host cell genome. As can be seen by the "iffy" nature of this discussion, the mechanisms by which specific agents cause cancer are uncertain at best. There is little doubt, however, about which agents cause cancer. We shall turn to this question next.

RADIATION

Both X rays and ultraviolet radiation can transform normal cells into tumor cells. When cells in culture are X rayed, they begin to grow in an unregulated manner. That is, these transformed cells do not stop growing or stop moving when they come into contact with other cells. Tumor cells thus lack contact inhibition of growth and movement. These properties, easily observed in culture, are examples of the criteria used to determine whether cells have become cancerous. Such cells can also form tumors when inoculated into animals. We shall come back later to these and many other cellular aspects of cancer.

Substantial evidence has accumulated that radiation causes cancer in humans. For example, the survivors of the atom bomb blasts in Japan have a high incidence of leukemia, a cancer of the blood-forming tissues. The frequency of leukemia is proportional to the exposure to radiation during the blasts. Also, a high incidence of leukemia is observed in people who have received large doses of radiation for treating conditions other than leukemia. Radiologists who used some of the early X-ray equipment, which allowed substantial radiation leakage, also experienced a high incidence of leukemia. A variety of studies have shown that ultraviolet radiation from the sun is implicated in human skin cancer. Radiation damage of the chromosomes is probably a direct cause of such cancers.

VIRUS

Although virus particles are associated with some human tumors, it is difficult to prove that viruses are a major cause of human cancer because of the limited availability of human tissue for experimental studies. There is little doubt, however, that viruses can cause cancer in animals. This suggests that viral genes are also a causative factor in human cancer. When we talk about viruses, we must consider two major types: DNA viruses and RNA viruses. The DNA virus consists of a core of DNA surrounded by a protein coat. The RNA virus, on the other hand, consists of a core of RNA, though the coat is protein. Lipid is associated with the coats of some virus particles. Let us briefly consider the DNA tumor viruses first, then the RNA tumor viruses.

Two extensively studied DNA tumor viruses are polyoma virus and simian virus 40 (SV 40). These viruses typically attach to the cell surface and inject their DNA core into the cell. Once inside the cell, the viral DNA either begins to produce many additional viral particles, killing the cell, or the viral DNA incorporates itself into the host cell's genome (Figure 16–1). When the latter event occurs, the cell may become transformed into a cancer cell. Viral genes in the host genome may induce cancer by coding for new proteins that alter the behavior of the cells, or the insertion itself may damage the host DNA so that processes such as growth regulation are impaired.

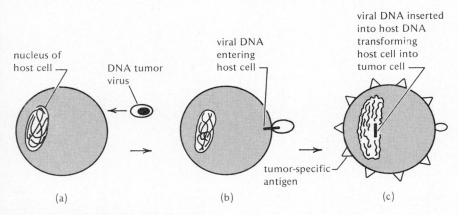

Figure 16–1. Viral transformation.

RNA viruses have also been implicated as causing cancer. A well-studied example is Rous sarcoma virus (RSV), which induces tumors when extracted from chicken tumors and inoculated into tumor-free birds. This can be demonstrated in culture by infecting normal chick cells with RSV and watching them enter the cancerous state. The transformation is signalled by a lack of contact inhibition of growth and movement in the cells (Figure 16–2). This conclusion is confirmed by the ability of these cells to form tumors when inoculated into chicks.

The means by which RNA tumor viruses transform cells into cancer cells has been the subject of intensive recent investigations. The RNA virus apparently attaches to the cell, then injects its RNA into the cell. Viral DNA is synthesized on the viral RNA template by the enzyme reverse transcriptase (RNA-dependent DNA polymerase). (This enzyme upsets the once accepted dogma that DNA can be synthesized only on a DNA template.) This viral DNA can now insert itself into the host DNA, just as the DNA from DNA virus does. It can induce cancer by coding for proteins that alter cell behavior (such as new cell surface proteins) or by altering the control of cell division.

A recent theory (*oncogene* theory) proposes that all cancer is caused by viral genes that were incorporated into the host species' genome long ago and are passed on from generation to generation. Chemicals, radiation, and other carcinogens "turn on" or activate these genes. Thus, all of us have genes that code for proteins that can transform our cells into cancer cells. Whether or not we develop cancer depends on whether or not these genes become activated, and whether or not our bodies can fight off the cells that do become transformed. This theory does offer a unified explanation of all cancer. It explains how many seemingly different agents can cause cancer (by

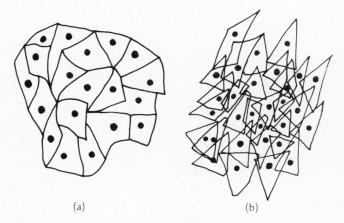

(a)   (b)

Figure 16–2. Cultures of normal and transformed cells. (a) Normal cells show contact inhibition. (b) Polyoma virus transformed cells overlap each other, showing decreased contact inhibition.

activating the cancer genes) and how certain individuals can have a genetic predisposition for developing certain tumors. Such a predisposition can be explained in terms of the numbers of these cancer-producing "oncogenes" in different families, and in terms of the ease with which these genes can be activated.

Is there any evidence to support the oncogene theory? The evidence so far is suggestive, but not unequivocal. Many agents that cause cancer, such as methylcholanthrene or radiation, not only transform normal cells into cancer cells, but also induce the appearance of type C RNA virus particles (type C is a classification of virus structural type) in some of these cells. A virus-specific protein antigen will also appear in these cells. These results suggest that cancer-causing agents can promote the expression of viral genes in cells that contained no visible virus. Such experiments, however, do not prove that the viral genome is the cause of cancer. The appearance of viral particles in treated cells could be a secondary effect of the X-ray or chemical treatment.

It is not clear whether all cells contain cancer-causing viral genes in their genomes. Some studies with identical twins suggest that cells in all individuals do not necessarily contain oncogenes. Type C DNA synthesized from type C RNA virus was hybridized with DNA from identical twins, one of whom had leukemia. Only DNA from the twin who had leukemia hybridized with the viral DNA. This suggests that the leukemia patient had viral DNA genes, while the healthy twin did not. This suggests that the viral genes were not present in the fertilized egg, but were picked up or formed by only one twin at some later time.

It can be said that certain viruses cause cancer, although it should be stressed that identifying virus particles in tumor cells does not prove that the virus caused the tumor. It remains to be seen whether viruses are an important cause of cancer in humans. Certain viruses called *herpes* viruses are associated with many human diseases, such as cold sores, mononucleosis, and certain genital infections. Similar viruses have also been identified in human tumors, such as Burkitt's lymphoma. Virus isolated from these human tumors can cause tumors in animals. It still remains to be determined, however, whether the herpes virus was the actual cause of the human tumor. Such human cancer experiments are obviously difficult—we cannot inject human beings with virus to determine whether the virus causes cancer. Rather good evidence, however, is being obtained from carefully performed animal experiments.

How do viruses cause cancer? The literature is full of interesting studies that give insight into this question. One of the more recent of these was performed by Sefton and colleagues in 1981. They found that cells transformed by Rous sarcoma virus produce an enzyme, a protein kinase, that apparently phosphorylates certain amino acid residues (tyrosines) in the protein that anchors microfilaments to the plasma membrane. This protein (called *vinculin*) also plays a role in helping cells attach to the substratum.

Phosphorylation of the tyrosines in vinculin may disrupt microfilament organization and cause the alterations in cell shape and adhesiveness that accompany the transformation by Rous sarcoma virus. Such studies are beginning to shed light on the cellular changes that accompany viral transformation at the biochemical level.

CHEMICALS

Cancer-causing substances are called carcinogens. Many chemical carcinogens have been identified, and we shall review some here. These chemicals may cause cancer by altering genetic information, as discussed previously. Proof that a given chemical causes cancer is obtained by exposing cells to these chemicals in culture or in living animals. If a chemical causes cancer in mouse cells when administered at a carefully determined concentration, it will probably cause cancer in human cells as well, because all mammalian cells are quite similar. The chemical concentration used in the living animal experiments are based upon the weight of the animal *versus* the weight of a human being and the lifespan of the animal *versus* that of a human being. In other words, substances are given to animals in concentrations similar to those that would build up in the human body over a long period of time. Thus, when we hear that animals are sometimes given large doses of chemicals in these tests or that animal tests are just not relevant to human cancer, we should remember that most chemicals that cause cancer in animals have been found to cause cancer in humans. Mammalian cells are very similar from species to species.

Animal tests, however, are not the only means of determining whether a chemical can cause cancer. Many data are available from human studies examining the incidence of cancer in specific occupational groups, in smokers, and in other groupings of people. These studies have shown that cigarette tar, asbestos, benzene, nickel, vinyl chloride, 2-naphthylamine, hematite, chromium, cadmium oxide, cadmium sulphate, bis(chloromethyl) ether, benzidine, auramine, arsenic compounds, 4-aminobiphenyl, soot, and tars can cause cancer in humans (Figure 16-3). Many other substances, such as chloroform and nitrosamines, have also been implicated in causing human cancer. The American Cancer Society has estimated that fifty percent of cancer deaths could be eliminated if people stopped smoking cigarettes. This dramatic statement is a guess, but an educated one. It points to the enormous number of cancer cases that are caused by things we eat, breathe, or touch. Some cancer specialists suggest that eighty-five percent of all cancers may be induced by environmental factors!

Before leaving the topic of chemicals that cause cancer, let us say a word or two about the difficulty of testing potential carcinogens. Not all cancer-causing agents cause cancer immediately. Some cause cancer only after cells have been exposed to other substances. Some agents, for instance, work in a two-step process. One substance sets up a precancerous condition, while

Chemical	Target organs(s)	Route of exposure
4-Aminobiphenyl	bladder	inhalation, oral
Arsenic compounds	skin, ? lung	oral, inhalation
Asbestos (crocidolite, amosite, chrysolite, and anthophyllite)	lung and chest cavity gastrointestinal tract	inhalation, oral
Auramine	bladder	oral, inhalation, skin
Benzene	bone marrow	inhalation, skin
Benzidine	bladder	inhalation, oral, skin
Bis(chloromethyl)ether	lung	inhalation
Cadmium oxide and sulphate	prostate, ? lung	inhalation, oral
Chromium (chromate-producing industries)	lung	inhalation
Hematite (mining)	lung	inhalation
2-Naphthylamine	bladder	inhalation, oral
Nickel (nickel refining)	nasal cavity, lung	inhalation
Soot and tars	lung skin (scrotum)	skin contact
Vinyl chloride	liver, brain, lung	inhalation, skin

Note: This table does not include many chemicals that produce cancers in animals, but whose effects on humans are not definitely known. From *Cancer and the Worker*, N.Y. Acad. Sci., 1977, p. 12.

Figure 16–3. Industrial chemicals believed to cause (or suspected of causing) cancer in humans.

another substance causes the actual tumor to form. Some studies show that even months or years after the first agent was applied to the skin of a mouse, the second agent will induce a tumor there. If the first agent was not applied, the second does not promote tumor development. These findings illustrate how difficult it is to determine whether an agent is a carcinogen. Many potential carcinogens may go undetected because the right second agent was not used.

Sodium nitrite is another example of the problem. Sodium nitrite, a widely used preservative in certain meat products, is not a known carcinogen. However, *nitrosamine*, a product of the reaction between sodium nitrite and amine, is a potent carcinogen. Does the human body form nitrosamine after ingesting sodium nitrite? The question is still controversial. Some studies indicate that nitrosamine is actually formed. It is difficult, however, to determine if the amount formed is hazardous. Recent studies indicate that nitrosamine is formed when bacon containing sodium nitrite is cooked. As a result, federal regulatory agencies are now developing stringent guidelines for the use of sodium nitrite in bacon and meat products that are heated at high temperatures. We can conclude that suspected carcinogens are often not eliminated from the environment until solid data are available. Decisions to regulate hazardous substances are difficult and often based upon economic considerations, as well as health-related factors. Increasing efforts are being made to control carcinogens in the environment; so we can end this section on a somewhat optimistic note.

TUMOR TYPES

Tumors are abnormal growths. Some tumors grow slowly, do not spread, and do not endanger the organism. Warts and moles are examples of such tumors. Such growths are termed benign tumors. Benign tumors, however, can change into tumors that can invade body tissues, often ending in death. Tumors that spread are termed malignant tumors. They spread by two major means. First, the tumor cells can crawl to nearby tissues. Such movement is called invasion. Tumors also spread by invading blood vessels, and the malignant cells are carried by the blood to distant sites in the body. Movement in the bloodstream is called *metastasis*. We shall consider the mechanisms of tumor spreading in the section on cellular aspects.

Many terms are used to classify tumors. *Carcinomas* are tumors derived from epithelial tissue. *Sarcomas* are tumors derived from connective tissue. *Leukemia* is a tumor of the blood-forming tissues that produces a large number of certain white blood cells. *Melanomas* are derived from pigment cells; *teratomas* are tumors originating from germ cells. Teratomas often develop into embryo-like growths consisting of many types of differentiated tissues. Some investigators believe that teratomas are abnormally activated eggs or primordial germ cells that begin to develop like normal embryos, then go awry. These investigators suggest that some of these tumors could form normal embryos if they grew in the right place in the body. In any case, teratomas are so like embryos that they offer magnificent material for studies on the relationship between embryonic cells and cancer cells, the nature of the benign and malignant states, and the nature of differentiation. These remarkable tumors reveal a great deal about these key areas.

DIAGNOSIS AND TREATMENT

Before turning to the cellular aspects of cancer, let us say a few words about the diagnosis and treatment of cancer.

Many tumors are curable if treated before they spread. Even after spreading, appropriate treatment can cure many tumors. It is much more difficult, however, to treat a tumor that has spread because one must deal with numerous secondary tumors at sites often distant from the original or primary tumor. Thus, successful treatment of many cancers boils down to diagnosing tumors early in their development.

Certain cancers are now almost always curable, even though they were major killers not too long ago. One example is cervical cancer in human females. This form of cancer, once a major killer, is now often completely curable because of early diagnosis. The key diagnostic tool is the Pap test. In this test, a smear of cervical cells is examined with a microscope. Cancer cells, precancerous cells, and normal cells can be easily distinguished. Because of this test, cervical cancer can often be cured with freezing or surgery before any spread has occurred—or even before a real tumor develops.

Other tumors, however, are difficult to detect early. For example, some lung tumors and breast tumors may grow for up to ten years before they are detectable. Present methods of detecting these tumors usually involve the use of diagnostic X-rays. These techniques, however, often cannot detect a tumor smaller than about one centimeter in diameter. A one-centimeter tumor already contains about one billion cells. Some tumors take about ten years to develop to the one centimeter stage; by then, the tumor may have already spread. Thus, some types of cancer that currently kill many people, such as lung and breast cancer, may not be detectable until they have been growing for ten years. Detection of such tumors will improve with improving technology in the near future. Such improved diagnosis should produce an increased cure rate.

Cancer is treated by many well-established methods, as well as some new experimental ones. The well-established methods include surgery, radiation therapy, and chemotherapy. The newer experimental methods include immunotherapy and bone marrow transplants.

Surgery is the treatment of choice if the tumor can be removed without excessive danger to the individual. Surgery can cure cancers that have not spread to other parts of the body.

Localized radiation is the treatment of choice in some tumors. For example, cancer of the voice box is often cured with radiation treatment with little speech impairment. Although very effective, surgery often drastically impairs speech function. If carefully controlled, localized radiation therapy can cure cancer. Under other conditions, radiation can also cause cancer.

Chemotherapy is the method of choice in treating many nonlocalized cancers, such as leukemia and tumors that have spread. Chemotherapy uses chemicals that interfere with the processes of living cells. Some chemicals mimic metabolites needed for the synthesizing of DNA or RNA. When these substances are used by the cell, the cell produces defective nucleic acid, leading to cell death. Some drugs inhibit protein synthesis; others inhibit respiration. The drug dosage must be carefully regulated, because these agents are also harmful to normal cells. However, because tumor cells usually are in a constant state of growth, such drugs become incorporated into these cells more easily than into normal cells. Individuals undergoing chemotherapy, however, often display side effects (hair loss, nausea) as a result of the toxicity of these drugs. Still, many cancers, including certain leukemias, have been treated effectively with chemotherapy. A decade or two ago, leukemia was considered not curable; now many people have been free of the disease for five, ten, fifteen, or more years as a result of modern chemotherapy.

Immunotherapy is an experimental technique that uses the body's natural defenses against tumor cells. Cancer cells often display new surface antigens called tumor-specific antigens. These antigens can be recognized by the body's immune system, just as the body can recognize invading bacteria. The white blood cells of the body then attack the invading bacteria and destroy them. White blood cells can also recognize tumor cells and destroy them. For some reason, however, the body's immune system does not function properly in some individuals. A poorly functioning immune system could render an individual more susceptible to cancer or to persistent forms of infectious diseases. (Patients with AIDS—acquired immune deficiency syndrome—have a greatly increased chance of developing Kaposi's sarcoma, an otherwise very rare form of cancer.) Immunotherapy is based upon the idea that cancer can be controlled by stimulating the immune system, much as the Salk vaccine protects against polio. Various laboratories around the world are experimenting with immunotherapy as a treatment for cancer. Immunotherapy is usually tried on patients who have little chance of recovery using conventional treatments. These patients are inoculated with a substance that acts as a generalized stimulus of the immune system, or with a combination of such a substance with living or dead tumor cells or with tumor cell-surface antigen material. The substance usually used as a generalized stimulus is BCG (Bacillus of Calmette and Guerin), a weakened strain of tuberculosis bacteria that has been used to treat tuberculosis. This material activates the immune system. Vaccines are also made with living or dead tumor cells or with parts of tumor cells. Immunotherapy has great potential as a cancer treatment. It is, however, very difficult to carry out. Different tumors have different tumor-specific antigens, and it is very difficult to obtain large quantities of pure human tumor cells for the vaccines. A different tumor vaccine would probably be needed for treating each specific tumor.

The last form of treatment discussed here is that of bone marrow transplants. This experimental method is used to treat certain terminal cases of leukemia. Leukemia is a disease in which certain white blood cells are produced in great number by the bone marrow. If the diseased bone marrow is destroyed, the leukemia should be cured. Unfortunately, white blood cells are also essential for fighting disease. A patient without white blood cells is vulnerable to any of the thousands of diseases normally kept in check by white blood cells. The bone marrow transplant technique was developed to get around this problem. The patient's bone marrow is first destroyed with strong whole-body radiation and chemicals. Healthy bone marrow is then removed with a syringe from the bones of an identical twin or from those of a person with an identical tissue type. The healthy marrow is inoculated into the bones of the patient, colonizing them. If all goes well, it begins to produce normal white blood cells.

This is a very drastic treatment. The patient is subjected to high radiation doses that can have severe side effects. Also, such transplants can be made only between identical twins or between persons who have identical tissue types, with similar antigens on the surfaces of the body cells. If the tissue types are not identical, the "graft *versus* host" reaction may occur. The new white blood cells produced by the marrow transplant recognize the rest of the body as foreign and begin to destroy it. A heart or kidney transplant is rejected by a similar process. In this case, however, the host's white blood cells destroy the transplanted kidney or heart.

Many cancers are curable by conventional techniques. Experimental treatments and better diagnostic methods offer new hope for the future. We shall now turn to the cellular aspects of cancer.

CELLULAR ASPECTS: CANCER CELLS VERSUS EMBRYONIC CELLS

TUMOR SPREAD AND EMBRYONIC CELL MIGRATION

One of the major problems in dealing with malignant tumors is that cells separate from these tumors and spread to other parts of the body. In some ways, this spread resembles the migration of embryonic cells. Primordial germ cells, for instance, move from other regions in the embryo to the gonad area. Neural crest cells home into many areas of the body. Optic nerve cells migrate to specific regions of the optic tectum.

The same mechanisms that control embryonic cell migration may also control tumor spread. Earlier, we discussed the experiments of Barbera, Marchase, and Roth, who showed that specific adhesive recognition may control homing of optic nerves from the retina to specific regions of the optic tectum (Chapters 10 and 12). These experiments involved rotating dorsal or ventral retina cells with dorsal or ventral chunks of optic tectum. Similarly, Nicolson and Winkelhake showed that cells from primary tumors may form secondary tumors at specific sites because the tumor cells stick best there. These experiments were done by rotating cells from tumors, such as malignant melanomas, with body tissue cells. The melanoma cells stick better to lung cells than to other body cells (Figure 16-4). The lung is where most secondary tumors develop from the line of spreading melanoma used in the experiments (Figure 16-5). Thus, adhesive recognition controls the sites of secondary tumor formation just as it controls the homing of embryonic cells to specific regions of the embryo.

Another factor involved in tumor spread is the initial separation of cells from the primary tumor. Coman used microneedles to separate cell pairs and showed that tumor cells separate more easily than cells in the normal tissues from which the tumors are derived (Figure 16-6). Thus, tumors show reduced cell adhesion (or more correctly, tissue cohesion). Experiments by Oppenheimer's and Roseman's groups suggest that the reduced adhesiveness

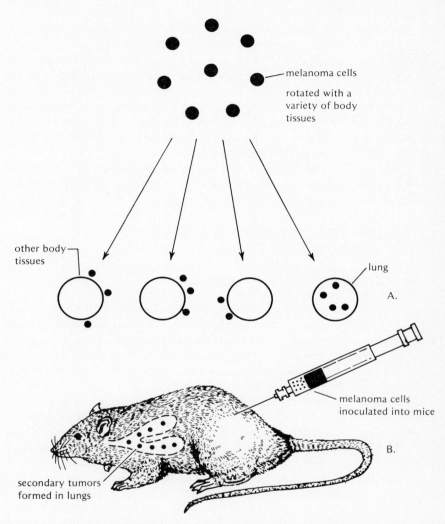

Figure 16-4. Melanoma tumor cells home to the lung and also stick best to lung tissue. (a) Melanoma cells stuck best to lung tissue when rotated with a variety of body tissues. (b) Melanoma cells form secondary tumors most often in the lung. Based on experiments by Nicolson and Winkelhake, *Nature* 255 (1975): 230-32.

of some tumor cells may result from a decreased ability of the cells to synthesize, store, or transport L-glutamine, an amino acid that is required to form the complex carbohydrates needed for cell adhesion in the tumor lines studied (mouse ascites tumors) (Figures 16-7, 16- 8).

Another important factor is the increased production of proteases, enzymes that catalyze the hydrolysis of proteins, by tumor cells. Increased protease at the cell surface may assist the tumors in spreading and invading

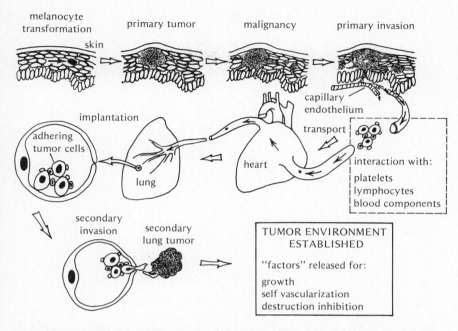

Figure 16–5. Metastatic spread of skin melanoma to lung. From Nicolson et al., *Cell and Tissue Interactions* (New York: Raven Press, 1977), pp. 225–41.

other regions of the body. Most tumor cells also exhibit mobile cell-surface receptor sites. This can be demonstrated by using fluorescent tagged lectins that bind to cell-surface receptor sites that contain carbohydrate. Using the tagged molecules, the receptors can be moved to one pole of the cell, just as Roberson, Neri, and Oppenheimer did for migratory embryonic cells (Chapter 6). Thus, tumor cells and certain embryonic cells both possess cell surfaces

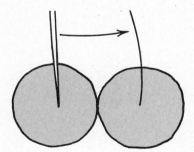

Figure 16–6. Cell pairs separated with microneedles. Tumor cell pairs required less separating force than normal cell pairs of the same tissue type. Based upon experiments by Coman (1944).

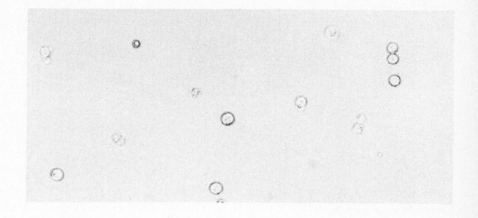

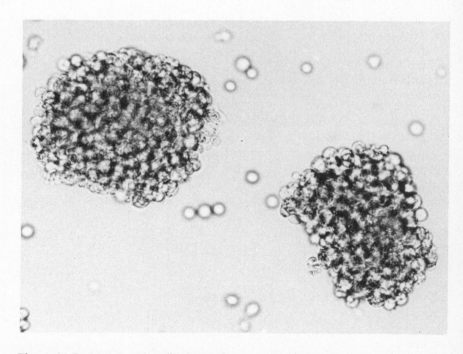

Figure 16-7. Mouse ascites (body cavity) tumor cells rotated in different media. Media without L-glutamine. Media with L-glutamine. Mouse ascites tumor cells stick together (aggregate) when rotated in media containing L-glutamine. This suggests that ascites tumor cells have difficulty in synthesizing, storing, or transporting L-glutamine. Most normal cells do not require added L-glutamine to adhere under these conditions. From Oppenheimer et al., *Proc. Natl. Acad. Sci. U.S.* 63 (1969): 1395-1402.

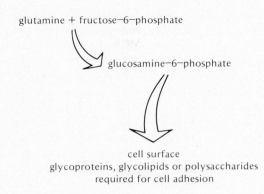

glutamine + fructose–6–phosphate

glucosamine–6–phosphate

cell surface
glycoproteins, glycolipids or polysaccharides
required for cell adhesion

Figure 16–8. Synthesis of cell surface molecules required for the adhesion of mouse ascites tumor cells. This reaction is based upon many experiments with mouse ascites tumors. These pathways may be of general importance in controlling cell-cell adhesion.

that do not restrict the movement of specific receptors. These mobile cell surfaces may help tumor and embryonic cells to migrate.

Many normal cells contain cytoskeletons made up of organized bundles of microfilaments and microtubules. This can be observed by using fluorescent antibodies that bind to the actin of microfilaments and to the tubulin of microtubules. Most tumor cells, however, display a diffuse fluorescence pattern when stained with these antibodies. This suggests that although microfilament and microtubule subunits are present in tumor cells, the subunits are not organized into distinct microfilaments and microtubules. Microfilaments and microtubules may prevent cell-surface receptor-site mobility in normal cells by anchoring some of the cell-surface sites from within the cell. The lack of an organized cytoskeleton in tumor cells may allow greater cell-surface receptor-site mobility and thereby permit increased cell migration ability and continued growth.

We already mentioned that tumor cells show less contact inhibition of growth and movement than do normal cells. Cell-cell contact in normal tissues stops cell movement and growth. In tumors, spread may occur partly because contact no longer inhibits cell growth and movement. Contact inhibition may be controlled by cellular cyclic AMP. Virus-transformed tumor cells often contain less cyclic AMP than nontransformed cells. Some experiments show that contact inhibition of growth and movement can be restored by treating tumor cells with derivatives of cyclic AMP. Cyclic AMP may induce contact-inhibition by influencing the cytoskeleton. The relationship between the cell surface, cyclic AMP, the cytoskeleton, and cell growth and migration is a stimulating topic that is probably of major importance to the cancer problem. One possible hypothesis regarding the nature of cell transformation is that genetic changes cause cell surface changes, which in turn lower the levels of cellular cyclic AMP (probably by lowering the level of the membrane-bound enzyme, adenyl cyclase). Low cyclic AMP levels may then induce many of the other cellular changes associated with malignancy. Major breakthroughs in our understanding of tumor cell transformation will probably come from studies in this area.

Many theories have been proposed to explain the altered behavior of cancer cells, including unregulated growth and spread. Only a few of these have been described in this chapter. Let us conclude by mentioning one last theory that was proposed by Holley. He suggests that a major factor in the unregulated growth of tumor cells is their increased nutrient uptake from the medium or from surrounding regions of the tumor. He suggests that if cells get a constant supply of the sugars, amino acids, vitamins, and other substances needed for growth, they will continue to grow. This hypothesis is supported by findings that many types of tumor cells take up significantly more nutrients than their normal counterparts. The plasma membrane of tumor cells, therefore, is more permeable to nutrients. Whether this altered permeability is a cause of the transformed state or just a secondary effect remains to be determined.

EMBRYO-LIKE TUMORS OR TUMOR-LIKE EMBRYOS?

We have already mentioned that teratomas are tumors derived from germ cells. These tumors are often indistinguishable from early embryos, and in fact develop rather like normal embryos if they are maintained in the proper place in the body. Teratomas and other tumors may also carry specific embryonic antigens on the surfaces of their cells. This suggests that some of the same genes that are active in tumors are active in embryos. Normal adult cells seldom display these antigens.

Mouse teratomas can be isolated from the testes of some strains of young mice and transplanted to other mice or maintained in culture. When teratoma cells are inoculated into the body cavities of mice, they develop into free-floating clusters called "embryoid bodies." These clusters resemble normal 6-day-old mouse embryos. If grown in the eye cavity of mice they will differentiate into many tissues. If maintained in the body cavity, however, embryoid bodies continue to proliferate rapidly. A variety of experiments indicate that these teratomas are derived from germ cells. Mintz and Illmansee isolated teratoma cells and injected them into early mouse embryo blastocysts from another strain of mice. These blastocysts were then implanted into the uteri of other female mice. The embryos developed into normal, tumor-free adult mice. Some of the adult tissues were derived from the normal mouse embryo cells and others from the teratoma cells (Figure 16-9). The teratoma cells had genotypic markers; so did the normal blastocyst cells. The markers included strain-specific enzyme patterns, hemoglobin, immunoglobulins, and pigment. Thus, normal mice developed, in part, from cancer cells! The important thing to remember about this experi-

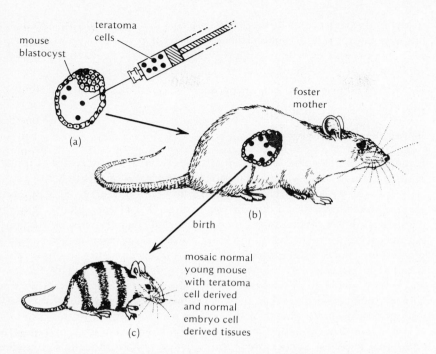

Figure 16-9. Development of normal mouse from teratoma cells and normal mouse embryo cells. (a) Teratoma cells (black) are inoculated into a normal mouse blastocyst (white). (b) The embryo (including the teratoma cells) is grown in the uterus of a female mouse. (c) The mouse developing from the combination consists of normal tissues, some derived from teratoma cells (black) and some from normal embryo cells (white). Based on experiments by Mintz and Illmensee.

ment is that teratoma tumor cells can differentiate into normal adult tissues. Here we see the ultimate relationship between normal embryo cells and certain tumor cells. Both can divide and remain relatively undifferentiated under some conditions; whereas under other conditions they can differentiate into normal adult tissues.

Normal six-day-old mouse embryo cells will develop into teratomas if implanted in the testes of adult mice! Older embryo cells, however, do not form tumors under these conditions. Here again we see that young embryo cells are similar to certain tumor cells. Both can grow in an unregulated way or can differentiate into normal adult tissue, depending upon their environment.

Teratomas are being used in many studies dealing with the malignant and benign states and the conditions required for cell growth and cell differentiation. Teratoma cells are excellent for these studies because they can be grown in enormous numbers in the body cavity of mice or in tissue culture. These cells remain malignant if maintained in the free-floating form in the body cavity, but differentiate into normal tissues if implanted into embryo blastocysts or under the skin. For these reasons, teratomas will probably provide us with much information about the relationship between the maligant and benign states and between cell growth and cell differentiation.

RECENT DEVELOPMENTS IN CANCER BIOLOGY

It is appropriate to mention some recent developments in the area of cancer biology so that we can have an up-to-date overview of this field.

Can cancer be prevented? Cancer prevention will in all likelihood receive much more attention in the future. We already know that the number one killer cancer, lung cancer, can largely be prevented by eliminating the carcinogens taken into the lungs during cigarette smoking. Not only would lung cancer be greatly reduced if carcinogenic cigarette smoke were eliminated, but cancers of the mouth, esophagus, pancreas, larynx, bladder, and kidney would also be substantially reduced. For example, the American Cancer Society estimates that in males, 97% of deaths due to lung cancer and 99% of deaths due to cancer of the larynx are attributable to smoking.

A variety of so-called anticarcinogens have been studied that offer additional hope for cancer prevention. Let us take a moment to examine some recent studies.

One study (reported in 1980) indicated that two combinations of drugs (retinoic acid plus the steroid dexamethasone, and retinoic acid plus the protease inhibitor TLCK) inhibited the promotion of skin tumors by specific carcinogens. Other studies showed that carcinogen-induced mammary cancer was prevented in rats by a drug (2-bromo-alpha-ergocryptine) that suppressed the secretion of the hormone prolactin if this drug was administered along with the retinoid retinyl acetate.

Another study showed that vitamin C and vitamin E reduced the level of mutagens (such as the N-nitroso derivatives of intestinal lipids) in the feces of humans to 25% of control values (as measured in individuals not fed vitamins A or E). The two vitamins apparently prevent the formation of carcinogens by acting as antioxidants. A study also showed that vitamin C inhibited the malignant transformation of mouse fibroblast cells by the carcinogen methylcholanthrene in culture.

Vitamin A and its analogs (retinoids) apparently prevent carcinogen-induced cancers in rats. For example, in one experiment, rats were given three biweekly doses of the carcinogen N-methyl-N-nitrosourea to the bladder. Some of these rats were then treated with low or high doses of the vitamin A analog 13-cis-retinoic acid by adding the retinoid to the diet for eight months. The animals were killed and the surface of the bladder of the rats was carefully examined. The rats fed the high dose of retinoic acid had many fewer and less extensive bladder cancers than the other animals. This is just one of many studies that show that vitamin A analogs effectively prevent cancers induced by carcinogens.

Other studies show that vitamin A can completely prevent lung cancers in hamsters. The mode of action of these compounds is not known, but some studies suggest that these substances prevent the carcinogen from binding to cellular DNA. In one study DNA from the tracheal epithelium of animals deficient in vitamin A bound much more of the carcinogen benzo(a)pyrene than did DNA from animals with normal vitamin A levels. Vitamin A also stimulates the specific differentiation of several cell types. How vitamin A and its analogs suppress cancer induction is still not well understood. Other studies even suggest that vitamin A may enhance the carcinogenicity of some compounds. Much is still to be learned about the relationship between vitamin A and carcinogen activity.

A variety of substances other than vitamin A inhibit the action of carcinogens. Vitamin C (ascorbic acid) may inhibit the development of some cancers. Certain studies suggest that vitamin C increases the survival time of patients with some terminal cancers. Vitamin C can block the formation of carcinogenic nitrosamines (and other N-nitroso compounds). Nitrosamines can form in the acid conditions of the stomach by a reaction between the widely used food preservative sodium nitrite and specific amines. Vitamin C binds to nitrite and thus reduces the amount of nitrite available for forming nitrosamines. Cancers induced in animals fed with amines and sodium nitrite are reduced when the animals are simultaneously fed ascorbic acid.

High levels of riboflavin in the diet of rats fed carcinogenic azo dyes substantially reduced the induction of liver cancers. The azo dyes rapidly decrease the riboflavin content of the liver. High levels of added riboflavin presumably prevent the depletion of liver riboflavin and protect the liver from the carcinogen.

Spontaneous mammary tumors, leukemia, and lung cancers in mice are reduced if their caloric intake is restricted. Restricting calories also reduced skin cancer induction by dibenzanthracene to less than half its value in control animals. Azo dyes induced fewer liver cancers in animals fed with saturated lipids than in animals fed with unsaturated fat. How these dietary factors reduce cancer induction is not well understood.

Some interesting results show that the hormonal status of animals affects their susceptibility to carcinogens. The pituitary, thyroid or adrenal gland was removed from test rats. Significantly fewer liver cancers formed when the rats were treated with a variety of carcinogens. If animals are given high levels of specific hormones, liver cancer induction by such carcinogens as 2-acetylaminofluorene is greatly increased. Normal hormones probably activate the oxygenase enzymes that, in turn, activate carcinogens. If the hormone levels drop, oxygenase activity would decrease and with it, carcinogen activation. Hormones might also increase the rate of cell division. This could make the cells more susceptible to transformation.

In the future, increased efforts will be made to develop anticarcinogens. In addition, more emphasis will probably be placed upon the role of diet in cancer causation. Some forms of cancer can probably be prevented by altering what we eat and drink. For example, recent studies by Maugh link chlorinated organic contaminants in drinking water with rectal, bladder, and colon cancer. The chlorine in chlorinated drinking water forms such contaminants by interacting with the organic compounds and bromine present in the water to form trace amounts of halomethanes.

At least seven carcinogens have been discovered that are produced by fungi that contaminate foods. Aflatoxin is produced by the fungus *Aspergillus flavus*, a common contaminant of groundnuts and rice stored in hot and humid areas. Aflatoxin is a prime example of a liver carcinogen that can be eliminated by the proper treatment and storage of food. Nitrosamines, another group of potent carcinogens, form by the interaction of the food preservative sodium nitrite with amines, a breakdown product of protein. Some of the sodium nitrite we take in comes from natural sources, such as vegetables and our own bacteria. However, the exposure of our cells to nitrosamines would probably be reduced if sodium nitrite were substantially reduced in or eliminated from preserved food products.

Finally, studies suggest that diets high in fiber reduce the incidence of cancers of the colon and rectum. A diet high in bulk promotes frequent bowel movements, reducing the time in which carcinogens present in the large intestine contents remain in contact with the lining of the bowel, where cancers begin. Large amounts of bulk also reduce the concentration of carcinogens in the bowel. Diets high in fat are correlated with high incidences of colon-rectal cancer. Time will tell whether substantial decreases in cancer rates can be achieved by altering the diet. For example, can the incidence of colon-rectal cancer be reduced by increasing dietary fiber, decreasing dietary fat, and increasing the dietary levels of some of the anticarcinogens mentioned previously? This area of research makes sense and possibly will lead to preventive measures against some common forms of human cancer.

TYROSINE PHOSPHORYLATION AND GROWTH STIMULATION

Epidermal growth factor, insulin, platelet-derived growth factor, and some cancer-causing viruses all have something in common. All of them phosphorylate tyrosine groups in and on the cell. When the growth factors bind to cellular receptor sites, the receptors start phosphorylating tyrosines. When some cancer viruses (such as Rous sarcoma virus) transform cells, the concentration of phosphorylated tyrosines increase many-fold. This increase is apparently due to tyrosine kinases (enzymes that phosphorylate tyrosines) that may be the transforming proteins of the viruses. We shall soon learn whether tyrosine phosphorylation is of general importance in cellular transformation and loss of growth control and, if so, how the phosphorylated product alters the cell's growth characteristics.

EXPRESSION OF ONCOGENES DURING NORMAL EMBRYONIC DEVELOPMENT

Viral oncogenes (apparently originating in the normal vertebrate genome) are required for inducing specific virally initiated cancers, including sarcomas, carcinomas, lymphomas, and leukemias. Recent work by Müller and colleagues shows that, during normal mouse prenatal and early postnatal development, specific tissues at specific stages express normal cellular genes identical to the oncogenes of FBJ mouse osteosarcoma virus, Abelson mouse leukemia virus, and Harvey sarcoma virus. For example, rapidly proliferating placental tissue like the trophoblast, which can invade the uterine epithelium and surrounding tissue, significantly expresses a specific oncogene base sequence. This suggests that many cancers result from the expression of cellular oncogenes that played an earlier role in such normal developmental processes as cell proliferation, trophoblast invasion, and differentiation.

HUMAN CANCER GENES

Now and then we hear the word "breakthrough" in cancer research. It is appropriate to discuss here a piece of work that may be one of the most important findings to come along in decades.

Several studies have shown that DNA fragments isolated from some spontaneous and chemically induced human, mouse, and other cancers can transform normal cells into cancer cells. DNA fragments from a human bladder carcinoma were cloned by inserting these genes into bacterial plasmids. The cloned human carcinoma genes were then analyzed by their ability to bind other DNA sequences. The results showed that the bladder carcinoma gene has a base sequence similar to that of a specific gene carried by mouse sarcoma virus. A human lung carcinoma gene was found to be similar to another gene carried by mouse sarcoma virus. Other studies showed that a specific phosphoprotein (called p21) encoded by a mouse sarcoma virus gene is present at high levels in cells transformed by the human bladder carcinoma gene. Thus, this gene is similar in base sequence to a transforming gene carried by a mouse sarcoma virus. In addition, cells transformed by the bladder gene express an RNA that hybridizes with both the bladder gene and the mouse sarcoma virus gene. Such transformed cells contain elevated levels of the protein encoded by the mouse sarcoma virus gene.

A fragment of DNA similar to the mouse sarcoma virus transforming gene can be isolated from normal human cells. This DNA will transform cells if it is activated by linking it to a fragment of the mouse sarcoma virus gene. Cells transformed by this human gene contained within their DNA the human DNA fragment linked to the viral DNA. The transformed cells express the p21 protein. The p21 protein is apparently the transforming agent of a mouse sarcoma virus. Normal cells contain low levels of p21 and of mRNA coding for this protein. These results support the notion that cancer is a disease caused by increased amounts of normal gene products.

The genetic basis of cancer is becoming clear. Continued efforts along the lines described in the final section of this chapter will probably soon pinpoint the exact mechanism(s) involved in human carcinogenesis.

"JUMPING" CANCER GENES

We have discussed the recent isolation of activated human cancer genes that have base sequences similar to those of certain viral oncogenes. We noted that some of these genes code for specific proteins, such as p21, that may cause cellular transformation. Recent work has shown that these transforming genes may code for a protein found in normal cells, but that the transformed cells produce it in greatly increased quantities. In other work, segments of human bladder cancer genes were systematically exchanged with normal cellular genes that transformed cells when appropriately activated. These recombinant genes were then transferred to cultured cells. Analysis showed that a specific base—a guanine—in the normal gene was replaced by a thymine in the transforming gene. As a result, valine was incorporated in the p21 protein encoded by the oncogene instead of glycine. In some cases, therefore, a specific one-base alteration may enhance the transforming ability of the gene product.

Recent work has also shown that genes with oncogene-like base sequences are involved in the translocations associated with specific cancers such as Burkitt's lymphoma. One example is the *myc* gene, which is similar to a gene carried by cancer-causing viruses. In about half of the Burkitt's lymphoma cells examined, the *myc* gene was translocated to a position close to an antibody-forming base sequence. Translocation of cancer-causing genes may activate them (or alter them) to bring about cell transformation. This may be how some cancers are induced. Decades of work have shown that many cancers are associated with specific chromosome abnormalities, including translocations. This new work may clarify the relationship between visible chromosomal changes and cancer.

Just as antibody-forming genes become activated by translocation, so too may cancer genes. "Jumping" genes may be the mechanism for certain beneficial events, such as antibody synthesis, and other cellular differentiations. However, they may also cause such pathological processes as cancer.

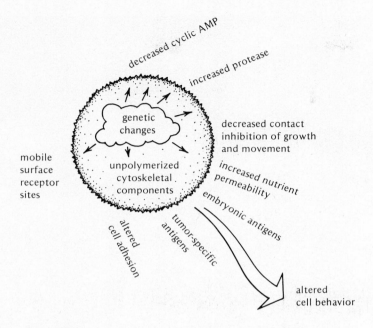

decreased cyclic AMP

increased protease

genetic changes

decreased contact inhibition of growth and movement

mobile surface receptor sites

unpolymerized cytoskeletal components

increased nutrient permeability

embryonic antigens

tumor-specific antigens

altered cell adhesion

altered cell behavior

Figure 16-10. Some cellular changes associated with cancer cells.

The rapid advance of these fields in the recent past suggests that we are on the brink of a major advance in our understanding of differentiation and cancer.

SUMMARY

Many agents can cause cancer. Such agents probably act on the genes. Genetic changes alter various cellular properties that, in turn, cause altered cell behavior (Figures 16-10, 16-11, 16-12). Embryonic cells clearly resemble tumor cells in many respects. Let us keep in mind, however, that some tumors do not possess embryonic cell characteristics, and it is unlikely that the activation of embryonic genes is the only reason for cancer. Such a conclusion is supported by the appearance of tumor-specific antigens (as well as embryonic antigens) on the surfaces of tumor cells. These antigens are usually not found on embryonic cells. Significant similarities, however, exist between many types of tumor and embryonic cells, as illustrated by teratoma experiments, cell-surface receptor-site mobilities, and other characteristics described in this chapter. Clearly, the study of embryology and the study of cancer are inextricably intertwined. Information from one area can lead to a better understanding of the other.

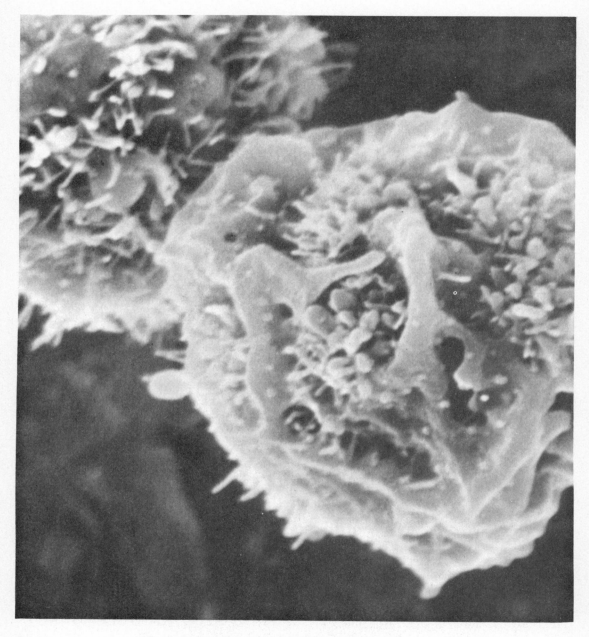

Figure 16-11. Mouse ascites tumor cells. Note the active cell surfaces.

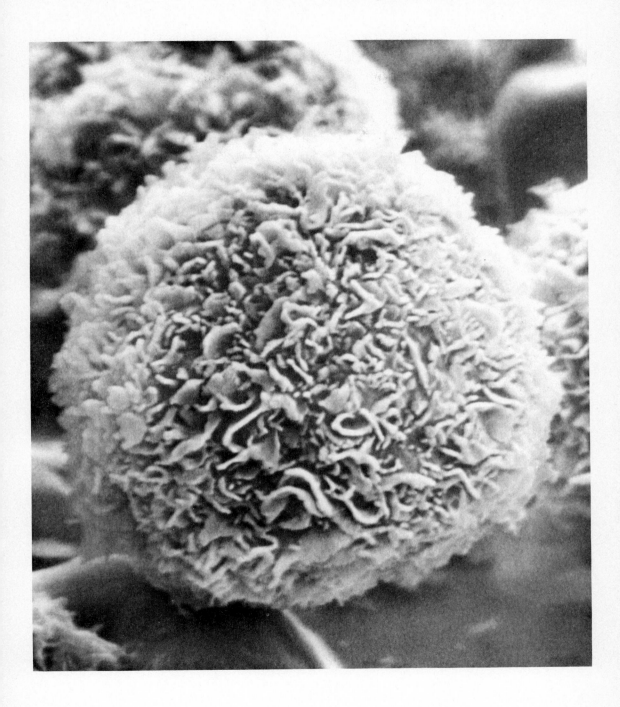

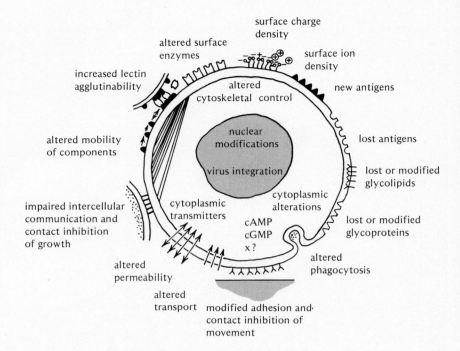

surface charge
density

altered surface
enzymes

surface ion
density

increased lectin
agglutinability

altered
cytoskeletal control

new antigens

nuclear
modifications

altered mobility
of components

lost antigens

virus integration

lost or modified
glycolipids

impaired intercellular
communication and
contact inhibition
of growth

cytoplasmic
transmitters

cytoplasmic
alterations

cAMP
cGMP
x ?

lost or modified
glycoproteins

altered
phagocytosis

altered
permeability

altered
transport

modified adhesion and·
contact inhibition of
movement

Figure 16-12. Some changes found after a transformation to the cancerous state.
From G. Nicolson, *Biochem. Biophys. Acta*, 458 (1976): 1–72. The reader is referred to this
article for further discussion on this topic.

READINGS AND REFERENCES

Baltimore, D. 1975. Tumor viruses. *Cold Spring Harbor Symp. Quant. Biol.* 39: 1187–1200.
Baltimore, D. 1976. Viruses, polymerases and cancer. *Science* 192: 632.
Chang, E. H., Furth, M. E., Scolnick, E. M., and Lowy, D. R. 1982. Tumorigenic
    transformation of mammalian cells induced by a normal human gene homologous to
    the oncogene of Harvey murine sarcoma virus. *Nature* 297: 479.
Coman, D. R. 1944. Decreased mutual adhesiveness, a property of cells from squamous
    cell carcinomas. *Cancer Res.* 4: 625–29.
German, J., ed. 1974. *Chromosomes and Cancer.* New York: John Wiley.
Heidelberger, C. 1975. Chemical carcinogenesis. *Ann. Rev. Biochem.* 44: 79.
Holley, R. W. 1972. A unifying hypothesis concerning the nature of malignant growth.
    Proc. Natl. Acad. Sci. U.S. 69: 2840.
Kolata, G. 1983. Is tyrosine the key to growth control? *Science* 219: 377–78.
Koprowski, H., ed. 1977. *Neoplastic Transformation: Mechanisms and Consequences.* Elmsford,
    NY: Pergamon.
Krontiris, T. G., and Cooper, G. M. 1981. Transforming activity of human tumor DNAs.
    Proc. Natl. Acad. Sci. U.S. 78: 1181–84.
Markert, C. L. 1968. Neoplasia: A disease of cell differentiation. *Cancer Res.* 28: 1908–19.
Marx, J. L. 1982. The case of the misplaced gene. *Science* 28 : 983–85.

Mintz, B., and Illmensee, K., 1976. Totipotency and normal differentiation of single teratocarcinoma cells cloned by injection into blastocysts. *Proc. Natl. Acad. Sci. U.S.* 73: 549–553.

Müller, R., Slamon, D. J., Tremblay, J. M., Cline, M., and Verma, I. M. 1982. Differential expression of cellular oncogenes during pre-and postnatal development of the mouse. *Nature* 299: 640–44.

Maugh, T. H., II. 1981. New studies link chlorination and cancer. *Science* 211: 694.

Nicolson, G. L. 1976. Transmembrane control of the receptors on normal and tumor cells. II. Surface changes associated with transformation and malignancy. *Biochim. Biophys. Acta* 458: 1–72.

Nicolson, G. L. 1978. Cell and tissue interactions leading to malignant tumor spread (metastasis). *Amer. Zool.* 18: 71–80.

Oppenheimer, S. B., Edidin, M., Orr, C. W., and Roseman, S. 1969. An L-glutainime requirement for intercellular adhesion. *Proc. Natl. Acad. Sci. U.S.* 63: 1395–1402.

Oppenheimer, S. B. 1982. *Cancer: A Biological and Clinical Introduction*. Boston, Mass.: Allyn and Bacon.

Parada, L. F., Tabin, C. J., Shih, C.. and Weinberg, R. A. 1982. Human bladder carcinoma oncogene is homologue of Harvey sarcoma virus Ras gene. *Nature* 297: 474.

Robbin, R., Chow, I. N., and Black, P. H. Proteolytic enzymes, cell surface changes and viral transformation. *Adv. Cancer Res.* 22: 203–60.

Reddy, B. S., Cohen, L. A., McCoy, G. D., Hill, P., Weisburger, J. H., and Wynder, E. L. 1980. Nutrition and its relationship to cancer. *Adv. Can. Res.* 32: 237–345.

Reif, A. E. 1981. The causes of cancer. *American Scientist* 69: 437–47.

Rigby, P. W. J. 1981. The detection of cellular transforming genes. *Nature* 290: 186–87.

Rigby, P. W. J. 1982. The oncogene circle closes. *Nature* 297: 451.

Sefton, B. M., Hunter, T., Ball, E. H., and Singer, S. J. 1981. Vinculin: A cytoskeletal target of the transforming protein of Rous sarcoma virus. *Cell* 24: 165–74.

Sheppard, J. R. 1971. Restoration of inhibited growth to transformed cells by dibutyryl adenosine 3′, 5′-cyclic monophosphate. *Proc. Natl. Acad. Sci. U.S.* 68: 1316–20.

Shih, C., Padhy, L. C., Murray, M., and Weinberg, R. A. 1981. Transforming genes of carcinomas and neuroblastomas introduced into mouse fibroblasts. *Nature* 290: 261–64.

Sporn, M. B. 1980. Combination chemoprevention of cancer. *Nature* 287: 107.

Stevens, L. C. 1975. Spontaneous parthenogenesis and teratocarcinogenesis in mice. *33rd. Symposium, The Society of Developmental Biology*. New York: Academic Press.

Tabin, C. J., Bradley, S. M., Bargmann, C. I., Weinberg, R. A., Papageorge, A. G., Scolnick, E. M., Dhar, R., Lowy, D. R., and Chang, E. H. 1982. Mechanism of activation of a human oncogene. *Nature* 300: 143–49.

Weber, K., Lazarides, E., Goldman, R. E., Vogel, A., and Pollack, R. 1975. Localization and distribution of actin fibers in normal, transformed and revertant cells. *Cold Spring Harbor Symp. Quant. Biol.* 39: 363–69.

17

AGING AND DEVELOPMENTAL BIOLOGY

The biological basis of aging is not well understood, although more and more is being learned. The average life expectancy in this country might be increased 20 years by the end of this decade. The dream of a truly extended healthful life may become a reality in the not too distant future. We shall briefly examine some of the new information and ideas concerning aging. Studies of aging and studies of embryology and developmental biology go hand in hand. Early embryonic development produces the differentiated adult organism, which in turn, forms the aged individual. Some investigators consider aging to be a part of a normal developmental continuum based on events like those governing the earlier phases of development.

Many hypotheses have been developed to explain aging, but none have been proven. Let us examine a few of them in order to gain some insight into developments in this area. We shall then examine some intriguing experiments in which investigators have successfully retarded the aging process.

HYPOTHESES OF AGING

Two general hypotheses of aging are:

1. Aging is caused primarily by factors originating outside of cells (extrinsic factors).

2. Aging is caused primarily by mechanisms within cells (intrinsic factors) that form a biological clock that runs rather independently of external factors.

Many combinations of these ideas have also been formulated. We shall see that a combination of both hypotheses may be the best explanation of aging at this time. What is the evidence that aging is primarily caused by agents originating outside of cells?

AGE-RELATED DISEASES

One might wish to distinguish between aging and age-related diseases. However, several extrinsic factors are known that cause certain age-related diseases (at least in part). Specific cancers, for instance, are caused by environmental agents. Radiation can cause leukemia and skin cancer; chemical agents (cigarette tars, asbestos, nickel, vinyl chloride, benzene, and arsenic compounds) can cause various types of cancer. Presumably, these agents act by causing mutations in DNA. Hypertension is associated with excess salt intake. Mercury exposure can lead to neurological degeneration. Excess cadmium intake promotes kidney diseases.

The general characteristics of aging, however, include a decline in physiological functions such as heart efficiency, body temperature stability under stress, kidney filtration rate, and breathing capacity. These characteristics have not yet been clearly linked to the presence or absence of specific extrinsic agents. The line between disease and general decline, however, may not be clear. Further work may show that long-term exposure to certain environmental agents is related to the general decline associated with aging.

MEMBRANE DAMAGE AND AGING

Many hypotheses have been developed to explain how cells become damaged over the years, leading to aging. One group of such models suggests that aging is the result of a leakage of hydrolytic enzymes from lysosomes (Figure 17-1). Lysosomes are membrane-bound bodies that contain these enzymes. Their normal function is to digest particles inside of cell vesicles. Clearly, if enzymes leak from lysosomes, the cell may begin to digest itself. The membrane surrounding lysosomes might be disrupted by a variety of agents, including free-radical compounds. Free-radical compounds are generated in metabolism or by radiation. They are unstable and contain unpaired electrons. Free radicals can react with the lipid of lysosomal membranes, damaging the membranes and allowing lysosomal enzyme to leak out. Future work should tell whether such enzyme leakage is important in the aging process.

REDUCTION OF AGING RATE BY MODIFICATION OF DIET

Several interesting experiments suggest that some aspects of aging are related to environmental factors. The rate of aging in rats is significantly reduced by a long-term restriction of dietary caloric intake (Figure 17-2). Immediately after weaning, the animals were placed either on a normal diet or on a diet that was nutritionally complete, but with just enough calories for slow growth. The animals on the latter diet remained stunted in size and

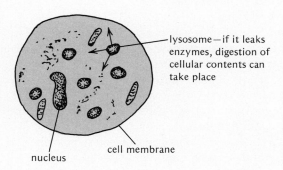

Figure 17-1. Leakage of hydrolytic enzymes from damaged lysosome.

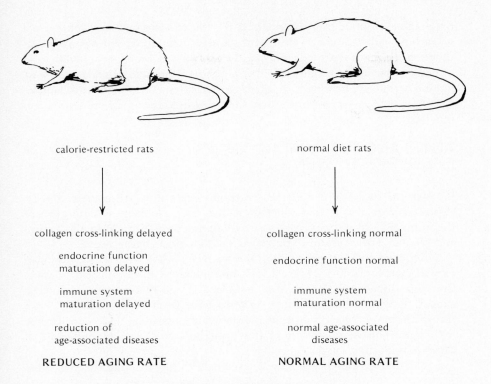

calorie-restricted rats

normal diet rats

collagen cross-linking delayed

collagen cross-linking normal

endocrine function
maturation delayed

endocrine function normal

immune system
maturation delayed

immune system
maturation normal

reduction of
age-associated diseases

normal age-associated
diseases

REDUCED AGING RATE

NORMAL AGING RATE

Figure 17-2. Some consequences of restricted calorie intake in rats.

juvenile in physiological functions, but resumed normal growth if returned to normal diet. More importantly, the animals on the restricted diets often lived 33% longer than the animals on normal diets. The cancer and respiratory diseases normally associated with aging were clearly reduced in the calorie-restricted individuals.

COLLAGEN CROSS-LINKING AND AGING

The experiments just described suggest that externally originating factors are involved in aging. The results, however, are subject to many interpretations. The calorie-restricted individuals displayed delayed cross-linking of the protein collagen. Increased collagen cross-linking is typically associated with aging. In fact, tissue age is often determined by measuring the degree of collagen cross-linking. Collagen is found in and around blood vessel walls and in bone, cartilage, skin, and tendon. Indeed, collagen is present in just about all body tissues.

Some investigators believe that increased collagen cross-linking makes tissue more rigid or stiff and restricts the diffusion of nutrients, gases, and wastes through the tissues. Both of these effects might impair physiological function (for instance, by decreasing the elasticity of blood vessel walls) and directly cause aging (Figure 17-3). The known increase in collagen cross-linking with age, however, might be a side effect of another mechanism. There is little direct evidence that the cross-linking impairs physiological function.

AGING AND THE IMMUNE AND ENDOCRINE SYSTEMS

In addition to delayed collagen cross-linking, calorie-restricted animals exhibit delayed maturation of hormonal functions and delayed maturation of the immune system. Several theories of aging consider the immune system and the endocrine system to be key regulators of the aging process.

Reduced functioning of the immune system has been proposed as an explanation for aging (Figure 17-4). Individuals with diminished immune function probably have increased incidence of age-related diseases such as cancer. Patients whose immune systems have been suppressed for organ transplant therapy have a much higher incidence of cancer than is normal for their age. The immune system functions best in young individuals and declines with increasing age.

The thymus gland may play a "pacemaker" role in aging. This immune

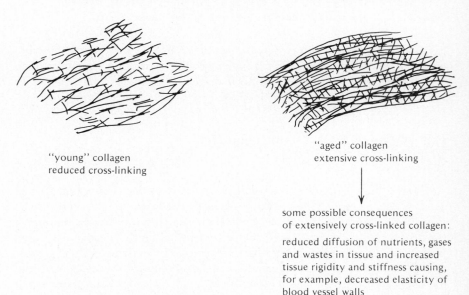

"young" collagen
reduced cross-linking

"aged" collagen
extensive cross-linking

some possible consequences
of extensively cross-linked collagen:

reduced diffusion of nutrients, gases
and wastes in tissue and increased
tissue rigidity and stiffness causing,
for example, decreased elasticity of
blood vessel walls

Figure 17-3. Possible consequences of extensively cross-linked collagen.

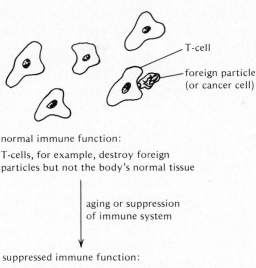

normal immune function:

T-cells, for example, destroy foreign
particles but not the body's normal tissue

aging or suppression
of immune system

suppressed immune function:

T-cells have decreased ability to destroy
foreign particles and cancer cells and
may attack the body's own tissues
(autoimmune disease)

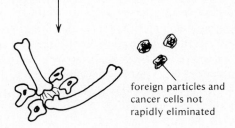

foreign particles and
cancer cells not
rapidly eliminated

Figure 17–4. Functioning of normal and suppressed immune systems. Suppressed immune system functioning is associated with aging.

system organ reaches its maximum size early in life and becomes smaller with age. In one experiment, the thymus gland was removed from young mice. The mice acquired "runting disease," a disease in which the immune system attacks normal tissue (*autoimmune* response), causing tissue degeneration. Autoimmune diseases are related to aging. Rheumatoid arthritis, for example, is caused by such a mechanism. Decreased immune function is also related to the increased susceptibility to infection and cancer that appears in old age. Delay in the maturation of the immune system, therefore, may be part of the explanation for the extended life of calorie-restricted animals.

We mentioned that the calorie-restricted rats also displayed delayed hormonal maturation. Hormones are important in aging. In one experiment, the pituitary gland was removed from an animal and the animal was maintained with a minimal hormone supplement. The animal showed a reduction in some age-related processes, such as collagen cross-linking. Anterior pituitary function is indeed depressed in calorie-restricted animals.

At the beginning of this discussion, we suggested that reduced aging in calorie-restricted rats was evidence for the extrinsic-factor model of aging. Obviously, calorie intake represents an externally generated agent. Now, after examining some of the consequences of calorie restriction, we can see that these same experiments could lead to other conclusions. For example, if the thymus gland is indeed a "pacemaker" that controls aging, then the normal development and decline of this gland would result from a normal developmental program. This program, however, could be modified by externally generated factors such as calorie intake. This suggests that aging is not controlled by internal "clocks" alone or by extrinsic factors alone, but by an interaction between internal "clocks" and extrinsic factors. Extrinsic factors might turn the internal "clocks" on or off, just as a variety of agents turn genes on and off during differentiation.

INTERNAL AGING CLOCKS

What other evidence favors the intrinsic or internal "clock" mechanism of aging? Many previously inactive genes become activated late in life. For example, genes that specify traits such as adult-onset polycystic kidney disease, hereditary glaucoma, hairy ears, and Huntington's chorea usually become active only in the later years of life (Figure 17-5). This suggests that the conditions associated with aging are programmed into our genetic information, but that they are repressed in early life. Aging, therefore, might represent the later part of a developmental timetable that includes embryonic development, maturation, and aging. Again, externally generated agents probably play a role in turning such "aging genes" on (or off).

Additional information regarding the built-in cellular "clocks" has been obtained through studies of isolated cells in culture. Mammalian cells in artificial culture generally go through three growth phases:

1. Establishment, or adaptation to the culture conditions.

2. Exponential growth in which the cells grow rapidly.

3. Cessation of growth and eventual degeneration.

Adult onset diabetes	Hairy ear rims
Adult onset polycystic kidneys	Huntington's chorea
Pereoneal atrophy (dominant)	Hereditary glaucoma
Dominant muscular distrophy	Dystrophia myotonica

Figure 17-5. Some genetic conditions that appear to become active during later life.

Cells from old donors usually go through fewer doublings than cells from fetal tissue. However, cells from old donors can exist quite long in culture, longer than the life span of the donor would suggest. These experiments, however, do suggest that normal cells contain some sort of "clock" that controls (at least in part) growth potential and growth cessation. Again, however, environmental conditions can modify the number of doublings that cells can undergo in culture. For example, hydrocortisone increases the number of doublings, but seldom by more than 25%. This 25% is in the same general range as the life extension caused by calorie restriction in rats (although cell doubling time should not be equated with the overall aging of organisms). These results suggest that a "clock" exists that can be modified, to some extent, by external agents.

GENETICS OF AGING

Aging may be, in part, caused by the buildup of cell products that form as the result of accumulated mutations in the genetic material. Such products could include defective enzymes that control DNA copying, which would result in the formation of abnormal proteins.

Some diseases caused by gene mutations can accelerate aging. These diseases include Werner's syndrome and the Hutchinson-Gilford progeria syndrome. In Werner's syndrome, the hair may turn gray in the teens. Cataracts, adult-onset diabetes, skin lesions, and sarcomas are common. In the Hutchinson-Gilford syndrome, the individual usually lives only until age 11 and always dies before age 30. Its symptoms include the early loss of hair, retarded growth, enlarged stiff joints, and hardening of the arteries. Such diseases suggest that genetic mutations can indeed cause age-related declines in physiological function.

The life span of a given species is well defined within a certain range of years or days. The life span of parents and children in a given family are also quite similar. Identical twins usually show quite similar life spans. These observations show that the genetic makeup of an individual controls his or her general life expectancy. The degree to which life span can be modified by altering extrinsic factors (such as calorie intake) remains one of the most intriguing questions in the area of aging today. Perhaps the genetically controlled life expectancy can be substantially altered. The experiments on altering calorie intake in rodents suggests that such modification may be feasible for human beings.

READINGS AND REFERENCES

Ham, R. G., and Veomett, M. J. 1980. *Mechanisms of Development*. St. Louis: C. V. Mosby.
Herman, E. 1976. In A. V. Everitt and J. A. Burgess, eds., *Hypothalamus, Pituitary and Aging*. Springfield, IL: Charles C Thomas.

Kohn, R. R. 1978. *Principles of Mammalian Aging*, 2nd ed. Englewood Cliffs, NJ: Prentice-Hall.

Makinodan, T. 1977. In C. E. Finch and L. Hayflick, eds., *Handbook of the Biology of Aging*. New York: Litton.

Sacher, G. A. 1977. In C. E. Finch and L. Hayflick, eds., *Handbook of the Biology of Aging*. New York: Litton.

Smith, L., Klug, T., and Adelman, R., eds. 1980. *Biochemistry of Aging*. National Institutes of Health Publication no. 81-1895.

Stern, C. 1973. *Principles of Human Genetics*, 3rd ed. San Francisco: W. H. Freeman.

GLOSSARY

This glossary consists mainly of terms used in the text. It is not intended as a complete listing of anatomical or embryological terms. More complete discussions of these terms in the context of developmental biology are given in the text. An unabridged dictionary should be consulted for definitions in a context broader than embryology.

acinus (pl. acini): Minute saclike structure.

acrosome: The anterior tip of a mature sperm cell. It aids the sperm in penetrating the outer egg coats and in establishing a connection with the egg cytoplasm.

actin: Protein subunit of microfilaments.

actinomycin D: An RNA synthesis inhibitor.

adhesion-promoting factors: Molecules that cause cells to stick together.

adrenal gland: Endocrine gland on or near the kidney.

adult hemoglobin: Hemoglobin found in adults.

albumen: White of an egg such as that of birds. A type of protein.

allantois: Extraembryonic saclike extension of the hindgut of amniotes. It aids in excretion and respiration.

amnion: Inner membrane surrounding the embryo.

amniotes: Vertebrates possessing an amnion during development.

amniotic cavity: Space between the amnion and the embryo proper.

amphibia: Class of vertebrates intermediate in many characteristics between the fishes and reptiles. Amphibians live part of the time in water and part on land.

amphioxus: Cephalochordates. Small, fishlike creatures that are often called lancelets, or lancet fish.

anamniotes: Vertebrates that lack an amnion during development.

anaphase: A stage in mitosis in which the chromatids of each chromosome separate and move to opposite poles. In meiosis, the stage in which paired chromosomes separate.

androgens: Male sex hormones.

angiosperm: Member of a class of plants that carry their seeds in a closed seed vessel.

animal hemisphere: Region of the egg where the nucleus resides.

anneal (reanneal): Reassociation of complementary nucleic acid strands upon slow cooling.

anther: Pollen-producing part of a stamen.

antibody: Protein that immunizes the body against a specific foreign antigen.

antifertilizin: Sperm cell-surface molecule that binds to egg fertilizin.

antigen: Substance that stimulates the production of antibodies and reacts with them.

anucleolate: Without nucleoli.

anura: Amphibia without tails. Frogs and toads.

anus: Posterior opening of digestive tube.

aorta: A main trunk of the arterial system.

apical ectodermal ridge: Thickened epidermis at the tip of limb buds.

archenteron: Primitive or embryonic digestive tube.

archenteron roof: Prospective notochord.

area opaca: Peripheral zone of the chick blastoderm that is attached to the yolk underneath.

area pellucida: Relatively transparent central region of chick blastoderm underlaid by the subgerminal space.

atrium: Heart chamber. Sometimes also applied to a chamber in other organs.

autonomic: Self-controlling and independent of outside influences.

auxin: A plant-growth hormone, indolacetic acid.

Aves: The class of birds.

axon: The outgrowth of a nerve cell that conducts impulses away from the cell body.

B-cell (bursa lymphocyte): Lymphocyte active in humoral immunity.

benign tumor: Slow-growing, nonspreading growth that does not harm the host.

bilateral cleavage: Cleavage pattern that forms an embryo with a right and left side that are mirror images of each other.

bile: Fluid secreted by the liver.

bindin: Protein isolated from sperm acrosomal granules. It may help the sperm recognize and adhere to the surface of the egg cell.

bipolar neuron: Nerve cell with an axon at one end and a dendrite at the other.

blastema: Primitive aggregation of cells from which an organ develops.

blastocoel: Cavity present in the blastula-stage embryo.

blastocyst: Blastula-like stage of the mammalian embryo.

blastocyst cavity: Blastocoel-like cavity in the mammalian blastocyst.

blastoderm: Primitive cellular plate of early embryos.

blastodisc: A plate of cytoplasm at the animal pole of the bird egg.

blastomere: One of the cells that results from the cleavage of the zygote.

blastopore: Opening into the archenteron from outside the embryo.

blastula: Embryonic stage in which the embryo is a hollow ball with a cavity. In some embryos, the blastula stage resembles a cap of cells rather than a ball.

block to polyspermy: Preventing more than one sperm from entering the egg.

bone marrow transplant: Replacing an individual's diseased bone marrow with bone marrow from a compatible donor, after destroying the host individual's marrow with radiation and chemicals.

Bowman's capsule: Expanded and invaginated portion of kidney tubule that contacts the glomerulus.

BUdR: 5-bromodeoxyuridine, a thymidine analog.

bursa of Fabricius: Gland involved in the immune system of birds.

calyx: Outermost series of leaflike parts in a flower.

Capsella: Genus of weedy flowering plant, known as shepherd's purse.

carbohydrate binding proteins: Lectins; proteins that can combine with carbohydrates.

carcinogen: A cancer-causing agent.

carcinoma: Tumor derived from epithelial tissue.

cartilage: Gristle-like skeletal tissue.

cDNA: Complementary DNA; DNA made using RNA as a template.

cell–cell recognition: Factors that allow specific cells to interact and not other cells.

cell motility: Cell movement.

cellular immunity: Direct attack by lymphocytes on a foreign cell.

centriole: Minute cytoplasmic granules surrounded by a zone of gelated cytoplasm. An organelle containing microtubules located at the spindle poles in dividing cells or embedded at the cell periphery. It forms the basal portion of a cilium or flagellum.

centrolecithal eggs: Eggs with yolk concentrated in the interior, such as insect eggs.

cerebellum: Brain region derived from the roof of the metencephalon.

cerebral hemispheres: Brain regions derived from the roof of the telencephalon.

chalaza: The tissue supporting the yolk in a bird's egg.

chemotaxis: Movement toward increasing concentration of a chemical.

chondroblast: Embryonic cartilage-forming cell.

chondrocyte: Cartilage cell.

Chordata: Phylum of animals with a permanent or transient notochord.

choroid coat: An outer coat of the eyeball that develops from mesenchyme.

chorion: A surface coat outside the plasma membrane of the eggs of fishes and tunicates. The term is also used to describe an extraembryonic membrane made of somatopleure that surrounds some embryos.

chromatid: One of the two daughter strands of a duplicated chromosome.

chromosome: A deeply staining rod or thread in the cell nucleus that contains the genes.

ciliary body: Structure involved in controlling the shape of the eye lens.

cilium: A whiplike locomotor organelle produced by a centriole.

cleavage: Division of the fertilized egg.

cleft: A split (in embryonic tissue).

cleidoic eggs: Eggs such as those of birds that are insulated from the environment by albumen, membranes, and shell.

clitoris: Organ composed of erectile tissue; the homologue in females of the male penis.

cloaca: Combined urogenital and rectal chamber.

clonal selection hypothesis: A theory of the production of antibodies. Cells that produce a given antibody all derive from a single cell; contact with an antigen causes that specific cell to divide, producing daughter cells that recognize that antigen.

coelom: Body cavity.

colchicine: Drug that disrupts microtubules.

collagen: Fibrous protein present in connective tissue.

competence: Ability to form a structure.

competition-hybridization experiment: Experiment in which cold (nonradioactive) RNA competes with radioactive RNA for binding sites on a DNA molecule. Such experiments are used to suggest base-sequence similarities and differences among nucleic acid molecules.

contact guidance: Cell growth in response to mechanical components or to oriented components of the substratum.

contact inhibition: The inhibition of growth or movement of cells by contact with other cells.

contractility: The ability to draw together into a more compact form.

conus arteriosus: Truncus arteriosus. Anterior-most portion of the heart.

coracoid process: Bony structure extending from the shoulder blade to the breastbone.

cornea: Transparent front covering of the eye.

corolla: The circle of flower leaves, usually colored, forming the inner floral envelope.

corona radiata: Follicle cell layer surrounding an ovulated mammalian oocyte.

corpus luteum: Endocrine capsule formed from an ovulated follicle.

cortex: Outer portion of an organ or part.

cortical granule: Structure in the surface cytoplasm, just below the plasma membrane of eggs, that functions in the fertilization reaction and in the formation of a fertilization membrane.

CoT: Concentration × time, in moles per liter × seconds.

cotyledon: Embryonic leaf.

crossing over: The exchange of segments of genetic material between the chromatids of homologous chromosomes in the prophase of meiosis.

crystallins: Proteins in the lens of the eye.

cycloheximide: A protein synthesis inhibitor.

cytochalasin: Drug that disrupts microfilaments. It also has other effects on cellular processes.

cytotrophoblast: Inner or cellular layer of the trophoblast.

dendrite: A branched, receiving process of a nerve cell.

dermatome: An embryonic skin segment. Used specifically to denote the outer region of the somite that gives rise to the dorsal dermis of the skin.

dermis: Inner or lower layer of skin.

diabetes: Disease resulting from a lack of insulin production.

diencephalon: Posterior portion of forebrain.

digit: Finger or toe.

diploid: The normal number of chromosomes in all body cells except the gametes. Chromosomes in matching pairs.

diplotene: The stage in the prophase of meiosis, following pachytene, in which the chromosomes are distinctly double and begin to separate from one another, except at the chiasmata.

distal: Away from the point of attachment or origin.

DNA tumor virus: Tumor-causing virus with a DNA core.

dorsal: Relating to the top or back.

dorsal lip of the blastopore: Prospective notochord; the first region to enter the embryo during gastrulation.

Down's syndrome: Trisomy 21 or Mongolism. Down's individuals have three instead of two number 21 chromosomes.

ductus deferens: The sperm duct.

duodenum: The first division of the small intestine.

ectoderm: Outer germ layer of the embryo.

egg: the mature female gamete, normally haploid.

embryology: The study of the origin and development of the embryo.

embryonic field: An embryonic area that has the ability to regulate itself. It possesses the four properties given on pages 230–233.

embryonic hemoglobin: Hemoglobin found in embryos. It differs in structure from adult hemoglobin.

embryonic induction: Stimulation of differentiation in one tissue as the result of interaction with another tissue.

embryo sac: Female gametophyte in flowering plants.

endocardium: Inner lining of heart cavities.

endoderm: Innermost germ layer of the embryo.

endometrium: Lining of the uterus.

epiblast: Upper embryonic region of the blastoderm.

epidermis: The outer epithelial portion of the skin.

epididymis: First convoluted portion of the excretory duct of the testis. It passes from above downward along the posterior border of the testis.

epimere; somite. Dorsal region of mesoderm on each side of the neural tube, consisting of myotome, dermatome, and sclerotome.

equal cleavage: Division that produces daughter cells of equal size.

erythropoietin: Glycoprotein hormone that stimulates the synthesis of globin messenger RNA.

esophagus: Portion of gut between pharynx and stomach.

estrogen: Hormone that influences estrus or produces changes in the sexual characteristics of female mammals.

estrus: Period of sexual receptivity in the female.

eukaryotic cell: Cell of higher organisms (those above the bacteria and blue-green algae).

extraembryonic coelom: Cavity outside the embryo proper. It is continuous with the body cavity (coelom) and is surrounded by extraembryonic membranes.

eye cups: Cup-like structures formed by the invagination of the optic vesicles.

F-actin: Actin filament.

Fallopian tube: Oviduct.

fate map: Projection on the embryo of what specific areas will become at later stages.

femur: Thigh bone or upper leg bone.

fertilization: Union of egg and spermatozoan.

fertilizin: Acid mucopolysaccharide in egg jelly that binds sperm.

fetus: Embryo of an animal before birth; embryo in the uterus at later stages in development.

fibroblast: Primitive mesenchymal cell that gives rise to connective tissue.

fibula: External and smaller of the two bones of the lower leg.

filopodia: Long cellular projections.

flagellin: A globular protein that makes up flagella.

flagellum: Long, threadlike structure that protrudes from the cell body.

fluid-mosaic model of the cell surface: Theory that describes the cell surface as composed of a lipid bilayer with mobile proteins attached to and imbedded in the lipid.

follicle: Vesicular body in the ovary that contains the developing oocyte.

follicle cells: Cells that surround oocytes to nourish and protect them.

follicle-stimulating hormone: A hormone produced by the anterior pituitary gland that controls reproductive events, particularly the maturation of the oocyte and its follicle.

freeze-fracture technique: Cell microscopy technique that involves splitting the membrane down the middle and peeling away one of the lipid bilayers. The surface of the cleaved membrane is then coated with heavy metal, molding the contours of the fractured surface. The replica is examined with an electron microscope.

Fucus: A brown seaweed.

FUdR: 5-fluorodeoxyuridine, an inhibitor of DNA synthesis.

gall bladder: Sac-like vessel associated with the liver. It stores bile.

gamete: An egg cell or sperm cell.

gametophyte: The haploid generation in flowering plants. Pollen or embryo sac.

gamete: An egg cell or sperm cell.

ganglion: Aggregation of nerve cell bodies along the course of a nerve.

gastrula: Embryonic stage following the blastula. It consists of a layered sac.

gastrulation: Transformation of a blastula into a layered embryo, the gastrula.

gene: The hereditary unit in chromosomes. It consists of a DNA segment containing a base sequence that codes for a polypeptide chain.

gene cloning: Production of many copies of a given gene.

germinal epithelium: Surface of genital ridge; outer part of primitive gonad.

globin: Polypeptide chains of hemoglobin.

glomerulus: Tuft of capillary loops at the beginning of each kidney tubule.

glomus: Network of fine blood vessels associated with the ciliated funnels of the pronephric tubules.

glucagon: Pancreatic hormone involved in regulating blood sugar.

glycogen: A complex carbohydrate (polysaccharide) that stores energy-rich carbohydrate units in animals. Animal starch.

glycosaminoglycans: Sugar polymers consisting of uronic acids and amino acids.

glycosyl transferase: Enzyme that catalyzes the transfer of single sugar residues from nucleotide sugars to the growing ends of carbohydrate chains.

glycosyl transferase–carbohydrate acceptor model: A cell adhesion model. Cell-cell adhesion is mediated by glycosyl transferases on one cell that bind to carbohydrate chains on an adjacent cell.

gonad: A reproductive gland; the ovary or testis.

gonadotropic: Refers to a hormone that promotes gonad growth or activity.

Graafian follicle: Mature mammalian ovarian follicle.

gray crescent: A grayish surface cytoplasmic region of the amphibian embryo. Exposed soon after fertilization, it gives rise to the dorsal lip of the blastopore. Prospective notochord.

gymnosperm: Member of a class of plants whose ovules and seeds are not enclosed in a case. Certain evergreens are gymnosperms.

haploid: Half the chromosome number of normal body cells; one chromosome of each pair. Mature sperm and egg cells have a haploid chromosome number. At fertilization, the normal chromosome number (diploid) is restored in the zygote.

heme: Nonprotein, iron-containing chemical group that binds oxygen.

hemoglobin: Red blood cell protein (globin) combined with a nonprotein heme group that binds oxygen.

Hensen's node: Anterior thickened end of the primitive streak.

histone: A major type of chromosomal protein.

holoblastic cleavage: Total cleavage. Divisions pass through the entire zygote.

homologous chromosome pair: Chromosomes of identical size, shape, and gene content that derive from the mother and father.

humerus: Upper arm bone.

humoral immunity: Immunity through the synthesis of free soluble antibodies that combine with foreign antigens.

hyaline layer: Clear layer present, for example, at the surface of sea urchin embryos. It is derived from cortical granule material.

hybridization experiments: Experiments in nucleic acid technology. Comparisons of nucleic acid base sequences are made by measuring the binding of one nucleic acid strand with another.

hypoblast: Lower embryonic region of blastoderm.

hypomere: The most ventral subdivision of the mesoderm. Consists of the somatic and splanchnic mesoderm.

hypothalamus: Ventral portion of the thalamus of the diencephalon of the brain.

ileum: Lowest division of the small intestine.

immunoglobulin: An antibody.

incomplete cleavage: Cleavage in which divisions do not pass through the entire zygote.

indifferent stage of sexual development: Developmental stage at which the gonad has not yet differentiated into a testis or ovary.

induction: The process whereby an inducing tissue interacts with a responding tissue, causing the responding tissue to differentiate.

inner cell mass: Inner part of the mammalian embryo that forms the embryo proper.

instructive information: Genetic information provided by an inducing tissue to a responding tissue.

insulin: Pancreatic protein hormone functioning in carbohydrate metabolism.

integral membrane protein: Protein that is very strongly associated with or deeply imbedded in the lipid bilayer of the cell membrane.

intermediate mesoderm: Mesomere. The region of mesoderm between the epimere and hypomere.

intestine: The digestive tube passing from the stomach to the anus.

invagination: Folding or buckling in a set of cells so that an outer surface becomes an inner surface, as in gastrulation.

invasion: The spreading of tumor tissue by cells crawling to nearby tissues.

iris: Pigmented disk-like diaphragm of the eye. It controls the size of the pupil.

islets of Langerhans: Insulin-producing cells of the pancreas.

isolecithal egg (oligolecithal egg): An egg that possesses a small amount of rather evenly distributed yolk.

isozymes: Enzymes that exist in multiple molecular forms.

jejunum: Middle division of the small intestine.

karyotype: Chromosome size, shape, and number.

keratin: A hard, relatively insoluble protein. It is present largely in cutaneous (skin) structures.

kinetin: A plant hormone.

Klinefelter's syndrome: A condition resulting from an XXY karyotype in mammals.

lactate dehydrogenase (LDH): Enzyme that catalyzes the interconversion of pyruvate and lactate.

lampbrush chromosomes: Extended diplotene chromosomes that appear during oogenesis in some amphibian species. These chromosomes develop loops perpendicular to the long axis. The loops actively synthesize RNA.

lateral: On the side.

lateral lips of the blastopore: Side lips of the blastopore.

lectin: Carbohydrate-binding protein.

lens-forming ectoderm: Prospective lens.

leptotene: Stage in the prophase of meiosis immediately preceding homologous chromosome pairing (synapsis). In leptotene, the chromosomes appear as fine threadlike structures.

leukemia: Tumor of blood-forming tissues in which white blood cell number greatly increases.

leukocyte: A white blood corpuscle.

Leydig cells: Interstitial cells of the testis.

liquor folliculi: Liquid in ovarian follicle.

liver: Largest glandular organ in vertebrates. It secretes bile and is active in metabolism.

luteinizing hormone (LH): A hormone produced by the anterior pituitary gland that controls certain reproductive events. In the female, it plays an important role in causing ovulation; in the male, it stimulates the interstitial cells of the testis to synthesize male sex hormones.

lymph node: A gland-like body involved in the immune system.

lymphocyte: A white blood cell formed in lymph node tissue. It is involved in the immune response.

lysins: Enzymes present in acrosomal granules. They aid sperm in penetrating the outer egg coats.

macromeres: Large cleavage blastomeres.

macrophage: A large blood cell that engulfs and destroys other cells.

malignant tumor: Spreading tumor that can kill the host.

Malpighian body: Bowman's capsule plus its glomerulus; the functional renal unit.

mammalia: Highest class of living organisms. It includes all the vertebrates that suckle their young.

marsupialia: An order of mammals. Marsupials have an abdominal pouch for nurturing the young.

medulla: Inner portion of an organ or part. In the brain, the region that forms from the myelencephalon.

meiosis: Divisions in the germ cell line that result in haploid gametes.

melanin: Dark pigment in skin, hair, and eyes.

melanoma: Tumor derived from pigment cells.

melanophore: Cell containing melanin.

menstruation: Periodic discharge of blood, necrotic tissue, and glandular secretions from the uterus. A part of the reproductive cycle in human females.

meristem: Undifferentiated plant tissue that gives rise to stem and root tissue.

meroblastic cleavage: Cleavage that involves only a small cytoplasmic area at the animal pole of the egg.

mesencephalon: Midbrain.

mesenchymal factor: Factor that stimulates cell division and differentiation in pancreas epithelium.

mesenchyme: Embryonic connective tissue.

mesendoderm: Combination of prospective mesoderm and prospective endoderm.

mesentery: Double layer of peritoneum enclosing an organ.

mesocardium: Mesentery (two-layered membrane) that supports the heart.

mesoderm: Middle germ layer of the embryo.

mesomeres: Medium-sized cleavage blastomeres. Also, the intermediate mesoderm, the region between the epimere and hypomere.

mesonephric duct: Duct connecting the mesonephric tubules and the cloaca.

mesonephric tubules: Kidney tubules of adult fish and amphibians.

mesonephros: Second stage in the development of the amniote kidney; the functional kidney of adult fish and amphibians.

messenger RNA: RNA molecules containing the genetic information needed for synthesizing a polypeptide chain.

metamorphosis: Marked change in form and structure in an animal's development from embryo to adult.

metanephros: Last stage in the development of the amniote kidney.

metaphase: The stage of mitosis or meiosis during which the chromosomes lie in the central plane of the spindle.

metastasis: Tumor spread via the bloodstream.

metencephalon: Anterior portion of the hindbrain.

microfilaments: Rod-like elements composed of the protein actin.

micromeres: Small cleavage blastomeres.

micropinocytosis: Intake of very small fluid droplets by cells.

microtubules: Cylindrical units composed of globular subunits of the protein tubulin.

microvilli: Fine projections or extensions of the surface of cells.

mitochondrion: Cytoplasmic organelle involved in energy metabolism.

Mongolism: See Down's syndrome.

morphogenesis: Development of form and structure.

morphology: The study of the form or structure of organisms.

morula: Solid ball of blastomeres (about the 16-cell stage) that precedes the blastula stage.

Müllerian duct: Oviduct.

mutant: An organism or gene that has undergone a chromosomal change (mutation) that can be hereditary.

myelencephalon: Posterior portion of the hindbrain.

myoblast: Cell that gives rise to muscle.

myocardium: Muscle layer of the heart.

myofibril: Contractile muscle fiber.

myosin: A fibrous protein composed of two identical subunits wound around each other.

myotome: Prospective or actual muscle segment. Specifically used to denote the inner part of the somite that forms back muscles.

myotube: Muscle tubes containing many muscle nuclei in a common cytoplasm.

nephrocoel: Internal cavity of the nephrotome.

nephrostome: Funnel-shaped opening by which a kidney tubule communicates with the nephrocoel.

nephrotome: Segmented portion of intermediate mesoderm that gives rise to the kidney tubule.

nerve growth factor: Protein that stimulates nerve outgrowth from ganglia.

nests of oogonia: Clusters of oogonia surrounded by follicle cells.

neural crest: Tissue that forms from the region of ectoderm (the neural folds) between the prospective epidermis and the prospective neural tube.

neural folds: Elevated ridges of the neural plate.

neural groove: Trough formed when the neural plate bends upward.

neural plate: Embryonic region that becomes the nervous system.

neural retina: Inner sensory layer of the eye cup.

neural tube: Tube that forms the nervous system. It is derived from the neural plate.

neuroblast: Embryonic nerve cell.

neuron: Cellular unit of the nervous system.

neurulation: The stage or process of neural tube formation.

nonhistone protein: Proteins associated with DNA. Most are phosphorylated and acidic.

nonnotochordal mesoderm: Mesoderm (middle germ layer) that gives rise to derivatives other than the notochord.

notochord: Fibrocellular rod that constitutes the primitive skeletal axis.

nucleolus: Dense spherical granule in the cell nucleus. It is composed of ribosomal RNA and protein.

nucleosomes: Bead-like particles that make up eukaryotic chromosomes.

nucleus: Large body within the cell that contains the chromosomes. The nucleus also usually contains one or more nucleoli.

nurse cells: Cells in organisms such as insects that nourish the developing oocyte.

olfactory: Relating to the sense of smell.

oligolecithal (isolecithal) egg: An egg that possesses a small amount of rather evenly distributed yolk.

ontogeny: Development of the individual.

oncogene theory: Theory that all cells of animals contain cancer genes (oncogenes) that can transform normal cells into cancer cells if activated.

oocyte: Stage in the maturation of the female gamete.

oogenesis: Development of egg cells.

oogonia: Undifferentiated diploid cells that give rise to oocytes and eggs by meiosis.

opistonephros: Adult kidney in anamniotes.

optic chiasma: Place where the nerve fibers from both eyes cross.

optic tecta: Visual interpretation centers in the midbrain.

optic vesicle: Outpocketing of the diencephalon of the brain that forms the retina of the eye.

organizer: Spemann and Mangold's term for the dorsal lip of the blastopore.

organogenesis: The formation of organs.

ostium tubae: Funnel-shaped openings of the oviducts.

otic: Relating to the ear.

oviduct: Tube that transports the egg from the ovary.

ovule: In plants, the body in the ovary that gives rise to the seed.

ovum: Egg.

pachytene: The stage in meiotic prophase that immediately follows the pairing of homologous chromosomes (synapsis). The chromosome threads become shortened and thickened, and crossing over occurs.

pancreas: Abdominal digestive and endocrine gland.

parathyroid: Endocrine gland adjacent to the thyroid.

parthenogenesis: Development without fertilization.

penis: Male organ of copulation.

pericardial: Surrounding the heart.

perichondrium: The fibrous membrane that covers cartilage.

peripheral membrane protein: Protein loosely associated with the cell membrane.

peritoneum: Lining of the body cavity and covering of the organs.

permissive information: Information that turns on genes in a responding tissue.

petal: A division of the corolla of a flower.

pharynx: Anterior portion of the foregut.

phloem: Plant tissue that conducts sap.

physiological polyspermy: A method of fertilization whereby many sperm enter the egg, and all but one disintegrate after entry.

pigmented retina (pigmented layer of the eye): Outer layer of eye cup.

pinocytosis: Intake of fluid droplets by cells.

pistil: Female reproductive system of flowering plants.

pituitary: Endocrine gland at the base of the forebrain that secretes several hormones.

placenta: Organ of physiological communication between the mammalian mother and fetus.

placode: Thickened plate-like area of epithelium.

plasma cell: A mature antibody-secreting B lymphocyte.

polar body: Tiny nonfunctioning cell with little cytoplasm, produced during each meiotic division during oogenesis.

polar lobe: A cytoplasmic extrusion formed in some spirally cleaving embryos.

pollen: Male gametophyte in flowering plants.

polyploid: Having chromosomes in more than two sets: triploid, tetraploid, etc.

polysome: Group of ribosomes active in protein synthesis.

polyspermy: Penetration of an egg by more than one sperm.

pouches: Sac-like structures that form in embryos.

primary mesenchyme cells: Cells (derived from micromeres) that form the skeletal elements in sea urchins and related embryos.

primary oocytes: Cells that form as a result of growth of oogonia. These cells undergo the first meiotic division.

primary sex cords: Inner part of primitive gonad.

primary spermatocytes: Cells that form as the result of growth of spermatogonia. They undergo the first meiotic division.

primary tumor: Tumor in a host that develops by transforming normal cells into tumor cells.

primates: Highest order of mammals, including humans, monkeys, and lemurs.

primitive groove: Indentation in the midline of the primitive streak.

primitive streak: Thickening in the surface of some embryos at the beginning of gastrulation.

primordial germ cells: Primitive cells that give rise to gametes.

proamnion: Crescent-shaped area around the head of early bird embryos.

prokaryotic cell: Primitive bacterium or blue-green algae cell.

proctodeum: Terminal portion of the rectum formed in the embryo by an ectodermal invagination.

progesterone: A hormone of the corpus luteum that induces changes in the uterine lining, preparing it for the embryo.

pronephric duct: The duct connecting the pronephric tubules and the cloaca.

pronephric tubules: The functional units of the kidneys of aquatic larvae.

pronephros: First stage in the development of amniote kidney; the functional kidney of fish and amphibian embryos.

pronucleus: The nucleus of the egg or sperm before they fuse in fertilization.

prophase: Early phase of nuclear division characterized by the condensation of the chromosomes and their movement toward the equator of the spindle.

prosencephalon: Forebrain.

prospective (presumptive) region: A region in an early embryo that will become a specific structure in the more advanced embryo.

prostate: Gland that surrounds the beginning of the urethra in the male. In mammals, it produces a secretion that activates sperm in the seminal fluid.

protease: Enzyme that catalyzes the hydrolysis of protein.

protein kinase: An enzyme that catalyzes the phosphorylation of soluble yolk (phosvitin) into an insoluble form that can be stored in the egg.

proteoglycans: Glycosaminoglycans linked to proteins.

protoplast: Wall-free plant cell.

proximal: Nearest the trunk or point of origin.

pupil: Opening in the front of the eye that allows light to enter.

puromycin: A protein synthesis inhibitor.

radial cleavage: Cleavage that divides the embryo into an upper and lower tier of cells. The upper cells lie exactly over the lower cells. The cells are uniformly distributed around the polar axis of the egg.

radius: Forearm bone.

rectum: Terminal portion of the large intestine, connecting with the anus.

regeneration blastema: Undifferentiated cells together with their epidermal covering that accumulate at the surface of a wound.

regional inducing specificity: The ability of a specific region of a structure to induce a specific differentiation in a responding tissue.

renal: Pertaining to the kidney.

reptilia: Class of vertebrates that includes the alligators, crocodiles, lizards, turtles, and snakes.

rete cord cells: Cells that take part in testis formation.

rete testis: Network of tubules in the testis, formed from rete cord cells.

retina: Light-sensitive layer in the eye.

reverse transcriptase: An enzyme that catalyzes DNA synthesis using RNA as a template.

rhizoid: Root-like plant outgrowth that supports and nourishes the plant.

rhombencephalon: Hindbrain.

ribosomes: Particles on which protein synthesis occurs. The active protein synthesizing system is the polysome, which consists of a group of ribosomes.

RNA tumor virus: A tumor-causing virus with an RNA core.

rods and cones: Photoreceptor cells in the retina of the eye.

root apical meristem: Embryonic tissue at the tip of plant roots. It can give rise to adult tissue.

sarcoma: Tumor derived from connective tissue.

sclera: Tough outer coat of eyeball. It develops from mesenchyme.

sclerotome: The somite region that gives rise to the vertebral column.

scrotum: Sac containing the testes.

secondary mesenchyme cells: Cells that form the major portion of the mesoderm in the sea urchin embryo.

secondary oocyte: Product of the first meiotic division of the primary oocyte. It contains the haploid number of chromosomes.

secondary sex cords: Clusters of cells containing the oogonia; nests of oogonia.

secondary spermatocyte: Product of the first meiotic division of the primary spermatocyte. It contains the haploid number of chromosomes.

secondary tumor: Tumor that develops from the spread of a primary tumor.

selective adhesiveness: Specific adhesion between certain cells but not others.

selective gene amplification: Selective copying of certain genes. Thousands of DNA copies of specific genes are sometimes formed.

self-regulation: The ability of a structure to direct or control its own development.

seminiferous tubules: Sperm-forming tubules in the testis.

sepal: An individual leaf of the calyx.

Sertoli cells: Cells in the testis that support and nourish the developing sperm.

shoot apical meristem: Embryonic tissue at the tip of plant shoots. It can give rise to adult tissue.

sinus venosus: Heart chamber that receives venous blood.

somatic mesoderm: Hypomere mesoderm that is in close contact with ectoderm.

somatopleure: The combination of somatic mesoderm and ectoderm.

somite; epimere: Dorsal region of the mesoderm, consisting of myotome, dermatome, and sclerotome.

sorting out: The redistribution of cells in a mixed clump so that like cells group together.

species-specific differentiation of DNA: In this text, refers to specific base sequences of DNA that are similar in a species but vary among different species.

spermatid: Product of the second meiotic division of the secondary spermatocyte. Spermatids have the haploid chromosome number and differentiate into mature sperm cells.

spermatocyte: Stage in the maturation of male gametes that precedes the spermatid stage.

spermatogenesis: Development of sperm cells.

spermatogonia: Diploid cells in the testis that give rise to spermatocytes and sperm by meiosis.

spermatozoan: The fully mature haploid sperm cell that differentiates from the spermatid; the male gamete.

spermiogenesis: Transformation of spermatids into spermatozoa.

splanchnic mesoderm: Hypomere mesoderm that is in close contact with endoderm.

splanchnopleure: The combination of splanchnic mesoderm and endoderm.

spiral cleavage: Cleavage pattern resulting from oblique spindle orientations. The daughter cells in the upper tiers lie over the junctions between the lower cells.

stage-specific competence: The competence of a given embryonic stage to respond to an inducer.

stage-specific differentiation: The presence of some specific characteristics (such as RNAs and proteins) in some embryonic stages but not in others.

stamen: Male reproductive system of a flowering plant.

stigma: The part of a flower's pistil that receives the pollen.

stomach: Portion of gut between the esophagus and the small intestine.

stomodeum: Ectodermal invagination that forms the mouth cavity.

style: The elongation of a flower's carpel or ovary that bears the stigma.

sublethal cytolysis: Theory that embryonic induction can be caused by cell damage that does not bring embryonic death.

superficial cleavage: Incomplete cleavage in centrolecithal eggs. The cleavage does not extend all the way through the egg.

synapsis: Pairing of homologous chromosomes of maternal and paternal origin during the prophase of meiosis.

T-cell (thymus lymphocyte): Lymphocyte that plays a major role in cellular immunity.

telencephalon: Anterior portion of the forebrain.

teleosts: Modern bony fishes.

telolecithal: An egg whose yolk is concentrated at one pole. The yolk of a *moderately* telolecithal egg is substantially more concentrated in the vegetal hemisphere; the yolk of a *strongly* telolecithal egg is all in the vegetal hemisphere, and all the nonyolky cytoplasm sits as a cap atop the yolk.

telophase: The last stage in meiosis and mitosis during which the chromosomes reorganized themselves into two new nuclei.

teratoma: Tumor derived from reproductive cells. It often develops embryo-like growths.

tetrad: A complex of four chromatids, two from each of two paired homologous chromosomes. Tetrads are present during the prophase of meiosis.

thalamus: Side walls of the diencephalon.

thymus: Ductless gland in the neck. It is involved in the immune system.

thyroid: A gland or cartilage in the neck.

tibia: The inner and thicker of the two bones of the lower leg.

tissue-specific differentiation: Refers to specific characteristics (such as specific RNAs and proteins) taken on by some tissues and not others.

total (holoblastic) cleavage: Divisions that pass through the entire zygote.

trachea: Air tube extending from the larynx to the bronchi.

transcription: The synthesis of RNA from a DNA template.

transfer RNA: RNA molecule that transports amino acids to their proper places on the messenger RNA molecule during translation.

translation: Synthesis of proteins using genetic information coded in messenger RNA.

trisomy 21: See Down's syndrome.

trophoblast: Outer layer of the mammal embryo.

truncus arteriosus: Conus arteriosus. Anterior-most portion of heart.

trypsin: Proteolytic enzyme.

tryptophan pyrrolase: Enzyme that opens the indole ring of tryptophan.

tumor: Abnormal growth.

tumor-specific antigen: Antigen found primarily on or in tumor cells.

tunica: An enveloping layer.

ulna: The inner and larger of the two forearm bones.

unequal cleavage: Cleavage that results in two cells of unequal size.

ureter: Tube that conducts urine from the kidney to the bladder.

ureteric bud: Beginning of the formation of the ureter.

urethra: Canal leading from the bladder to the outside of the body. Used to discharge urine.

urodela: Amphibia with tails, such as salamanders and newts.

uterus: Womb. Hollow muscular organ in the mammalian mother in which the embryo develops into the fetus.

vagina: Genital canal in the female extending from the uterus to the vulva.

vas deferens: Sperm duct; ductus deferens.

vegetal hemisphere: The half of the egg in which yolk accumulates. The egg nucleus resides in the other (animal) hemisphere.

ventral: Relating to the belly.

ventral lip of the blastobore: Lower or belly region adjoining the blastopore; the last region to enter the blastopore during gastrulation.

ventricle: A cavity in the brain or heart.

virus: Infective agent, usually composed of a nucleic acid core and a protein coat.

vitelline membane: A surface coat outside the plasma membrane of some eggs.

vulva: The entrance to the vagina; the external genitalia of the female.

Wolffian duct: Mesonephric duct.

wrist: Part of arm between the hand and forearm.

Xenopus: African clawed toad (a frog).

xylem: In higher plants, the portion of a vascular bundle made up of woody tissue. It transports water.

yolk: Food reserve in the egg.

yolk plug: The center of the circular blastopore in amphibian embryos. It consists of yolky endoderm cells.

yolk sac: Bag-like extraembryonic membrane extending from the midgut of bird, fish, and reptile embryos.

zona pellucida: A surface coat outside the plasma membrane of mammalian eggs.

zone of polarizing activity (ZPA): An area of limb bud mesoderm near the posterior junction of the limb bud with the body. It plays a key role in determining limb orientation.

zygote: Fertilized egg.

zygotene: The stage in the prophase of meiosis in which the chromosomes arrange themselves in homologous pairs.

zymogen: Storage (inactive) form of an enzyme.

INDEX